Praise for *Python Data Science Handbook*, Second Edition

There are many data science books out there today but I find Jake VanderPlas's book to be exceptional. He takes a subject that is very broad and complex and breaks it down in a way that makes it easy to understand with great writing and exercises that get you using the concepts quickly.

—*Celeste Stinger, Site Reliability Engineer*

Jake VanderPlas's expertise and passion for sharing knowledge are undeniable. This freshly updated edition offers clear, easy-to-follow examples that will help you successfully set up and use essential data science and machine learning tools. If you're ready to dive into core techniques for using Python-based tools to gain real insights from your data, this is the book for you!

—*Anne Bonner, Founder and CEO, Content Simplicity*

Python Data Science Handbook has been a favorite of mine for years for recommending to data science students. The second edition improves on an already amazing book complete with compelling Jupyter notebooks that allow you to execute your favorite data science recipe while you read along.

—*Noah Gift, Duke Executive in Residence and Founder of Pragmatic AI Labs*

This updated edition is a great introduction to the libraries that make Python a top language for data science and scientific computing, presented in an accessible style with great examples throughout.

—*Allen Downey, author of* Think Python *and* Think Bayes

Python Data Science Handbook is an excellent guide for readers learning the Python data science stack. With complete practical examples written in an approachable manner, the reader will undoubtedly learn how to effectively store, manipulate, and gain insight from a dataset.

—*William Jamir Silva, Senior Software Engineer, Adjust GmbH*

Jake VanderPlas has a history of breaking down core Python concepts and tooling for those learning data science, and in the second edition of *Python Data Science Handbook* he has done it once again. In this book, he provides an overview of all the tools one would need to get started as well as the background on why certain things are the way they are, and he does so in an accessible way.

—*Jackie Kazil, Creator of the Mesa Library and Data Science Leader*

第2版

Python数据科学手册（影印版）

Python Data Science Handbook

杰克·万托布拉斯（Jake VanderPlas）著

Beijing · Boston · Farnham · Sebastopol · Tokyo

O'Reilly Media, Inc.授权东南大学出版社出版

南京 东南大学出版社

图书在版编目(CIP)数据

Python 数据科学手册 = Python Data Science Handbook, 2nd Edition：第 2 版：影印版：英文 / (美) 杰克·万托布拉斯 (Jake VanderPlas) 著. —南京：东南大学出版社，2023.3

ISBN 978 - 7 - 5766 - 0658 - 4

Ⅰ. ①P… Ⅱ.①杰… Ⅲ.①软件工具-程序设计-手册-英文 Ⅳ.①TP311.561-62

中国国家版本馆 CIP 数据核字(2023)第 002297 号

图字：10 - 2022 - 478 号

Python 数据科学手册　第 2 版（影印版）

著　　者：杰克·万托布拉斯 (Jake VanderPlas)
责任编辑：张　烨　　封面设计：Karen Montgomery，张　健　　责任印制：周荣虎
出版发行：东南大学出版社
社　　址：南京四牌楼 2 号　邮编：210096　电话：025-83793330
网　　址：http://www.seupress.com
电子邮件：press@ seupress.com
经　　销：全国各地新华书店
印　　刷：常州市武进第三印刷有限公司
开　　本：787mm×1000mm　1/16
印　　张：37
字　　数：725 千
版　　次：2023 年 3 月第 1 版
印　　次：2023 年 3 月第 1 次印刷
书　　号：ISBN 978 - 7 - 5766 - 0658 - 4
定　　价：148.00 元

本社图书若有印装质量问题，请直接与营销部联系。电话(传真)：025 - 83791830

Table of Contents

Part IV.   Visualization with Matplotlib

Part V.   Machine Learning

Preface

What Is Data Science?

This is a book about doing data science with Python, which immediately begs the question: what is *data science*? It's a surprisingly hard definition to nail down, especially given how ubiquitous the term has become. Vocal critics have variously dismissed it as a superfluous label (after all, what science doesn't involve data?) or a simple buzzword that only exists to salt resumes and catch the eye of overzealous tech recruiters.

In my mind, these critiques miss something important. Data science, despite its hype-laden veneer, is perhaps the best label we have for the cross-disciplinary set of skills that are becoming increasingly important in many applications across industry and academia. This *cross-disciplinary* piece is key: in my mind, the best existing definition of data science is illustrated by Drew Conway's Data Science Venn Diagram, first published on his blog in September 2010 (Figure P-1).

While some of the intersection labels are a bit tongue-in-cheek, this diagram captures the essence of what I think people mean when they say "data science": it is fundamentally an interdisciplinary subject. Data science comprises three distinct and overlapping areas: the skills of a *statistician* who knows how to model and summarize datasets (which are growing ever larger); the skills of a *computer scientist* who can design and use algorithms to efficiently store, process, and visualize this data; and the *domain expertise*—what we might think of as "classical" training in a subject—necessary both to formulate the right questions and to put their answers in context.

With this in mind, I would encourage you to think of data science not as a new domain of knowledge to learn, but a new set of skills that you can apply within your current area of expertise. Whether you are reporting election results, forecasting stock returns, optimizing online ad clicks, identifying microorganisms in microscope photos, seeking new classes of astronomical objects, or working with data in any other field, the goal of this book is to give you the ability to ask and answer new questions about your chosen subject area.

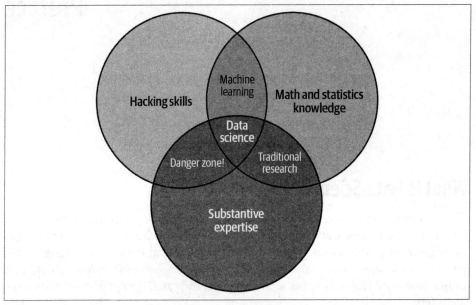

Figure P-1. Drew Conway's Data Science Venn Diagram (source: Drew Conway (https://oreil.ly/PkOOw), used with permission)

Who Is This Book For?

In my teaching both at the University of Washington and at various tech-focused conferences and meetups, one of the most common questions I have heard is this: "How should I learn Python?" The people asking are generally technically minded students, developers, or researchers, often with an already strong background in writing code and using computational and numerical tools. Most of these folks don't want to learn Python per se, but want to learn the language with the aim of using it as a tool for data-intensive and computational science. While a large patchwork of videos, blog posts, and tutorials for this audience is available online, I've long been frustrated by the lack of a single good answer to this question; that is what inspired this book.

The book is not meant to be an introduction to Python or to programming in general; I assume the reader has familiarity with the Python language, including defining functions, assigning variables, calling methods of objects, controlling the flow of a program, and other basic tasks. Instead, it is meant to help Python users learn to use Python's data science stack—libraries such as those mentioned in the following section, and related tools—to effectively store, manipulate, and gain insight from data.

Why Python?

Python has emerged over the past couple of decades as a first-class tool for scientific computing tasks, including the analysis and visualization of large datasets. This may have come as a surprise to early proponents of the Python language: the language itself was not specifically designed with data analysis or scientific computing in mind. The usefulness of Python for data science stems primarily from the large and active ecosystem of third-party packages: *NumPy* for manipulation of homogeneous array-based data, *Pandas* for manipulation of heterogeneous and labeled data, *SciPy* for common scientific computing tasks, *Matplotlib* for publication-quality visualizations, *IPython* for interactive execution and sharing of code, *Scikit-Learn* for machine learning, and many more tools that will be mentioned in the following pages.

If you are looking for a guide to the Python language itself, I would suggest the sister project to this book, *A Whirlwind Tour of Python* (*https://oreil.ly/jFtWj*). This short report provides a tour of the essential features of the Python language, aimed at data scientists who are already familiar with one or more other programming languages.

Outline of the Book

Each numbered part of this book focuses on a particular package or tool that contributes a fundamental piece of the Python data science story, and is broken into short self-contained chapters that each discuss a single concept:

- Part I, "Jupyter: Beyond Normal Python" introduces IPython and Jupyter. These packages provide the computational environment in which many Python-using data scientists work.

- Part II, "Introduction to NumPy" focuses on the NumPy library, which provides the `ndarray` for efficient storage and manipulation of dense data arrays in Python.

- Part III, "Data Manipulation with Pandas" introduces the Pandas library, which provides the `DataFrame` for efficient storage and manipulation of labeled/columnar data in Python.

- Part IV, "Visualization with Matplotlib" concentrates on Matplotlib, a library that provides capabilities for a flexible range of data visualizations in Python.

- Part V, "Machine Learning" focuses on the Scikit-Learn library, which provides efficient and clean Python implementations of the most important and established machine learning algorithms.

The PyData world is certainly much larger than these six packages, and it's growing every day. With this in mind, I make every attempt throughout this book to provide references to other interesting efforts, projects, and packages that are pushing the boundaries of what can be done in Python. Nevertheless, the packages I concentrate on are currently fundamental to much of the work being done in the Python data science space, and I expect they will remain important even as the ecosystem continues growing around them.

Installation Considerations

Installing Python and the suite of libraries that enable scientific computing is straightforward. This section will outline some of the things to keep in mind when setting up your computer.

Though there are various ways to install Python, the one I would suggest for use in data science is the Anaconda distribution, which works similarly whether you use Windows, Linux, or macOS. The Anaconda distribution comes in two flavors:

- Miniconda (*https://oreil.ly/dH7wJ*) gives you the Python interpreter itself, along with a command-line tool called *conda* which operates as a cross-platform package manager geared toward Python packages, similar in spirit to the apt or yum tools that Linux users might be familiar with.

- Anaconda (*https://oreil.ly/ndxjm*) includes both Python and conda, and additionally bundles a suite of other preinstalled packages geared toward scientific computing. Because of the size of this bundle, expect the installation to consume several gigabytes of disk space.

Any of the packages included with Anaconda can also be installed manually on top of Miniconda; for this reason I suggest starting with Miniconda.

To get started, download and install the Miniconda package—make sure to choose a version with Python 3—and then install the core packages used in this book:

```
[~]$ conda install numpy pandas scikit-learn matplotlib seaborn jupyter
```

Throughout the text, we will also make use of other more specialized tools in Python's scientific ecosystem; installation is usually as easy as typing **conda install** *package name*. If you ever come across packages that are not available in the default conda channel, be sure to check out *conda-forge* (*https://oreil.ly/CCvwQ*), a broad, community-driven repository of conda packages.

For more information on conda, including information about creating and using conda environments (which I would *highly* recommend), refer to conda's online documentation (*https://oreil.ly/MkqPw*).

Conventions Used in This Book

The following typographical conventions are used in this book:

Italic

Indicates new terms, URLs, email addresses, filenames, and file extensions.

`Constant width`

Used for program listings, as well as within paragraphs to refer to program elements such as variable or function names, databases, data types, environment variables, statements, and keywords.

`Constant width bold`

Shows commands or other text that should be typed literally by the user.

`Constant width italic`

Shows text that should be replaced with user-supplied values or by values determined by context.

 This element signifies a general note.

Using Code Examples

Supplemental material (code examples, figures, etc.) is available for download at *http://github.com/jakevdp/PythonDataScienceHandbook*.

If you have a technical question or a problem using the code examples, please send email to *bookquestions@oreilly.com*.

This book is here to help you get your job done. In general, if example code is offered with this book, you may use it in your programs and documentation. You do not need to contact us for permission unless you're reproducing a significant portion of the code. For example, writing a program that uses several chunks of code from this book does not require permission. Selling or distributing examples from O'Reilly books does require permission. Answering a question by citing this book and quoting example code does not require permission. Incorporating a significant amount of example code from this book into your product's documentation does require permission.

We appreciate, but generally do not require, attribution. An attribution usually includes the title, author, publisher, and ISBN. For example: *"Python Data Science*

Handbook, 2nd edition, by Jake VanderPlas (O'Reilly). Copyright 2023 Jake Vander-Plas, 978-1-098-12122-8."

If you feel your use of code examples falls outside fair use or the permission given above, feel free to contact us at *permissions@oreilly.com*.

O'Reilly Online Learning

 For more than 40 years, *O'Reilly Media* has provided technology and business training, knowledge, and insight to help companies succeed.

Our unique network of experts and innovators share their knowledge and expertise through books, articles, and our online learning platform. O'Reilly's online learning platform gives you on-demand access to live training courses, in-depth learning paths, interactive coding environments, and a vast collection of text and video from O'Reilly and 200+ other publishers. For more information, visit *https://oreilly.com*.

How to Contact Us

Please address comments and questions concerning this book to the publisher:

O'Reilly Media, Inc.
1005 Gravenstein Highway North
Sebastopol, CA 95472
800-998-9938 (in the United States or Canada)
707-829-0515 (international or local)
707-829-0104 (fax)

We have a web page for this book, where we list errata, examples, and any additional information. You can access this page at *https://oreil.ly/python-data-science-handbook*.

Email *bookquestions@oreilly.com* to comment or ask technical questions about this book.

For news and information about our books and courses, visit *https://oreilly.com*.

Find us on LinkedIn: *https://linkedin.com/company/oreilly-media*.

Follow us on Twitter: *https://twitter.com/oreillymedia*.

Watch us on YouTube: *https://youtube.com/oreillymedia*.

## PART I

Jupyter: Beyond Normal Python

There are many options for development environments for Python, and I'm often asked which one I use in my own work. My answer sometimes surprises people: my preferred environment is IPython (*http://ipython.org*) plus a text editor (in my case, Emacs or VSCode depending on my mood). Jupyter got its start as the IPython shell, which was created in 2001 by Fernando Perez as an enhanced Python interpreter and has since grown into a project aiming to provide, in Perez's words, "Tools for the entire life cycle of research computing." If Python is the engine of our data science task, you might think of Jupyter as the interactive control panel.

As well as being a useful interactive interface to Python, Jupyter also provides a number of useful syntactic additions to the language; we'll cover the most useful of these additions here. Perhaps the most familiar interface provided by the Jupyter project is the Jupyter Notebook, a browser-based environment that is useful for development, collaboration, sharing, and even publication of data science results. As an example of the usefulness of the notebook format, look no further than the page you are reading: the entire manuscript for this book was composed as a set of Jupyter notebooks.

This part of the book will start by stepping through some of the Jupyter and IPython features that are useful to the practice of data science, focusing especially on the syntax they offer beyond the standard features of Python. Next, we will go into a bit more depth on some of the more useful *magic commands* that can speed up common tasks in creating and using data science code. Finally, we will touch on some of the features of the notebook that make it useful for understanding data and sharing results.

## CHAPTER 1

Getting Started in IPython and Jupyter

In writing Python code for data science, I generally go between three modes of working: I use the IPython shell for trying out short sequences of commands, the Jupyter Notebook for longer interactive analysis and for sharing content with others, and interactive development environments (IDEs) like Emacs or VSCode for creating reusable Python packages. This chapter focuses on the first two modes: the IPython shell and the Jupyter Notebook. Use of an IDE for software development is an important third tool in the data scientist's repertoire, but we will not directly address that here.

Launching the IPython Shell

The text in this part, like most of this book, is not designed to be absorbed passively. I recommend that as you read through it, you follow along and experiment with the tools and syntax we cover: the muscle memory you build through doing this will be far more useful than the simple act of reading about it. Start by launching the IPython interpreter by typing **ipython** on the command line; alternatively, if you've installed a distribution like Anaconda or EPD, there may be a launcher specific to your system.

Once you do this, you should see a prompt like the following:

```
Python 3.9.2 (v3.9.2:1a79785e3e, Feb 19 2021, 09:06:10)
Type 'copyright', 'credits' or 'license' for more information
IPython 7.21.0 -- An enhanced Interactive Python. Type '?' for help.

In [1]:
```

With that, you're ready to follow along.

Launching the Jupyter Notebook

The Jupyter Notebook is a browser-based graphical interface to the IPython shell, and builds on it a rich set of dynamic display capabilities. As well as executing Python/IPython statements, notebooks allow the user to include formatted text, static and dynamic visualizations, mathematical equations, JavaScript widgets, and much more. Furthermore, these documents can be saved in a way that lets other people open them and execute the code on their own systems.

Though you'll view and edit Jupyter notebooks through your web browser window, they must connect to a running Python process in order to execute code. You can start this process (known as a "kernel") by running the following command in your system shell:

```
$ jupyter lab
```

This command launches a local web server that will be visible to your browser. It immediately spits out a log showing what it is doing; that log will look something like this:

```
$ jupyter lab
[ServerApp] Serving notebooks from local directory: /Users/jakevdp/ \
PythonDataScienceHandbook
[ServerApp] Jupyter Server 1.4.1 is running at:
[ServerApp] http://localhost:8888/lab?token=dd852649
[ServerApp] Use Control-C to stop this server and shut down all kernels
(twice to skip confirmation).
```

Upon issuing the command, your default browser should automatically open and navigate to the listed local URL; the exact address will depend on your system. If the browser does not open automatically, you can open a window and manually open this address (*http://localhost:8888/lab/* in this example).

Help and Documentation in IPython

If you read no other section in this chapter, read this one: I find the tools discussed here to be the most transformative contributions of IPython to my daily workflow.

When a technologically minded person is asked to help a friend, family member, or colleague with a computer problem, most of the time it's less a matter of knowing the answer than of knowing how to quickly find an unknown answer. In data science it's the same: searchable web resources such as online documentation, mailing list threads, and Stack Overflow answers contain a wealth of information, even (especially?) about topics you've found yourself searching on before. Being an effective practitioner of data science is less about memorizing the tool or command you should use for every possible situation, and more about learning to effectively find the information you don't know, whether through a web search engine or another means.

One of the most useful functions of IPython/Jupyter is to shorten the gap between the user and the type of documentation and search that will help them do their work effectively. While web searches still play a role in answering complicated questions, an amazing amount of information can be found through IPython alone. Some examples of the questions IPython can help answer in a few keystrokes include:

- How do I call this function? What arguments and options does it have?
- What does the source code of this Python object look like?
- What is in this package I imported?
- What attributes or methods does this object have?

Here we'll discuss the tools provided in the IPython shell and Jupyter Notebook to quickly access this information, namely the ? character to explore documentation, the ?? characters to explore source code, and the Tab key for autocompletion.

Accessing Documentation with ?

The Python language and its data science ecosystem are built with the user in mind, and one big part of that is access to documentation. Every Python object contains a reference to a string, known as a *docstring*, which in most cases will contain a concise summary of the object and how to use it. Python has a built-in help function that can access this information and print the results. For example, to see the documentation of the built-in len function, you can do the following:

```
In [1]: help(len)
Help on built-in function len in module builtins:

len(obj, /)
    Return the number of items in a container.
```

Depending on your interpreter, this information may be displayed as inline text or in a separate pop-up window.

Because finding help on an object is so common and useful, IPython and Jupyter introduce the ? character as a shorthand for accessing this documentation and other relevant information:

```
In [2]: len?
Signature: len(obj, /)
Docstring: Return the number of items in a container.
Type:      builtin_function_or_method
```

This notation works for just about anything, including object methods:

```
In [3]: L = [1, 2, 3]
In [4]: L.insert?
Signature: L.insert(index, object, /)
Docstring: Insert object before index.
Type:      builtin_function_or_method
```

or even objects themselves, with the documentation from their type:

```
In [5]: L?
Type:        list
String form: [1, 2, 3]
Length:      3
Docstring:
Built-in mutable sequence.

If no argument is given, the constructor creates a new empty list.
The argument must be an iterable if specified.
```

Importantly, this will even work for functions or other objects you create yourself! Here we'll define a small function with a docstring:

```
In [6]: def square(a):
   ....:     """Return the square of a."""
   ....:     return a ** 2
   ....:
```

Note that to create a docstring for our function, we simply placed a string literal in the first line. Because docstrings are usually multiple lines, by convention we used Python's triple-quote notation for multiline strings.

Now we'll use the ? to find this docstring:

```
In [7]: square?
Signature: square(a)
Docstring: Return the square of a.
File:      <ipython-input-6>
Type:      function
```

This quick access to documentation via docstrings is one reason you should get in the habit of always adding such inline documentation to the code you write.

Accessing Source Code with ??

Because the Python language is so easily readable, another level of insight can usually be gained by reading the source code of the object you're curious about. IPython and Jupyter provide a shortcut to the source code with the double question mark (??):

```
In [8]: square??
Signature: square(a)
Source:
def square(a):
```

```
    """Return the square of a."""
    return a ** 2
File:      <ipython-input-6>
Type:      function
```

For simple functions like this, the double question mark can give quick insight into the under-the-hood details.

If you play with this much, you'll notice that sometimes the ?? suffix doesn't display any source code: this is generally because the object in question is not implemented in Python, but in C or some other compiled extension language. If this is the case, the ?? suffix gives the same output as the ? suffix. You'll find this particularly with many of Python's built-in objects and types, including the len function from earlier:

```
In [9]: len??
Signature: len(obj, /)
Docstring: Return the number of items in a container.
Type:      builtin_function_or_method
```

Using ? and/or ?? is a powerful and quick way of finding information about what any Python function or module does.

Exploring Modules with Tab Completion

Another useful interface is the use of the Tab key for autocompletion and exploration of the contents of objects, modules, and namespaces. In the examples that follow, I'll use <TAB> to indicate when the Tab key should be pressed.

Tab completion of object contents

Every Python object has various attributes and methods associated with it. Like the help function mentioned earlier, Python has a built-in dir function that returns a list of these, but the tab-completion interface is much easier to use in practice. To see a list of all available attributes of an object, you can type the name of the object followed by a period (.) character and the Tab key:

```
In [10]: L.<TAB>
            append() count    insert   reverse
            clear    extend   pop      sort
            copy     index    remove
```

To narrow down the list, you can type the first character or several characters of the name, and the Tab key will find the matching attributes and methods:

```
In [10]: L.c<TAB>
            clear() count()
            copy()

In [10]: L.co<TAB>
             copy()  count()
```

If there is only a single option, pressing the Tab key will complete the line for you. For example, the following will instantly be replaced with L.count:

```
In [10]: L.cou<TAB>
```

Though Python has no strictly enforced distinction between public/external attributes and private/internal attributes, by convention a preceding underscore is used to denote the latter. For clarity, these private methods and special methods are omitted from the list by default, but it's possible to list them by explicitly typing the underscore:

```
In [10]: L._<TAB>
         __add__              __delattr__      __eq__
         __class__            __delitem__      __format__()
         __class_getitem__() __dir__()        __ge__                >
         __contains__         __doc__          __getattribute__
```

For brevity, I've only shown the first few columns of the output. Most of these are Python's special double-underscore methods (often nicknamed "dunder" methods).

Tab completion when importing

Tab completion is also useful when importing objects from packages. Here we'll use it to find all possible imports in the itertools package that start with co:

```
In [10]: from itertools import co<TAB>
         combinations()                        compress()
         combinations_with_replacement() count()
```

Similarly, you can use tab-completion to see which imports are available on your system (this will change depending on which third-party scripts and modules are visible to your Python session):

```
In [10]: import <TAB>
         abc                  anyio
         activate_this        appdirs
         aifc                 appnope          >
         antigravity          argon2

In [10]: import h<TAB>
         hashlib html
         heapq   http
         hmac
```

Beyond tab completion: Wildcard matching

Tab completion is useful if you know the first few characters of the name of the object or attribute you're looking for, but is little help if you'd like to match characters in the middle or at the end of the name. For this use case, IPython and Jupyter provide a means of wildcard matching for names using the * character.

For example, we can use this to list every object in the namespace whose name ends with `Warning`:

```
In [10]: *Warning?
BytesWarning                  RuntimeWarning
DeprecationWarning            SyntaxWarning
FutureWarning                 UnicodeWarning
ImportWarning                 UserWarning
PendingDeprecationWarning     Warning
ResourceWarning
```

Notice that the * character matches any string, including the empty string.

Similarly, suppose we are looking for a string method that contains the word `find` somewhere in its name. We can search for it this way:

```
In [11]: str.*find*?
str.find
str.rfind
```

I find this type of flexible wildcard search can be useful for finding a particular command when getting to know a new package or reacquainting myself with a familiar one.

Keyboard Shortcuts in the IPython Shell

If you spend any amount of time on a computer, you've probably found a use for keyboard shortcuts in your workflow. Most familiar perhaps are Cmd-c and Cmd-v (or Ctrl-c and Ctrl-v), used for copying and pasting in a wide variety of programs and systems. Power users tend to go even further: popular text editors like Emacs, Vim, and others provide users an incredible range of operations through intricate combinations of keystrokes.

The IPython shell doesn't go this far, but does provide a number of keyboard shortcuts for fast navigation while typing commands. While some of these shortcuts do work in the browser-based notebooks, this section is primarily about shortcuts in the IPython shell.

Once you get accustomed to these, they can be very useful for quickly performing certain commands without moving your hands from the "home" keyboard position. If you're an Emacs user or if you have experience with Linux-style shells, the following will be very familiar. I'll group these shortcuts into a few categories: *navigation shortcuts*, *text entry shortcuts*, *command history shortcuts*, and *miscellaneous shortcuts*.

Navigation Shortcuts

While the use of the left and right arrow keys to move backward and forward in the line is quite obvious, there are other options that don't require moving your hands from the "home" keyboard position:

Keystroke	Action
Ctrl-a	Move cursor to beginning of line
Ctrl-e	Move cursor to end of the line
Ctrl-b or the left arrow key	Move cursor back one character
Ctrl-f or the right arrow key	Move cursor forward one character

Text Entry Shortcuts

While everyone is familiar with using the Backspace key to delete the previous character, reaching for the key often requires some minor finger gymnastics, and it only deletes a single character at a time. In IPython there are several shortcuts for removing some portion of the text you're typing; the most immediately useful of these are the commands to delete entire lines of text. You'll know these have become second-nature if you find yourself using a combination of Ctrl-b and Ctrl-d instead of reaching for Backspace to delete the previous character!

Keystroke	Action
Backspace key	Delete previous character in line
Ctrl-d	Delete next character in line
Ctrl-k	Cut text from cursor to end of line
Ctrl-u	Cut text from beginning of line to cursor
Ctrl-y	Yank (i.e., paste) text that was previously cut
Ctrl-t	Transpose (i.e., switch) previous two characters

Command History Shortcuts

Perhaps the most impactful shortcuts discussed here are the ones IPython provides for navigating the command history. This command history goes beyond your current IPython session: your entire command history is stored in a SQLite database in your IPython profile directory.

The most straightforward way to access previous commands is by using the up and down arrow keys to step through the history, but other options exist as well:

Keystroke	Action
Ctrl-p (or the up arrow key)	Access previous command in history
Ctrl-n (or the down arrow key)	Access next command in history
Ctrl-r	Reverse-search through command history

The reverse-search option can be particularly useful. Recall that earlier we defined a function called `square`. Let's reverse-search our Python history from a new IPython shell and find this definition again. When you press Ctrl-r in the IPython terminal, you'll see the following prompt:

```
In [1]:
(reverse-i-search)`':
```

If you start typing characters at this prompt, IPython will autofill the most recent command, if any, that matches those characters:

```
In [1]:
(reverse-i-search)`sqa': square??
```

At any point, you can add more characters to refine the search, or press Ctrl-r again to search further for another command that matches the query. If you followed along earlier, pressing Ctrl-r twice more gives:

```
In [1]:
(reverse-i-search)`sqa': def square(a):
    """Return the square of a"""
    return a ** 2
```

Once you have found the command you're looking for, press Return and the search will end. You can then use the retrieved command and carry on with your session:

```
In [1]: def square(a):
    """Return the square of a"""
    return a ** 2

In [2]: square(2)
Out[2]: 4
```

Note that you can use Ctrl-p/Ctrl-n or the up/down arrow keys to search through your history in a similar way, but only by matching characters at the beginning of the line. That is, if you type **def** and then press Ctrl-p, it will find the most recent command (if any) in your history that begins with the characters def.

Miscellaneous Shortcuts

Finally, there are a few miscellaneous shortcuts that don't fit into any of the preceding categories, but are nevertheless useful to know:

Keystroke	Action
Ctrl-l	Clear terminal screen
Ctrl-c	Interrupt current Python command
Ctrl-d	Exit IPython session

The Ctrl-c shortcut in particular can be useful when you inadvertently start a very long-running job.

While some of the shortcuts discussed here may seem a bit obscure at first, they quickly become automatic with practice. Once you develop that muscle memory, I suspect you will even find yourself wishing they were available in other contexts.

## CHAPTER 2

Enhanced Interactive Features

Much of the power of IPython and Jupyter comes from the additional interactive tools they make available. This chapter will cover a number of those tools, including so-called magic commands, tools for exploring input and output history, and tools to interact with the shell.

IPython Magic Commands

The previous chapter showed how IPython lets you use and explore Python efficiently and interactively. Here we'll begin discussing some of the enhancements that IPython adds on top of the normal Python syntax. These are known in IPython as *magic commands*, and are prefixed by the % character. These magic commands are designed to succinctly solve various common problems in standard data analysis. Magic commands come in two flavors: *line magics*, which are denoted by a single % prefix and operate on a single line of input, and *cell magics*, which are denoted by a double %% prefix and operate on multiple lines of input. I'll demonstrate and discuss a few brief examples here, and come back to a more focused discussion of several useful magic commands later.

Running External Code: %run

As you begin developing more extensive code, you will likely find yourself working in IPython for interactive exploration, as well as a text editor to store code that you want to reuse. Rather than running this code in a new window, it can be convenient to run it within your IPython session. This can be done with the %run magic command.

For example, imagine you've created a *myscript.py* file with the following contents:

```
# file: myscript.py

def square(x):
    """square a number"""
    return x ** 2

for N in range(1, 4):
    print(f"{N} squared is {square(N)}")
```

You can execute this from your IPython session as follows:

```
In [1]: %run myscript.py
1 squared is 1
2 squared is 4
3 squared is 9
```

Note also that after you've run this script, any functions defined within it are available for use in your IPython session:

```
In [2]: square(5)
Out[2]: 25
```

There are several options to fine-tune how your code is run; you can see the documentation in the normal way, by typing **%run?** in the IPython interpreter.

Timing Code Execution: %timeit

Another example of a useful magic function is `%timeit`, which will automatically determine the execution time of the single-line Python statement that follows it. For example, we may want to check the performance of a list comprehension:

```
In [3]: %timeit L = [n ** 2 for n in range(1000)]
430 µs ± 3.21 µs per loop (mean ± std. dev. of 7 runs, 1000 loops each)
```

The benefit of `%timeit` is that for short commands it will automatically perform multiple runs in order to attain more robust results. For multiline statements, adding a second % sign will turn this into a cell magic that can handle multiple lines of input. For example, here's the equivalent construction with a `for` loop:

```
In [4]: %%timeit
   ...: L = []
   ...: for n in range(1000):
   ...:     L.append(n ** 2)
   ...:
484 µs ± 5.67 µs per loop (mean ± std. dev. of 7 runs, 1000 loops each)
```

We can immediately see that list comprehensions are about 10% faster than the equivalent `for` loop construction in this case. We'll explore `%timeit` and other approaches to timing and profiling code in "Profiling and Timing Code" on page 26.

Help on Magic Functions: ?, %magic, and %lsmagic

Like normal Python functions, IPython magic functions have docstrings, and this useful documentation can be accessed in the standard manner. So, for example, to read the documentation of the `%timeit` magic function, simply type this:

```
In [5]: %timeit?
```

Documentation for other functions can be accessed similarly. To access a general description of available magic functions, including some examples, you can type this:

```
In [6]: %magic
```

For a quick and simple list of all available magic functions, type this:

```
In [7]: %lsmagic
```

Finally, I'll mention that it is quite straightforward to define your own magic functions if you wish. I won't discuss it here, but if you are interested, see the references listed in "More IPython Resources" on page 31.

Input and Output History

Previously you saw that the IPython shell allows you to access previous commands with the up and down arrow keys, or equivalently the Ctrl-p/Ctrl-n shortcuts. Additionally, in both the shell and notebooks, IPython exposes several ways to obtain the output of previous commands, as well as string versions of the commands themselves. We'll explore those here.

IPython's In and Out Objects

By now I imagine you're becoming familiar with the `In [1]:/Out[1]:` style of prompts used by IPython. But it turns out that these are not just pretty decoration: they give a clue as to how you can access previous inputs and outputs in your current session. Suppose we start a session that looks like this:

```
In [1]: import math

In [2]: math.sin(2)
Out[2]: 0.9092974268256817

In [3]: math.cos(2)
Out[3]: -0.4161468365471424
```

We've imported the built-in `math` package, then computed the sine and the cosine of the number 2. These inputs and outputs are displayed in the shell with In/Out labels, but there's more—IPython actually creates some Python variables called `In` and `Out` that are automatically updated to reflect this history:

```
In [4]: In
Out[4]: ['', 'import math', 'math.sin(2)', 'math.cos(2)', 'In']

In [5]: Out
Out[5]:
{2: 0.9092974268256817,
 3: -0.4161468365471424,
 4: ['', 'import math', 'math.sin(2)', 'math.cos(2)', 'In', 'Out']}
```

The In object is a list, which keeps track of the commands in order (the first item in the list is a placeholder so that In [1] can refer to the first command):

```
In [6]: print(In[1])
import math
```

The Out object is not a list but a dictionary mapping input numbers to their outputs (if any):

```
In [7]: print(Out[2])
.9092974268256817
```

Note that not all operations have outputs: for example, import statements and print statements don't affect the output. The latter may be surprising, but makes sense if you consider that print is a function that returns None; for brevity, any command that returns None is not added to Out.

This can be useful when you want to interact with past results. For example, let's check the sum of sin(2) ** 2 and cos(2) ** 2 using the previously computed results:

```
In [8]: Out[2] ** 2 + Out[3] ** 2
Out[8]: 1.0
```

The result is 1.0, as we'd expect from the well-known trigonometric identity. In this case, using these previous results probably is not necessary, but it can become quite handy if you execute a very expensive computation and forget to assign the result to a variable.

Underscore Shortcuts and Previous Outputs

The standard Python shell contains just one simple shortcut for accessing previous output: the variable _ (i.e., a single underscore) is kept updated with the previous output. This works in IPython as well:

```
In [9]: print(_)
.0
```

But IPython takes this a bit further—you can use a double underscore to access the second-to-last output, and a triple underscore to access the third-to-last output (skipping any commands with no output):

```
In [10]: print(__)
-0.4161468365471424

In [11]: print(___)
.9092974268256817
```

IPython stops there: more than three underscores starts to get a bit hard to count, and at that point it's easier to refer to the output by line number.

There is one more shortcut I should mention, however—a shorthand for Out[X] is $\_X$ (i.e., a single underscore followed by the line number):

```
In [12]: Out[2]
Out[12]: 0.9092974268256817

In [13]: _2
Out[13]: 0.9092974268256817
```

Suppressing Output

Sometimes you might wish to suppress the output of a statement (this is perhaps most common with the plotting commands that we'll explore in Part IV). Or maybe the command you're executing produces a result that you'd prefer not to store in your output history, perhaps so that it can be deallocated when other references are removed. The easiest way to suppress the output of a command is to add a semicolon to the end of the line:

```
In [14]: math.sin(2) + math.cos(2);
```

The result is computed silently, and the output is neither displayed on the screen nor stored in the Out dictionary:

```
In [15]: 14 in Out
Out[15]: False
```

Related Magic Commands

For accessing a batch of previous inputs at once, the %history magic command is very helpful. Here is how you can print the first four inputs:

```
In [16]: %history -n 1-3
   1: import math
   2: math.sin(2)
   3: math.cos(2)
```

As usual, you can type %history? for more information and a description of options available (see Chapter 1 for details on the ? functionality). Other useful magic commands are %rerun, which will re-execute some portion of the command history, and %save, which saves some set of the command history to a file).

IPython and Shell Commands

When working interactively with the standard Python interpreter, one of the frustrations is the need to switch between multiple windows to access Python tools and system command-line tools. IPython bridges this gap and gives you a syntax for executing shell commands directly from within the IPython terminal. The magic happens with the exclamation point: anything appearing after ! on a line will be executed not by the Python kernel, but by the system command line.

The following discussion assumes you're on a Unix-like system, such as Linux or macOS. Some of the examples that follow will fail on Windows, which uses a different type of shell by default, though if you use the *Windows Subsystem for Linux* (*https://oreil.ly/H5MEE*) the examples here should run correctly. If you're unfamiliar with shell commands, I'd suggest reviewing the Unix shell tutorial (*https://oreil.ly/RrD2Y*) put together by the always excellent Software Carpentry Foundation.

Quick Introduction to the Shell

A full introduction to using the shell/terminal/command line is well beyond the scope of this chapter, but for the uninitiated I will offer a quick introduction here. The shell is a way to interact textually with your computer. Ever since the mid-1980s, when Microsoft and Apple introduced the first versions of their now ubiquitous graphical operating systems, most computer users have interacted with their operating systems through the familiar menu selections and drag-and-drop movements. But operating systems existed long before these graphical user interfaces, and were primarily controlled through sequences of text input: at the prompt, the user would type a command, and the computer would do what the user told it to. Those early prompt systems were the precursors of the shells and terminals that most data scientists still use today.

Someone unfamiliar with the shell might ask why you would bother with this, when many of the same results can be accomplished by simply clicking on icons and menus. A shell user might reply with another question: why hunt for icons and menu items when you can accomplish things much more easily by typing? While it might sound like a typical tech preference impasse, when moving beyond basic tasks it quickly becomes clear that the shell offers much more control of advanced tasks— though admittedly the learning curve can be intimidating.

As an example, here is a sample of a Linux/macOS shell session where a user explores, creates, and modifies directories and files on their system (osx:~ $ is the

prompt, and everything after the $ is the typed command; text that is preceded by a #
is meant just as description, rather than something you would actually type in):

```
osx:~ $ echo "hello world"          # echo is like Python's print function
hello world

osx:~ $ pwd                         # pwd = print working directory
/home/jake                          # This is the "path" that we're sitting in

osx:~ $ ls                          # ls = list working directory contents
notebooks   projects

osx:~ $ cd projects/                # cd = change directory

osx:projects $ pwd
/home/jake/projects

osx:projects $ ls
datasci_book    mpld3    myproject.txt

osx:projects $ mkdir myproject      # mkdir = make new directory

osx:projects $ cd myproject/

osx:myproject $ mv ../myproject.txt ./  # mv = move file. Here we're moving the
                                    # file myproject.txt from one directory
                                    # up (../) to the current directory (./).
osx:myproject $ ls
myproject.txt
```

Notice that all of this is just a compact way to do familiar operations (navigating a
directory structure, creating a directory, moving a file, etc.) by typing commands
rather than clicking icons and menus. With just a few commands (pwd, ls, cd, mkdir,
and cp) you can do many of the most common file operations, but it's when you go
beyond these basics that the shell approach becomes really powerful.

Shell Commands in IPython

Any standard shell command can be used directly in IPython by prefixing it with
the ! character. For example, the ls, pwd, and echo commands can be run as follows:

```
In [1]: !ls
myproject.txt

In [2]: !pwd
/home/jake/projects/myproject

In [3]: !echo "printing from the shell"
printing from the shell
```

Passing Values to and from the Shell

Shell commands can not only be called from IPython, but can also be made to interact with the IPython namespace. For example, you can save the output of any shell command to a Python list using the assignment operator, =:

```
In [4]: contents = !ls
```

```
In [5]: print(contents)
['myproject.txt']
```

```
In [6]: directory = !pwd
```

```
In [7]: print(directory)
['/Users/jakevdp/notebooks/tmp/myproject']
```

These results are not returned as lists, but as a special shell return type defined in IPython:

```
In [8]: type(directory)
IPython.utils.text.SList
```

This looks and acts a lot like a Python list but has additional functionality, such as the grep and fields methods and the s, n, and p properties that allow you to search, filter, and display the results in convenient ways. For more information on these, you can use IPython's built-in help features.

Communication in the other direction—passing Python variables into the shell—is possible using the {*varname*} syntax:

```
In [9]: message = "hello from Python"
```

```
In [10]: !echo {message}
hello from Python
```

The curly braces contain the variable name, which is replaced by the variable's contents in the shell command.

Shell-Related Magic Commands

If you play with IPython's shell commands for a while, you might notice that you cannot use !cd to navigate the filesystem:

```
In [11]: !pwd
/home/jake/projects/myproject
```

```
In [12]: !cd ..
```

```
In [13]: !pwd
/home/jake/projects/myproject
```

The reason is that shell commands in the notebook are executed in a temporary sub-shell that does not maintain state from command to command. If you'd like to change the working directory in a more enduring way, you can use the %cd magic command:

```
In [14]: %cd ..
/home/jake/projects
```

In fact, by default you can even use this without the % sign:

```
In [15]: cd myproject
/home/jake/projects/myproject
```

This is known as an *automagic* function, and the ability to execute such commands without an explicit % can be toggled with the %automagic magic function.

Besides %cd, other available shell-like magic functions are %cat, %cp, %env, %ls, %man, %mkdir, %more, %mv, %pwd, %rm, and %rmdir, any of which can be used without the % sign if automagic is on. This makes it so that you can almost treat the IPython prompt as if it's a normal shell:

```
In [16]: mkdir tmp

In [17]: ls
myproject.txt  tmp/

In [18]: cp myproject.txt tmp/

In [19]: ls tmp
myproject.txt

In [20]: rm -r tmp
```

This access to the shell from within the same terminal window as your Python session lets you more naturally combine Python and the shell in your workflows with fewer context switches.

## CHAPTER 3

Debugging and Profiling

In addition to the enhanced interactive tools discussed in the previous chapter, Jupyter provides a number of ways to explore and understand the code you are running, such as by tracking down bugs in the logic or unexpected slow execution. This chapter will discuss some of these tools.

Errors and Debugging

Code development and data analysis always require a bit of trial and error, and IPython contains tools to streamline this process. This section will briefly cover some options for controlling Python's exception reporting, followed by exploring tools for debugging errors in code.

Controlling Exceptions: %xmode

Most of the time when a Python script fails, it will raise an exception. When the interpreter hits one of these exceptions, information about the cause of the error can be found in the *traceback*, which can be accessed from within Python. With the %xmode magic function, IPython allows you to control the amount of information printed when the exception is raised. Consider the following code:

```
In [1]: def func1(a, b):
            return a / b

        def func2(x):
            a = x
            b = x - 1
            return func1(a, b)

In [2]: func2(1)
ZeroDivisionError                       Traceback (most recent call last)
```

```
<ipython-input-2-b2e110f6fc8f> in <module>()
----> 1 func2(1)

<ipython-input-1-d849e34d61fb> in func2(x)
      5         a = x
      6         b = x - 1
----> 7         return func1(a, b)

<ipython-input-1-d849e34d61fb> in func1(a, b)
      1 def func1(a, b):
----> 2         return a / b
      3
      4 def func2(x):
      5         a = x

ZeroDivisionError: division by zero
```

Calling func2 results in an error, and reading the printed trace lets us see exactly what happened. In the default mode, this trace includes several lines showing the context of each step that led to the error. Using the %xmode magic function (short for *exception mode*), we can change what information is printed.

%xmode takes a single argument, the mode, and there are three possibilities: Plain, Context, and Verbose. The default is Context, which gives output like that just shown. Plain is more compact and gives less information:

```
In [3]: %xmode Plain
Out[3]: Exception reporting mode: Plain

In [4]: func2(1)
Traceback (most recent call last):

  File "<ipython-input-4-b2e110f6fc8f>", line 1, in <module>
    func2(1)

  File "<ipython-input-1-d849e34d61fb>", line 7, in func2
    return func1(a, b)

  File "<ipython-input-1-d849e34d61fb>", line 2, in func1
    return a / b

ZeroDivisionError: division by zero
```

The Verbose mode adds some extra information, including the arguments to any functions that are called:

```
In [5]: %xmode Verbose
Out[5]: Exception reporting mode: Verbose

In [6]: func2(1)
ZeroDivisionError                         Traceback (most recent call last)
<ipython-input-6-b2e110f6fc8f> in <module>()
```

```
----> 1 func2(1)
        global func2 = <function func2 at 0x103729320>

<ipython-input-1-d849e34d61fb> in func2(x=1)
      5       a = x
      6       b = x - 1
----> 7       return func1(a, b)
        global func1 = <function func1 at 0x1037294d0>
        a = 1
        b = 0

<ipython-input-1-d849e34d61fb> in func1(a=1, b=0)
      1 def func1(a, b):
----> 2       return a / b
        a = 1
        b = 0
      3
      4 def func2(x):
      5       a = x

ZeroDivisionError: division by zero
```

This extra information can help you narrow in on why the exception is being raised. So why not use the Verbose mode all the time? As code gets complicated, this kind of traceback can get extremely long. Depending on the context, sometimes the brevity of Plain or Context mode is easier to work with.

Debugging: When Reading Tracebacks Is Not Enough

The standard Python tool for interactive debugging is pdb, the Python debugger. This debugger lets the user step through the code line by line in order to see what might be causing a more difficult error. The IPython-enhanced version of this is ipdb, the IPython debugger.

There are many ways to launch and use both these debuggers; we won't cover them fully here. Refer to the online documentation of these two utilities to learn more.

In IPython, perhaps the most convenient interface to debugging is the %debug magic command. If you call it after hitting an exception, it will automatically open an interactive debugging prompt at the point of the exception. The ipdb prompt lets you explore the current state of the stack, explore the available variables, and even run Python commands!

Let's look at the most recent exception, then do some basic tasks. We'll print the values of a and b, then type quit to quit the debugging session:

```
In [7]: %debug <ipython-input-1-d849e34d61fb>(2)func1()
      1 def func1(a, b):
----> 2       return a / b
      3
```

```
ipdb> print(a)
1
ipdb> print(b)
0
ipdb> quit
```

The interactive debugger allows much more than this, though—we can even step up and down through the stack and explore the values of variables there:

```
In [8]: %debug <ipython-input-1-d849e34d61fb>(2)func1()
      1 def func1(a, b):
----> 2     return a / b
      3

ipdb> up <ipython-input-1-d849e34d61fb>(7)func2()
      5     a = x
      6     b = x - 1
----> 7     return func1(a, b)

ipdb> print(x)
1
ipdb> up <ipython-input-6-b2e110f6fc8f>(1)<module>()
----> 1 func2(1)

ipdb> down <ipython-input-1-d849e34d61fb>(7)func2()
      5     a = x
      6     b = x - 1
----> 7     return func1(a, b)

ipdb> quit
```

This allows us to quickly find out not only what caused the error, but what function calls led up to the error.

If you'd like the debugger to launch automatically whenever an exception is raised, you can use the %pdb magic function to turn on this automatic behavior:

```
In [9]: %xmode Plain
        %pdb on
        func2(1)
Exception reporting mode: Plain
Automatic pdb calling has been turned ON
ZeroDivisionError: division by zero <ipython-input-1-d849e34d61fb>(2)func1()
      1 def func1(a, b):
----> 2     return a / b
      3

ipdb> print(b)
0
ipdb> quit
```

Finally, if you have a script that you'd like to run from the beginning in interactive mode, you can run it with the command %run -d, and use the next command to step through the lines of code interactively.

There are many more available commands for interactive debugging than I've shown here. Table 3-1 contains a description of some of the more common and useful ones.

Table 3-1. Partial list of debugging commands

Command	Description
l(ist)	Show the current location in the file
h(elp)	Show a list of commands, or find help on a specific command
q(uit)	Quit the debugger and the program
c(ontinue)	Quit the debugger, continue in the program
n(ext)	Go to the next step of the program
<enter>	Repeat the previous command
p(rint)	Print variables
s(tep)	Step into a subroutine
r(eturn)	Return out of a subroutine

For more information, use the help command in the debugger, or take a look at ipdb's online documentation (*https://oreil.ly/TVSAT*).

Profiling and Timing Code

In the process of developing code and creating data processing pipelines, there are often trade-offs you can make between various implementations. Early in developing your algorithm, it can be counterproductive to worry about such things. As Donald Knuth famously quipped, "We should forget about small efficiencies, say about 97% of the time: premature optimization is the root of all evil."

But once you have your code working, it can be useful to dig into its efficiency a bit. Sometimes it's useful to check the execution time of a given command or set of commands; other times it's useful to examine a multiline process and determine where the bottleneck lies in some complicated series of operations. IPython provides access to a wide array of functionality for this kind of timing and profiling of code. Here we'll discuss the following IPython magic commands:

%time
    Time the execution of a single statement

%timeit
    Time repeated execution of a single statement for more accuracy

```
%prun
```
Run code with the profiler

```
%lprun
```
Run code with the line-by-line profiler

```
%memit
```
Measure the memory use of a single statement

```
%mprun
```
Run code with the line-by-line memory profiler

The last four commands are not bundled with IPython; to use them, you'll need to get the line_profiler and memory_profiler extensions, which we will discuss in the following sections.

Timing Code Snippets: %timeit and %time

We saw the %timeit line magic and %%timeit cell magic in the introduction to magic functions in "IPython Magic Commands" on page 13; these can be used to time the repeated execution of snippets of code:

```
In [1]: %timeit sum(range(100))
1.53 µs ± 47.8 ns per loop (mean ± std. dev. of 7 runs, 1000000 loops each)
```

Note that because this operation is so fast, %timeit automatically does a large number of repetitions. For slower commands, %timeit will automatically adjust and perform fewer repetitions:

```
In [2]: %%timeit
        total = 0
        for i in range(1000):
            for j in range(1000):
                total += i * (-1) ** j
536 ms ± 15.9 ms per loop (mean ± std. dev. of 7 runs, 1 loop each)
```

Sometimes repeating an operation is not the best option. For example, if we have a list that we'd like to sort, we might be misled by a repeated operation; sorting a presorted list is much faster than sorting an unsorted list, so the repetition will skew the result:

```
In [3]: import random
        L = [random.random() for i in range(100000)]
        %timeit L.sort()
Out[3]: 1.71 ms ± 334 µs per loop (mean ± std. dev. of 7 runs, 1000 loops each)
```

For this, the %time magic function may be a better choice. It also is a good choice for longer-running commands, when short, system-related delays are unlikely to affect the result. Let's time the sorting of an unsorted and a presorted list:

```
In [4]: import random
        L = [random.random() for i in range(100000)]
        print("sorting an unsorted list:")
        %time L.sort()
Out[4]: sorting an unsorted list:
        CPU times: user 31.3 ms, sys: 686 µs, total: 32 ms
        Wall time: 33.3 ms
In [5]: print("sorting an already sorted list:")
        %time L.sort()
Out[5]: sorting an already sorted list:
        CPU times: user 5.19 ms, sys: 268 µs, total: 5.46 ms
        Wall time: 14.1 ms
```

Notice how much faster the presorted list is to sort, but notice also how much longer the timing takes with %time versus %timeit, even for the presorted list! This is a result of the fact that %timeit does some clever things under the hood to prevent system calls from interfering with the timing. For example, it prevents cleanup of unused Python objects (known as *garbage collection*) that might otherwise affect the timing. For this reason, %timeit results are usually noticeably faster than %time results.

For %time, as with %timeit, using the %% cell magic syntax allows timing of multiline scripts:

```
In [6]: %%time
        total = 0
        for i in range(1000):
            for j in range(1000):
                total += i * (-1) ** j
CPU times: user 655 ms, sys: 5.68 ms, total: 661 ms
Wall time: 710 ms
```

For more information on %time and %timeit, as well as their available options, use the IPython help functionality (e.g., type %time? at the IPython prompt).

Profiling Full Scripts: %prun

A program is made up of many single statements, and sometimes timing these statements in context is more important than timing them on their own. Python contains a built-in code profiler (which you can read about in the Python documentation), but IPython offers a much more convenient way to use this profiler, in the form of the magic function %prun.

By way of example, we'll define a simple function that does some calculations:

```
In [7]: def sum_of_lists(N):
            total = 0
            for i in range(5):
                L = [j ^ (j >> i) for j in range(N)]
                total += sum(L)
            return total
```

Now we can call %prun with a function call to see the profiled results:

```
In [8]: %prun sum_of_lists(1000000)
14 function calls in 0.932 seconds

Ordered by: internal time

ncalls  tottime  percall  cumtime  percall filename:lineno(function)
     5    0.808    0.162    0.808    0.162 <ipython-input-7-f105717832a2>:4(<listcomp>)
     5    0.066    0.013    0.066    0.013 {built-in method builtins.sum}
     1    0.044    0.044    0.918    0.918 <ipython-input-7-f105717832a2>:1
   > (sum_of_lists)
     1    0.014    0.014    0.932    0.932 <string>:1(<module>)
     1    0.000    0.000    0.932    0.932 {built-in method builtins.exec}
     1    0.000    0.000    0.000    0.000 {method 'disable' of '_lsprof.Profiler'
   > objects}
```

The result is a table that indicates, in order of total time on each function call, where the execution is spending the most time. In this case, the bulk of the execution time is in the list comprehension inside sum_of_lists. From here, we could start thinking about what changes we might make to improve the performance of the algorithm.

For more information on %prun, as well as its available options, use the IPython help functionality (i.e., type %prun? at the IPython prompt).

Line-by-Line Profiling with %lprun

The function-by-function profiling of %prun is useful, but sometimes it's more convenient to have a line-by-line profile report. This is not built into Python or IPython, but there is a line_profiler package available for installation that can do this. Start by using Python's packaging tool, pip, to install the line_profiler package:

```
$ pip install line_profiler
```

Next, you can use IPython to load the line_profiler IPython extension, offered as part of this package:

```
In [9]: %load_ext line_profiler
```

Now the %lprun command will do a line-by-line profiling of any function. In this case, we need to tell it explicitly which functions we're interested in profiling:

```
In [10]: %lprun -f sum_of_lists sum_of_lists(5000)
Timer unit: 1e-06 s

Total time: 0.014803 s
File: <ipython-input-7-f105717832a2>
```

```
Function: sum_of_lists at line 1

Line #      Hits         Time  Per Hit   % Time  Line Contents
==============================================================
     1                                           def sum_of_lists(N):
     2         1          6.0      6.0      0.0       total = 0
     3         6         13.0      2.2      0.1       for i in range(5):
     4         5      14242.0   2848.4     96.2           L = [j ^ (j >> i) for j
     5         5        541.0    108.2      3.7           total += sum(L)
     6         1          1.0      1.0      0.0       return total
```

The information at the top gives us the key to reading the results: the time is reported in microseconds, and we can see where the program is spending the most time. At this point, we may be able to use this information to modify aspects of the script and make it perform better for our desired use case.

For more information on %lprun, as well as its available options, use the IPython help functionality (i.e., type %lprun? at the IPython prompt).

Profiling Memory Use: %memit and %mprun

Another aspect of profiling is the amount of memory an operation uses. This can be evaluated with another IPython extension, the memory_profiler. As with the line_profiler, we start by pip-installing the extension:

```
$ pip install memory_profiler
```

Then we can use IPython to load it:

```
In [11]: %load_ext memory_profiler
```

The memory profiler extension contains two useful magic functions: %memit (which offers a memory-measuring equivalent of %timeit) and %mprun (which offers a memory-measuring equivalent of %lprun). The %memit magic function can be used rather simply:

```
In [12]: %memit sum_of_lists(1000000)
peak memory: 141.70 MiB, increment: 75.65 MiB
```

We see that this function uses about 140 MB of memory.

For a line-by-line description of memory use, we can use the %mprun magic function. Unfortunately, this works only for functions defined in separate modules rather than the notebook itself, so we'll start by using the %%file cell magic to create a simple module called mprun_demo.py, which contains our sum_of_lists function, with one addition that will make our memory profiling results more clear:

```
In [13]: %%file mprun_demo.py
         def sum_of_lists(N):
             total = 0
```

```
        for i in range(5):
            L = [j ^ (j >> i) for j in range(N)]
            total += sum(L)
            del L # remove reference to L
        return total
Overwriting mprun_demo.py
```

We can now import the new version of this function and run the memory line profiler:

```
In [14]: from mprun_demo import sum_of_lists
         %mprun -f sum_of_lists sum_of_lists(1000000)

Filename: /Users/jakevdp/github/jakevdp/PythonDataScienceHandbook/notebooks_v2/
> m prun_demo.py

Line #    Mem usage    Increment  Occurrences   Line Contents
=============================================================
     1     66.7 MiB     66.7 MiB           1   def sum_of_lists(N):
     2     66.7 MiB      0.0 MiB           1       total = 0
     3     75.1 MiB      8.4 MiB           6       for i in range(5):
     4    105.9 MiB     30.8 MiB     5000015           L = [j ^ (j >> i) for j
     5    109.8 MiB      3.8 MiB           5           total += sum(L)
     6     75.1 MiB    -34.6 MiB           5           del L # remove reference to L
     7     66.9 MiB     -8.2 MiB           1       return total
```

Here, the Increment column tells us how much each line affects the total memory budget: observe that when we create and delete the list L, we are adding about 30 MB of memory usage. This is on top of the background memory usage from the Python interpreter itself.

For more information on %memit and %mprun, as well as their available options, use the IPython help functionality (e.g., type %memit? at the IPython prompt).

More IPython Resources

In this set of chapters, we've just scratched the surface of using IPython to enable data science tasks. Much more information is available both in print and on the web, and here I'll list some other resources that you may find helpful.

Web Resources

The IPython website (http://ipython.org)
  The IPython website provides links to documentation, examples, tutorials, and a variety of other resources.

The nbviewer website (http://nbviewer.jupyter.org)
This site shows static renderings of any Jupyter notebook available on the internet. The front page features some example notebooks that you can browse to see what other folks are using IPython for!

A curated collection of Jupyter notebooks (https://github.com/jupyter/jupyter/wiki)
This ever-growing list of notebooks, powered by nbviewer, shows the depth and breadth of numerical analysis you can do with IPython. It includes everything from short examples and tutorials to full-blown courses and books composed in the notebook format!

Video tutorials
Searching the internet, you will find many video tutorials on IPython. I'd especially recommend seeking tutorials from the PyCon, SciPy, and PyData conferences by Fernando Perez and Brian Granger, two of the primary creators and maintainers of IPython and Jupyter.

Books

Python for Data Analysis (O'Reilly) (https://oreil.ly/ik2g7)
Wes McKinney's book includes a chapter that covers using IPython as a data scientist. Although much of the material overlaps what we've discussed here, another perspective is always helpful.

Learning IPython for Interactive Computing and Data Visualization (Packt)
This short book by Cyrille Rossant offers a good introduction to using IPython for data analysis.

IPython Interactive Computing and Visualization Cookbook (Packt)
Also by Cyrille Rossant, this book is a longer and more advanced treatment of using IPython for data science. Despite its name, it's not just about IPython; it also goes into some depth on a broad range of data science topics.

Finally, a reminder that you can find help on your own: IPython's ?-based help functionality (discussed in Chapter 1) can be useful if you use it well and use it often. As you go through the examples here and elsewhere, this can be used to familiarize yourself with all the tools that IPython has to offer.

## PART II

Introduction to NumPy

This part of the book, along with Part III, outlines techniques for effectively loading, storing, and manipulating in-memory data in Python. The topic is very broad: datasets can come from a wide range of sources and in a wide range of formats, including collections of documents, collections of images, collections of sound clips, collections of numerical measurements, or nearly anything else. Despite this apparent heterogeneity, many datasets can be represented fundamentally as arrays of numbers.

For example, images—particularly digital images—can be thought of as simply two-dimensional arrays of numbers representing pixel brightness across the area. Sound clips can be thought of as one-dimensional arrays of intensity versus time. Text can be converted in various ways into numerical representations, such as binary digits representing the frequency of certain words or pairs of words. No matter what the data is, the first step in making it analyzable will be to transform it into arrays of numbers. (We will discuss some specific examples of this process in Chapter 40.)

For this reason, efficient storage and manipulation of numerical arrays is absolutely fundamental to the process of doing data science. We'll now take a look at the specialized tools that Python has for handling such numerical arrays: the NumPy package and the Pandas package (discussed in Part III).

This part of the book will cover NumPy in detail. NumPy (short for *Numerical Python*) provides an efficient interface to store and operate on dense data buffers. In some ways, NumPy arrays are like Python's built-in list type, but NumPy arrays provide much more efficient storage and data operations as the arrays grow larger in size. NumPy arrays form the core of nearly the entire ecosystem of data science tools in

Python, so time spent learning to use NumPy effectively will be valuable no matter what aspect of data science interests you.

If you followed the advice in the Preface and installed the Anaconda stack, you already have NumPy installed and ready to go. If you're more the do-it-yourself type, you can go to NumPy.org (*http://www.numpy.org*) and follow the installation instructions found there. Once you do, you can import NumPy and double-check the version:

```
In [1]: import numpy
        numpy.__version__
Out[1]: '1.21.2'
```

For the pieces of the package discussed here, I'd recommend NumPy version 1.8 or later. By convention, you'll find that most people in the SciPy/PyData world will import NumPy using np as an alias:

```
In [2]: import numpy as np
```

Throughout this chapter, and indeed the rest of the book, you'll find that this is the way we will import and use NumPy.

Reminder About Built-in Documentation

As you read through this part of the book, don't forget that IPython gives you the ability to quickly explore the contents of a package (by using the Tab completion feature), as well as the documentation of various functions (using the ? character). For a refresher on these, revisit Chapter 1.

For example, to display all the contents of the NumPy namespace, you can type this:

```
In [3]: np.<TAB>
```

And to display NumPy's built-in documentation, you can use this:

```
In [4]: np?
```

Numpy offers more detailed documentation (*http://www.numpy.org*), along with tutorials and other resources.

## CHAPTER 4

Understanding Data Types in Python

Effective data-driven science and computation requires understanding how data is stored and manipulated. This chapter outlines and contrasts how arrays of data are handled in the Python language itself, and how NumPy improves on this. Understanding this difference is fundamental to understanding much of the material throughout the rest of the book.

Users of Python are often drawn in by its ease of use, one piece of which is dynamic typing. While a statically typed language like C or Java requires each variable to be explicitly declared, a dynamically typed language like Python skips this specification. For example, in C you might specify a particular operation as follows:

```
/* C code */
int result = 0;
for(int i=0; i<100; i++){
    result += i;
}
```

While in Python the equivalent operation could be written this way:

```
# Python code
result = 0
for i in range(100):
    result += i
```

Notice one main difference: in C, the data types of each variable are explicitly declared, while in Python the types are dynamically inferred. This means, for example, that we can assign any kind of data to any variable:

```
# Python code
x = 4
x = "four"
```

Here we've switched the contents of x from an integer to a string. The same thing in C would lead (depending on compiler settings) to a compilation error or other unintended consequences:

```
/* C code */
int x = 4;
x = "four";  // FAILS
```

This sort of flexibility is one element that makes Python and other dynamically typed languages convenient and easy to use. Understanding *how* this works is an important piece of learning to analyze data efficiently and effectively with Python. But what this type flexibility also points to is the fact that Python variables are more than just their values; they also contain extra information about the *type* of the value. We'll explore this more in the sections that follow.

A Python Integer Is More Than Just an Integer

The standard Python implementation is written in C. This means that every Python object is simply a cleverly disguised C structure, which contains not only its value, but other information as well. For example, when we define an integer in Python, such as x = 10000, x is not just a "raw" integer. It's actually a pointer to a compound C structure, which contains several values. Looking through the Python 3.10 source code, we find that the integer (long) type definition effectively looks like this (once the C macros are expanded):

```
struct _longobject {
    long ob_refcnt;
    PyTypeObject *ob_type;
    size_t ob_size;
    long ob_digit[1];
};
```

A single integer in Python 3.10 actually contains four pieces:

- `ob_refcnt`, a reference count that helps Python silently handle memory allocation and deallocation
- `ob_type`, which encodes the type of the variable
- `ob_size`, which specifies the size of the following data members
- `ob_digit`, which contains the actual integer value that we expect the Python variable to represent

This means that there is some overhead involved in storing an integer in Python as compared to a compiled language like C, as illustrated in Figure 4-1.

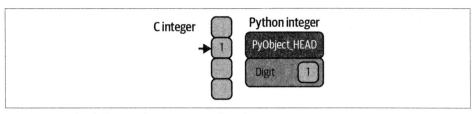

Figure 4-1. The difference between C and Python integers

Here, `PyObject_HEAD` is the part of the structure containing the reference count, type code, and other pieces mentioned before.

Notice the difference here: a C integer is essentially a label for a position in memory whose bytes encode an integer value. A Python integer is a pointer to a position in memory containing all the Python object information, including the bytes that contain the integer value. This extra information in the Python integer structure is what allows Python to be coded so freely and dynamically. All this additional information in Python types comes at a cost, however, which becomes especially apparent in structures that combine many of these objects.

A Python List Is More Than Just a List

Let's consider now what happens when we use a Python data structure that holds many Python objects. The standard mutable multielement container in Python is the list. We can create a list of integers as follows:

```
In [1]: L = list(range(10))
        L
Out[1]: [0, 1, 2, 3, 4, 5, 6, 7, 8, 9]

In [2]: type(L[0])
Out[2]: int
```

Or, similarly, a list of strings:

```
In [3]: L2 = [str(c) for c in L]
        L2
Out[3]: ['0', '1', '2', '3', '4', '5', '6', '7', '8', '9']

In [4]: type(L2[0])
Out[4]: str
```

Because of Python's dynamic typing, we can even create heterogeneous lists:

```
In [5]: L3 = [True, "2", 3.0, 4]
        [type(item) for item in L3]
Out[5]: [bool, str, float, int]
```

But this flexibility comes at a cost: to allow these flexible types, each item in the list must contain its own type, reference count, and other information. That is, each item is a complete Python object. In the special case that all variables are of the same type, much of this information is redundant, so it can be much more efficient to store the data in a fixed-type array. The difference between a dynamic-type list and a fixed-type (NumPy-style) array is illustrated in Figure 4-2.

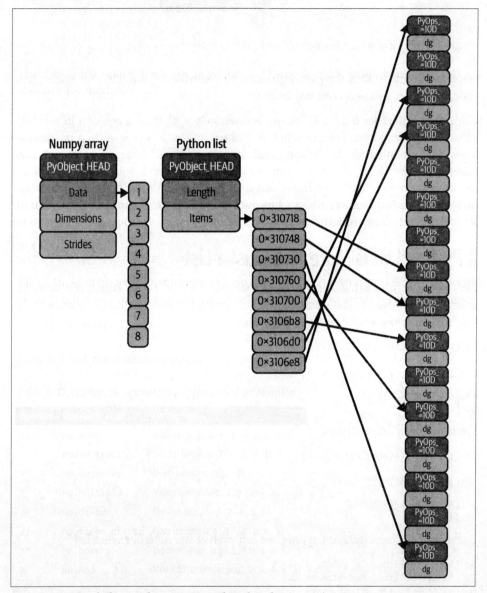

Figure 4-2. The difference between C and Python lists

At the implementation level, the array essentially contains a single pointer to one contiguous block of data. The Python list, on the other hand, contains a pointer to a block of pointers, each of which in turn points to a full Python object like the Python integer we saw earlier. Again, the advantage of the list is flexibility: because each list element is a full structure containing both data and type information, the list can be filled with data of any desired type. Fixed-type NumPy-style arrays lack this flexibility, but are much more efficient for storing and manipulating data.

Fixed-Type Arrays in Python

Python offers several different options for storing data in efficient, fixed-type data buffers. The built-in array module (available since Python 3.3) can be used to create dense arrays of a uniform type:

```
In [6]: import array
        L = list(range(10))
        A = array.array('i', L)
        A
Out[6]: array('i', [0, 1, 2, 3, 4, 5, 6, 7, 8, 9])
```

Here, 'i' is a type code indicating the contents are integers.

Much more useful, however, is the ndarray object of the NumPy package. While Python's array object provides efficient storage of array-based data, NumPy adds to this efficient *operations* on that data. We will explore these operations in later chapters; next, I'll show you a few different ways of creating a NumPy array.

Creating Arrays from Python Lists

We'll start with the standard NumPy import, under the alias np:

```
In [7]: import numpy as np
```

Now we can use np.array to create arrays from Python lists:

```
In [8]: # Integer array
        np.array([1, 4, 2, 5, 3])
Out[8]: array([1, 4, 2, 5, 3])
```

Remember that unlike Python lists, NumPy arrays can only contain data of the same type. If the types do not match, NumPy will upcast them according to its type promotion rules; here, integers are upcast to floating point:

```
In [9]: np.array([3.14, 4, 2, 3])
Out[9]: array([3.14, 4.  , 2.  , 3.  ])
```

If we want to explicitly set the data type of the resulting array, we can use the `dtype` keyword:

```
In [10]: np.array([1, 2, 3, 4], dtype=np.float32)
Out[10]: array([1., 2., 3., 4.], dtype=float32)
```

Finally, unlike Python lists, which are always one-dimensional sequences, NumPy arrays can be multidimensional. Here's one way of initializing a multidimensional array using a list of lists:

```
In [11]: # Nested lists result in multidimensional arrays
         np.array([range(i, i + 3) for i in [2, 4, 6]])
Out[11]: array([[2, 3, 4],
                [4, 5, 6],
                [6, 7, 8]])
```

The inner lists are treated as rows of the resulting two-dimensional array.

Creating Arrays from Scratch

Especially for larger arrays, it is more efficient to create arrays from scratch using routines built into NumPy. Here are several examples:

```
In [12]: # Create a length-10 integer array filled with 0s
         np.zeros(10, dtype=int)
Out[12]: array([0, 0, 0, 0, 0, 0, 0, 0, 0, 0])

In [13]: # Create a 3x5 floating-point array filled with 1s
         np.ones((3, 5), dtype=float)
Out[13]: array([[1., 1., 1., 1., 1.],
                [1., 1., 1., 1., 1.],
                [1., 1., 1., 1., 1.]])

In [14]: # Create a 3x5 array filled with 3.14
         np.full((3, 5), 3.14)
Out[14]: array([[3.14, 3.14, 3.14, 3.14, 3.14],
                [3.14, 3.14, 3.14, 3.14, 3.14],
                [3.14, 3.14, 3.14, 3.14, 3.14]])

In [15]: # Create an array filled with a linear sequence
         # starting at 0, ending at 20, stepping by 2
         # (this is similar to the built-in range function)
         np.arange(0, 20, 2)
Out[15]: array([ 0,  2,  4,  6,  8, 10, 12, 14, 16, 18])

In [16]: # Create an array of five values evenly spaced between 0 and 1
         np.linspace(0, 1, 5)
Out[16]: array([0.  , 0.25, 0.5 , 0.75, 1.  ])

In [17]: # Create a 3x3 array of uniformly distributed
         # pseudorandom values between 0 and 1
         np.random.random((3, 3))
Out[17]: array([[0.09610171, 0.88193001, 0.70548015],
```

```
                    [0.35885395, 0.91670468, 0.8721031 ],
                    [0.73237865, 0.09708562, 0.52506779]])

In [18]: # Create a 3x3 array of normally distributed pseudorandom
         # values with mean 0 and standard deviation 1
         np.random.normal(0, 1, (3, 3))
Out[18]: array([[-0.46652655, -0.59158776, -1.05392451],
                [-1.72634268,  0.03194069, -0.51048869],
                [ 1.41240208,  1.77734462, -0.43820037]])

In [19]: # Create a 3x3 array of pseudorandom integers in the interval [0, 10)
         np.random.randint(0, 10, (3, 3))
Out[19]: array([[4, 3, 8],
                [6, 5, 0],
                [1, 1, 4]])

In [20]: # Create a 3x3 identity matrix
         np.eye(3)
Out[20]: array([[1., 0., 0.],
                [0., 1., 0.],
                [0., 0., 1.]])

In [21]: # Create an uninitialized array of three integers; the values will be
         # whatever happens to already exist at that memory location
         np.empty(3)
Out[21]: array([1., 1., 1.])
```

NumPy Standard Data Types

NumPy arrays contain values of a single type, so it is important to have detailed knowledge of those types and their limitations. Because NumPy is built in C, the types will be familiar to users of C, Fortran, and other related languages.

The standard NumPy data types are listed in Table 4-1. Note that when constructing an array, they can be specified using a string:

```
np.zeros(10, dtype='int16')
```

Or using the associated NumPy object:

```
np.zeros(10, dtype=np.int16)
```

More advanced type specification is possible, such as specifying big- or little-endian numbers; for more information, refer to the NumPy documentation (*http://numpy.org*). NumPy also supports compound data types, which will be covered in Chapter 12.

Table 4-1. Standard NumPy data types

Data type	Description
bool_	Boolean (True or False) stored as a byte
int_	Default integer type (same as C long; normally either int64 or int32)
intc	Identical to C int (normally int32 or int64)
intp	Integer used for indexing (same as C ssize_t; normally either int32 or int64)
int8	Byte (−128 to 127)
int16	Integer (−32768 to 32767)
int32	Integer (−2147483648 to 2147483647)
int64	Integer (−9223372036854775808 to 9223372036854775807)
uint8	Unsigned integer (0 to 255)
uint16	Unsigned integer (0 to 65535)
uint32	Unsigned integer (0 to 4294967295)
uint64	Unsigned integer (0 to 18446744073709551615)
float_	Shorthand for float64
float16	Half-precision float: sign bit, 5 bits exponent, 10 bits mantissa
float32	Single-precision float: sign bit, 8 bits exponent, 23 bits mantissa
float64	Double-precision float: sign bit, 11 bits exponent, 52 bits mantissa
complex_	Shorthand for complex128
complex64	Complex number, represented by two 32-bit floats
complex128	Complex number, represented by two 64-bit floats

## CHAPTER 5

The Basics of NumPy Arrays

Data manipulation in Python is nearly synonymous with NumPy array manipulation: even newer tools like Pandas (Part III) are built around the NumPy array. This chapter will present several examples of using NumPy array manipulation to access data and subarrays, and to split, reshape, and join the arrays. While the types of operations shown here may seem a bit dry and pedantic, they comprise the building blocks of many other examples used throughout the book. Get to know them well!

We'll cover a few categories of basic array manipulations here:

Attributes of arrays
Determining the size, shape, memory consumption, and data types of arrays

Indexing of arrays
Getting and setting the values of individual array elements

Slicing of arrays
Getting and setting smaller subarrays within a larger array

Reshaping of arrays
Changing the shape of a given array

Joining and splitting of arrays
Combining multiple arrays into one, and splitting one array into many

NumPy Array Attributes

First let's discuss some useful array attributes. We'll start by defining random arrays of one, two, and three dimensions. We'll use NumPy's random number generator, which we will *seed* with a set value in order to ensure that the same random arrays are generated each time this code is run:

```
In [1]: import numpy as np
        rng = np.random.default_rng(seed=1701)  # seed for reproducibility

        x1 = rng.integers(10, size=6)  # one-dimensional array
        x2 = rng.integers(10, size=(3, 4))  # two-dimensional array
        x3 = rng.integers(10, size=(3, 4, 5))  # three-dimensional array
```

Each array has attributes including `ndim` (the number of dimensions), `shape` (the size of each dimension), `size` (the total size of the array), and `dtype` (the type of each element):

```
In [2]: print("x3 ndim: ", x3.ndim)
        print("x3 shape:", x3.shape)
        print("x3 size: ", x3.size)
        print("dtype:   ", x3.dtype)
Out[2]: x3 ndim:  3
        x3 shape: (3, 4, 5)
        x3 size:  60
        dtype:    int64
```

For more discussion of data types, see Chapter 4.

Array Indexing: Accessing Single Elements

If you are familiar with Python's standard list indexing, indexing in NumPy will feel quite familiar. In a one-dimensional array, the i_{th} value (counting from zero) can be accessed by specifying the desired index in square brackets, just as with Python lists:

```
In [3]: x1
Out[3]: array([9, 4, 0, 3, 8, 6])
```

```
In [4]: x1[0]
Out[4]: 9
```

```
In [5]: x1[4]
Out[5]: 8
```

To index from the end of the array, you can use negative indices:

```
In [6]: x1[-1]
Out[6]: 6
```

```
In [7]: x1[-2]
Out[7]: 8
```

In a multidimensional array, items can be accessed using a comma-separated (*row*, *column*) tuple:

```
In [8]: x2
Out[8]: array([[3, 1, 3, 7],
               [4, 0, 2, 3],
               [0, 0, 6, 9]])

In [9]: x2[0, 0]
Out[9]: 3

In [10]: x2[2, 0]
Out[10]: 0

In [11]: x2[2, -1]
Out[11]: 9
```

Values can also be modified using any of the preceding index notation:

```
In [12]: x2[0, 0] = 12
         x2
Out[12]: array([[12,  1,  3,  7],
                [ 4,  0,  2,  3],
                [ 0,  0,  6,  9]])
```

Keep in mind that, unlike Python lists, NumPy arrays have a fixed type. This means, for example, that if you attempt to insert a floating-point value into an integer array, the value will be silently truncated. Don't be caught unaware by this behavior!

```
In [13]: x1[0] = 3.14159  # this will be truncated!
         x1
Out[13]: array([3, 4, 0, 3, 8, 6])
```

Array Slicing: Accessing Subarrays

Just as we can use square brackets to access individual array elements, we can also use them to access subarrays with the *slice* notation, marked by the colon (:) character. The NumPy slicing syntax follows that of the standard Python list; to access a slice of an array x, use this:

```
x[start:stop:step]
```

If any of these are unspecified, they default to the values start=0, stop=<size of dimension>, step=1. Let's look at some examples of accessing subarrays in one dimension and in multiple dimensions.

One-Dimensional Subarrays

Here are some examples of accessing elements in one-dimensional subarrays:

```
In [14]: x1
Out[14]: array([3, 4, 0, 3, 8, 6])
```

```
In [15]: x1[:3]  # first three elements
Out[15]: array([3, 4, 0])

In [16]: x1[3:]  # elements after index 3
Out[16]: array([3, 8, 6])

In [17]: x1[1:4]  # middle subarray
Out[17]: array([4, 0, 3])

In [18]: x1[::2]  # every second element
Out[18]: array([3, 0, 8])

In [19]: x1[1::2]  # every second element, starting at index 1
Out[19]: array([4, 3, 6])
```

A potentially confusing case is when the step value is negative. In this case, the defaults for start and stop are swapped. This becomes a convenient way to reverse an array:

```
In [20]: x1[::-1]  # all elements, reversed
Out[20]: array([6, 8, 3, 0, 4, 3])

In [21]: x1[4::-2]  # every second element from index 4, reversed
Out[21]: array([8, 0, 3])
```

Multidimensional Subarrays

Multidimensional slices work in the same way, with multiple slices separated by commas. For example:

```
In [22]: x2
Out[22]: array([[12,  1,  3,  7],
               [ 4,  0,  2,  3],
               [ 0,  0,  6,  9]])

In [23]: x2[:2, :3]  # first two rows & three columns
Out[23]: array([[12,  1,  3],
               [ 4,  0,  2]])

In [24]: x2[:3, ::2]  # three rows, every second column
Out[24]: array([[12,  3],
               [ 4,  2],
               [ 0,  6]])

In [25]: x2[::-1, ::-1]  # all rows & columns, reversed
Out[25]: array([[ 9,  6,  0,  0],
               [ 3,  2,  0,  4],
               [ 7,  3,  1, 12]])
```

One commonly needed routine is accessing single rows or columns of an array. This can be done by combining indexing and slicing, using an empty slice marked by a single colon (:):

```
In [26]: x2[:, 0]  # first column of x2
Out[26]: array([12,  4,  0])
```

```
In [27]: x2[0, :]  # first row of x2
Out[27]: array([12,  1,  3,  7])
```

In the case of row access, the empty slice can be omitted for a more compact syntax:

```
In [28]: x2[0]  # equivalent to x2[0, :]
Out[28]: array([12,  1,  3,  7])
```

Subarrays as No-Copy Views

Unlike Python list slices, NumPy array slices are returned as *views* rather than *copies* of the array data. Consider our two-dimensional array from before:

```
In [29]: print(x2)
Out[29]: [[12  1  3  7]
          [ 4  0  2  3]
          [ 0  0  6  9]]
```

Let's extract a 2 × 2 subarray from this:

```
In [30]: x2_sub = x2[:2, :2]
         print(x2_sub)
Out[30]: [[12  1]
          [ 4  0]]
```

Now if we modify this subarray, we'll see that the original array is changed! Observe:

```
In [31]: x2_sub[0, 0] = 99
         print(x2_sub)
Out[31]: [[99  1]
          [ 4  0]]

In [32]: print(x2)
Out[32]: [[99  1  3  7]
          [ 4  0  2  3]
          [ 0  0  6  9]]
```

Some users may find this surprising, but it can be advantageous: for example, when working with large datasets, we can access and process pieces of these datasets without the need to copy the underlying data buffer.

Creating Copies of Arrays

Despite the features of array views, it's sometimes useful to instead explicitly copy the data within an array or a subarray. This is easiest to do with the copy method:

```
In [33]: x2_sub_copy = x2[:2, :2].copy()
         print(x2_sub_copy)
Out[33]: [[99  1]
          [ 4  0]]
```

If we now modify this subarray, the original array is not touched:

```
In [34]: x2_sub_copy[0, 0] = 42
         print(x2_sub_copy)
Out[34]: [[42  1]
          [ 4  0]]

In [35]: print(x2)
Out[35]: [[99  1  3  7]
          [ 4  0  2  3]
          [ 0  0  6  9]]
```

Reshaping of Arrays

Another useful type of operation is reshaping of arrays, which can be done with the reshape method. For example, if you want to put the numbers 1 through 9 in a 3×3 grid, you can do the following:

```
In [36]: grid = np.arange(1, 10).reshape(3, 3)
         print(grid)
Out[36]: [[1 2 3]
          [4 5 6]
          [7 8 9]]
```

Note that for this to work, the size of the initial array must match the size of the reshaped array, and in most cases the reshape method will return a no-copy view of the initial array.

A common reshaping operation is converting a one-dimensional array into a two-dimensional row or column matrix:

```
In [37]: x = np.array([1, 2, 3])
         x.reshape((1, 3))  # row vector via reshape
Out[37]: array([[1, 2, 3]])

In [38]: x.reshape((3, 1))  # column vector via reshape
Out[38]: array([[1],
                [2],
                [3]])
```

A convenient shorthand for this is to use np.newaxis in the slicing syntax:

```
In [39]: x[np.newaxis, :]  # row vector via newaxis
Out[39]: array([[1, 2, 3]])

In [40]: x[:, np.newaxis]  # column vector via newaxis
Out[40]: array([[1],
                [2],
                [3]])
```

This is a pattern that we will utilize often throughout the remainder of the book.

Array Concatenation and Splitting

All of the preceding routines worked on single arrays. NumPy also provides tools to combine multiple arrays into one, and to conversely split a single array into multiple arrays.

Concatenation of Arrays

Concatenation, or joining of two arrays in NumPy, is primarily accomplished using the routines np.concatenate, np.vstack, and np.hstack. np.concatenate takes a tuple or list of arrays as its first argument, as you can see here:

```
In [41]: x = np.array([1, 2, 3])
         y = np.array([3, 2, 1])
         np.concatenate([x, y])
Out[41]: array([1, 2, 3, 3, 2, 1])
```

You can also concatenate more than two arrays at once:

```
In [42]: z = np.array([99, 99, 99])
         print(np.concatenate([x, y, z]))
Out[42]: [ 1  2  3  3  2  1 99 99 99]
```

And it can be used for two-dimensional arrays:

```
In [43]: grid = np.array([[1, 2, 3],
                          [4, 5, 6]])
```

```
In [44]: # concatenate along the first axis
         np.concatenate([grid, grid])
Out[44]: array([[1, 2, 3],
                [4, 5, 6],
                [1, 2, 3],
                [4, 5, 6]])
```

```
In [45]: # concatenate along the second axis (zero-indexed)
         np.concatenate([grid, grid], axis=1)
Out[45]: array([[1, 2, 3, 1, 2, 3],
                [4, 5, 6, 4, 5, 6]])
```

For working with arrays of mixed dimensions, it can be clearer to use the np.vstack (vertical stack) and np.hstack (horizontal stack) functions:

```
In [46]: # vertically stack the arrays
         np.vstack([x, grid])
Out[46]: array([[1, 2, 3],
                [1, 2, 3],
                [4, 5, 6]])
```

```
In [47]: # horizontally stack the arrays
         y = np.array([[99],
                       [99]])
         np.hstack([grid, y])
```

```
Out[47]: array([[ 1,  2,  3, 99],
                [ 4,  5,  6, 99]])
```

Similarly, for higher-dimensional arrays, np.dstack will stack arrays along the third axis.

Splitting of Arrays

The opposite of concatenation is splitting, which is implemented by the functions np.split, np.hsplit, and np.vsplit. For each of these, we can pass a list of indices giving the split points:

```
In [48]: x = [1, 2, 3, 99, 99, 3, 2, 1]
         x1, x2, x3 = np.split(x, [3, 5])
         print(x1, x2, x3)
Out[48]: [1 2 3] [99 99] [3 2 1]
```

Notice that N split points leads to $N + 1$ subarrays. The related functions np.hsplit and np.vsplit are similar:

```
In [49]: grid = np.arange(16).reshape((4, 4))
         grid
Out[49]: array([[ 0,  1,  2,  3],
                [ 4,  5,  6,  7],
                [ 8,  9, 10, 11],
                [12, 13, 14, 15]])
In [50]: upper, lower = np.vsplit(grid, [2])
         print(upper)
         print(lower)
Out[50]: [[0 1 2 3]
          [4 5 6 7]]
         [[ 8  9 10 11]
          [12 13 14 15]]
In [51]: left, right = np.hsplit(grid, [2])
         print(left)
         print(right)
Out[51]: [[ 0  1]
          [ 4  5]
          [ 8  9]
          [12 13]]
         [[ 2  3]
          [ 6  7]
          [10 11]
          [14 15]]
```

Similarly, for higher-dimensional arrays, np.dsplit will split arrays along the third axis.

## CHAPTER 6

Computation on NumPy Arrays: Universal Functions

Up until now, we have been discussing some of the basic nuts and bolts of NumPy. In the next few chapters, we will dive into the reasons that NumPy is so important in the Python data science world: namely, because it provides an easy and flexible interface to optimize computation with arrays of data.

Computation on NumPy arrays can be very fast, or it can be very slow. The key to making it fast is to use vectorized operations, generally implemented through NumPy's *universal functions* (ufuncs). This chapter motivates the need for NumPy's ufuncs, which can be used to make repeated calculations on array elements much more efficient. It then introduces many of the most common and useful arithmetic ufuncs available in the NumPy package.

The Slowness of Loops

Python's default implementation (known as CPython) does some operations very slowly. This is partly due to the dynamic, interpreted nature of the language; types are flexible, so sequences of operations cannot be compiled down to efficient machine code as in languages like C and Fortran. Recently there have been various attempts to address this weakness: well-known examples are the PyPy project (*http://pypy.org*), a just-in-time compiled implementation of Python; the Cython project (*http://cython.org*), which converts Python code to compilable C code; and the Numba project (*http://numba.pydata.org*), which converts snippets of Python code to fast LLVM bytecode. Each of these has its strengths and weaknesses, but it is safe to say that none of the three approaches has yet surpassed the reach and popularity of the standard CPython engine.

The relative sluggishness of Python generally manifests itself in situations where many small operations are being repeated; for instance, looping over arrays to operate on each element. For example, imagine we have an array of values and we'd like to compute the reciprocal of each. A straightforward approach might look like this:

```
In [1]: import numpy as np
        rng = np.random.default_rng(seed=1701)

        def compute_reciprocals(values):
            output = np.empty(len(values))
            for i in range(len(values)):
                output[i] = 1.0 / values[i]
            return output

        values = rng.integers(1, 10, size=5)
        compute_reciprocals(values)
Out[1]: array([0.11111111, 0.25      , 1.        , 0.33333333, 0.125     ])
```

This implementation probably feels fairly natural to someone from, say, a C or Java background. But if we measure the execution time of this code for a large input, we see that this operation is very slow—perhaps surprisingly so! We'll benchmark this with IPython's %timeit magic (discussed in "Profiling and Timing Code" on page 26):

```
In [2]: big_array = rng.integers(1, 100, size=1000000)
        %timeit compute_reciprocals(big_array)
Out[2]: 2.61 s ± 192 ms per loop (mean ± std. dev. of 7 runs, 1 loop each)
```

It takes several seconds to compute these million operations and to store the result! When even cell phones have processing speeds measured in gigaflops (i.e., billions of numerical operations per second), this seems almost absurdly slow. It turns out that the bottleneck here is not the operations themselves, but the type checking and function dispatches that CPython must do at each cycle of the loop. Each time the reciprocal is computed, Python first examines the object's type and does a dynamic lookup of the correct function to use for that type. If we were working in compiled code instead, this type specification would be known before the code executed and the result could be computed much more efficiently.

Introducing Ufuncs

For many types of operations, NumPy provides a convenient interface into just this kind of statically typed, compiled routine. This is known as a *vectorized* operation. For simple operations like the element-wise division here, vectorization is as simple as using Python arithmetic operators directly on the array object. This vectorized approach is designed to push the loop into the compiled layer that underlies NumPy, leading to much faster execution.

Compare the results of the following two operations:

```
In [3]: print(compute_reciprocals(values))
        print(1.0 / values)
Out[3]: [0.11111111 0.25       1.         0.33333333 0.125     ]
        [0.11111111 0.25       1.         0.33333333 0.125     ]
```

Looking at the execution time for our big array, we see that it completes orders of magnitude faster than the Python loop:

```
In [4]: %timeit (1.0 / big_array)
Out[4]: 2.54 ms ± 383 µs per loop (mean ± std. dev. of 7 runs, 100 loops each)
```

Vectorized operations in NumPy are implemented via ufuncs, whose main purpose is to quickly execute repeated operations on values in NumPy arrays. Ufuncs are extremely flexible—before we saw an operation between a scalar and an array, but we can also operate between two arrays:

```
In [5]: np.arange(5) / np.arange(1, 6)
Out[5]: array([0.        , 0.5       , 0.66666667, 0.75      , 0.8       ])
```

And ufunc operations are not limited to one-dimensional arrays. They can act on multidimensional arrays as well:

```
In [6]: x = np.arange(9).reshape((3, 3))
        2 ** x
Out[6]: array([[  1,   2,   4],
               [  8,  16,  32],
               [ 64, 128, 256]])
```

Computations using vectorization through ufuncs are nearly always more efficient than their counterparts implemented using Python loops, especially as the arrays grow in size. Any time you see such a loop in a NumPy script, you should consider whether it can be replaced with a vectorized expression.

Exploring NumPy's Ufuncs

Ufuncs exist in two flavors: *unary ufuncs*, which operate on a single input, and *binary ufuncs*, which operate on two inputs. We'll see examples of both these types of functions here.

Array Arithmetic

NumPy's ufuncs feel very natural to use because they make use of Python's native arithmetic operators. The standard addition, subtraction, multiplication, and division can all be used:

```
In [7]: x = np.arange(4)
        print("x     =", x)
        print("x + 5 =", x + 5)
        print("x - 5 =", x - 5)
```

```
        print("x * 2  =", x * 2)
        print("x / 2  =", x / 2)
        print("x // 2 =", x // 2)  # floor division
Out[7]: x      = [0 1 2 3]
        x + 5  = [5 6 7 8]
        x - 5  = [-5 -4 -3 -2]
        x * 2  = [0 2 4 6]
        x / 2  = [0.  0.5 1.  1.5]
        x // 2 = [0 0 1 1]
```

There is also a unary ufunc for negation, a ** operator for exponentiation, and a % operator for modulus:

```
In [8]: print("-x       = ", -x)
        print("x ** 2 = ", x ** 2)
        print("x % 2  = ", x % 2)
Out[8]: -x      = [ 0 -1 -2 -3]
        x ** 2 = [0 1 4 9]
        x % 2  = [0 1 0 1]
```

In addition, these can be strung together however you wish, and the standard order of operations is respected:

```
In [9]: -(0.5*x + 1) ** 2
Out[9]: array([-1.  , -2.25, -4.  , -6.25])
```

All of these arithmetic operations are simply convenient wrappers around specific ufuncs built into NumPy. For example, the + operator is a wrapper for the add ufunc:

```
In [10]: np.add(x, 2)
Out[10]: array([2, 3, 4, 5])
```

Table 6-1 lists the arithmetic operators implemented in NumPy.

Table 6-1. Arithmetic operators implemented in NumPy

Operator	Equivalent ufunc	Description
+	np.add	Addition (e.g., 1 + 1 = 2)
-	np.subtract	Subtraction (e.g., 3 - 2 = 1)
-	np.negative	Unary negation (e.g., -2)
*	np.multiply	Multiplication (e.g., 2 * 3 = 6)
/	np.divide	Division (e.g., 3 / 2 = 1.5)
//	np.floor_divide	Floor division (e.g., 3 // 2 = 1)
**	np.power	Exponentiation (e.g., 2 ** 3 = 8)
%	np.mod	Modulus/remainder (e.g., 9 % 4 = 1)

Additionally, there are Boolean/bitwise operators; we will explore these in Chapter 9.

Absolute Value

Just as NumPy understands Python's built-in arithmetic operators, it also understands Python's built-in absolute value function:

```
In [11]: x = np.array([-2, -1, 0, 1, 2])
         abs(x)
Out[11]: array([2, 1, 0, 1, 2])
```

The corresponding NumPy ufunc is `np.absolute`, which is also available under the alias `np.abs`:

```
In [12]: np.absolute(x)
Out[12]: array([2, 1, 0, 1, 2])
```

```
In [13]: np.abs(x)
Out[13]: array([2, 1, 0, 1, 2])
```

This ufunc can also handle complex data, in which case it returns the magnitude:

```
In [14]: x = np.array([3 - 4j, 4 - 3j, 2 + 0j, 0 + 1j])
         np.abs(x)
Out[14]: array([5., 5., 2., 1.])
```

Trigonometric Functions

NumPy provides a large number of useful ufuncs, and some of the most useful for the data scientist are the trigonometric functions. We'll start by defining an array of angles:

```
In [15]: theta = np.linspace(0, np.pi, 3)
```

Now we can compute some trigonometric functions on these values:

```
In [16]: print("theta      = ", theta)
         print("sin(theta) = ", np.sin(theta))
         print("cos(theta) = ", np.cos(theta))
         print("tan(theta) = ", np.tan(theta))
Out[16]: theta      = [0.         1.57079633 3.14159265]
         sin(theta) = [0.0000000e+00 1.0000000e+00 1.2246468e-16]
         cos(theta) = [ 1.000000e+00  6.123234e-17 -1.000000e+00]
         tan(theta) = [ 0.00000000e+00  1.63312394e+16 -1.22464680e-16]
```

The values are computed to within machine precision, which is why values that should be zero do not always hit exactly zero. Inverse trigonometric functions are also available:

```
In [17]: x = [-1, 0, 1]
         print("x         = ", x)
         print("arcsin(x) = ", np.arcsin(x))
         print("arccos(x) = ", np.arccos(x))
         print("arctan(x) = ", np.arctan(x))
Out[17]: x         = [-1, 0, 1]
         arcsin(x) = [-1.57079633  0.          1.57079633]
```

```
arccos(x) =  [3.14159265 1.57079633 0.         ]
arctan(x) =  [-0.78539816  0.          0.78539816]
```

Exponents and Logarithms

Other common operations available in NumPy ufuncs are the exponentials:

```
In [18]: x = [1, 2, 3]
         print("x   =", x)
         print("e^x =", np.exp(x))
         print("2^x =", np.exp2(x))
         print("3^x =", np.power(3., x))
Out[18]: x   = [1, 2, 3]
         e^x = [ 2.71828183  7.3890561  20.08553692]
         2^x = [2. 4. 8.]
         3^x = [ 3.  9. 27.]
```

The inverse of the exponentials, the logarithms, are also available. The basic `np.log` gives the natural logarithm; if you prefer to compute the base-2 logarithm or the base-10 logarithm, these are available as well:

```
In [19]: x = [1, 2, 4, 10]
         print("x        =", x)
         print("ln(x)    =", np.log(x))
         print("log2(x)  =", np.log2(x))
         print("log10(x) =", np.log10(x))
Out[19]: x        = [1, 2, 4, 10]
         ln(x)    = [0.         0.69314718 1.38629436 2.30258509]
         log2(x)  = [0.         1.         2.         3.32192809]
         log10(x) = [0.         0.30103    0.60205999 1.        ]
```

There are also some specialized versions that are useful for maintaining precision with very small input:

```
In [20]: x = [0, 0.001, 0.01, 0.1]
         print("exp(x) - 1 =", np.expm1(x))
         print("log(1 + x) =", np.log1p(x))
Out[20]: exp(x) - 1 = [0.         0.0010005  0.01005017 0.10517092]
         log(1 + x) = [0.         0.0009995  0.00995033 0.09531018]
```

When x is very small, these functions give more precise values than if the raw `np.log` or `np.exp` were to be used.

Specialized Ufuncs

NumPy has many more ufuncs available, including for hyperbolic trigonometry, bitwise arithmetic, comparison operations, conversions from radians to degrees, rounding and remainders, and much more. A look through the NumPy documentation reveals a lot of interesting functionality.

Another excellent source for more specialized ufuncs is the submodule `scipy.spe cial`. If you want to compute some obscure mathematical function on your data, chances are it is implemented in `scipy.special`. There are far too many functions to list them all, but the following snippet shows a couple that might come up in a statistics context:

```
In [21]: from scipy import special
```

```
In [22]: # Gamma functions (generalized factorials) and related functions
         x = [1, 5, 10]
         print("gamma(x)      =", special.gamma(x))
         print("ln|gamma(x)|  =", special.gammaln(x))
         print("beta(x, 2)    =", special.beta(x, 2))
Out[22]: gamma(x)      = [1.0000e+00 2.4000e+01 3.6288e+05]
         ln|gamma(x)|  = [ 0.          3.17805383 12.80182748]
         beta(x, 2)    = [0.5         0.03333333 0.00909091]
```

```
In [23]: # Error function (integral of Gaussian),
         # its complement, and its inverse
         x = np.array([0, 0.3, 0.7, 1.0])
         print("erf(x)  =", special.erf(x))
         print("erfc(x) =", special.erfc(x))
         print("erfinv(x) =", special.erfinv(x))
Out[23]: erf(x)  = [0.          0.32862676 0.67780119 0.84270079]
         erfc(x) = [1.          0.67137324 0.32219881 0.15729921]
         erfinv(x) = [0.          0.27246271 0.73286908         inf]
```

There are many, many more ufuncs available in both NumPy and `scipy.special`. Because the documentation of these packages is available online, a web search along the lines of "gamma function python" will generally find the relevant information.

Advanced Ufunc Features

Many NumPy users make use of ufuncs without ever learning their full set of features. I'll outline a few specialized features of ufuncs here.

Specifying Output

For large calculations, it is sometimes useful to be able to specify the array where the result of the calculation will be stored. For all ufuncs, this can be done using the `out` argument of the function:

```
In [24]: x = np.arange(5)
         y = np.empty(5)
         np.multiply(x, 10, out=y)
         print(y)
Out[24]: [ 0. 10. 20. 30. 40.]
```

This can even be used with array views. For example, we can write the results of a computation to every other element of a specified array:

```
In [25]: y = np.zeros(10)
         np.power(2, x, out=y[::2])
         print(y)
Out[25]: [ 1.  0.  2.  0.  4.  0.  8.  0. 16.  0.]
```

If we had instead written y[::2] = 2 ** x, this would have resulted in the creation of a temporary array to hold the results of 2 ** x, followed by a second operation copying those values into the y array. This doesn't make much of a difference for such a small computation, but for very large arrays the memory savings from careful use of the out argument can be significant.

Aggregations

For binary ufuncs, aggregations can be computed directly from the object. For example, if we'd like to *reduce* an array with a particular operation, we can use the reduce method of any ufunc. A reduce repeatedly applies a given operation to the elements of an array until only a single result remains.

For example, calling reduce on the add ufunc returns the sum of all elements in the array:

```
In [26]: x = np.arange(1, 6)
         np.add.reduce(x)
Out[26]: 15
```

Similarly, calling reduce on the multiply ufunc results in the product of all array elements:

```
In [27]: np.multiply.reduce(x)
Out[27]: 120
```

If we'd like to store all the intermediate results of the computation, we can instead use accumulate:

```
In [28]: np.add.accumulate(x)
Out[28]: array([ 1,  3,  6, 10, 15])
```

```
In [29]: np.multiply.accumulate(x)
Out[29]: array([  1,   2,   6,  24, 120])
```

Note that for these particular cases, there are dedicated NumPy functions to compute the results (np.sum, np.prod, np.cumsum, np.cumprod), which we'll explore in Chapter 7.

Outer Products

Finally, any ufunc can compute the output of all pairs of two different inputs using the outer method. This allows you, in one line, to do things like create a multiplication table:

```
In [30]: x = np.arange(1, 6)
         np.multiply.outer(x, x)
Out[30]: array([[ 1,  2,  3,  4,  5],
                [ 2,  4,  6,  8, 10],
                [ 3,  6,  9, 12, 15],
                [ 4,  8, 12, 16, 20],
                [ 5, 10, 15, 20, 25]])
```

The ufunc.at and ufunc.reduceat methods are useful as well, and we will explore them in Chapter 10.

We will also encounter the ability of ufuncs to operate between arrays of different shapes and sizes, a set of operations known as *broadcasting*. This subject is important enough that we will devote a whole chapter to it (see Chapter 8).

Ufuncs: Learning More

More information on universal functions (including the full list of available functions) can be found on the NumPy (*http://www.numpy.org*) and SciPy (*http://www.scipy.org*) documentation websites.

Recall that you can also access information directly from within IPython by importing the packages and using IPython's tab completion and help (?) functionality, as described in Chapter 1.

## CHAPTER 7

Aggregations: min, max, and Everything in Between

A first step in exploring any dataset is often to compute various summary statistics. Perhaps the most common summary statistics are the mean and standard deviation, which allow you to summarize the "typical" values in a dataset, but other aggregations are useful as well (the sum, product, median, minimum and maximum, quantiles, etc.).

NumPy has fast built-in aggregation functions for working on arrays; we'll discuss and try out some of them here.

Summing the Values in an Array

As a quick example, consider computing the sum of all values in an array. Python itself can do this using the built-in sum function:

```
In [1]: import numpy as np
        rng = np.random.default_rng()

In [2]: L = rng.random(100)
        sum(L)
Out[2]: 52.76825337322368
```

The syntax is quite similar to that of NumPy's sum function, and the result is the same in the simplest case:

```
In [3]: np.sum(L)
Out[3]: 52.76825337322366
```

However, because it executes the operation in compiled code, NumPy's version of the operation is computed much more quickly:

```
In [4]: big_array = rng.random(1000000)
        %timeit sum(big_array)
        %timeit np.sum(big_array)
Out[4]: 89.9 ms ± 233 µs per loop (mean ± std. dev. of 7 runs, 10 loops each)
        521 µs ± 8.37 µs per loop (mean ± std. dev. of 7 runs, 1000 loops each)
```

Be careful, though: the sum function and the np.sum function are not identical, which can sometimes lead to confusion! In particular, their optional arguments have different meanings (sum(x, 1) initializes the sum at 1, while np.sum(x, 1) sums along axis 1), and np.sum is aware of multiple array dimensions, as we will see in the following section.

Minimum and Maximum

Similarly, Python has built-in min and max functions, used to find the minimum value and maximum value of any given array:

```
In [5]: min(big_array), max(big_array)
Out[5]: (2.0114398036064074e-07, 0.9999997912802653)
```

NumPy's corresponding functions have similar syntax, and again operate much more quickly:

```
In [6]: np.min(big_array), np.max(big_array)
Out[6]: (2.0114398036064074e-07, 0.9999997912802653)
```

```
In [7]: %timeit min(big_array)
        %timeit np.min(big_array)
Out[7]: 72 ms ± 177 µs per loop (mean ± std. dev. of 7 runs, 10 loops each)
        564 µs ± 3.11 µs per loop (mean ± std. dev. of 7 runs, 1000 loops each)
```

For min, max, sum, and several other NumPy aggregates, a shorter syntax is to use methods of the array object itself:

```
In [8]: print(big_array.min(), big_array.max(), big_array.sum())
Out[8]: 2.0114398036064074e-07 0.9999997912802653 499854.0273321711
```

Whenever possible, make sure that you are using the NumPy version of these aggregates when operating on NumPy arrays!

Multidimensional Aggregates

One common type of aggregation operation is an aggregate along a row or column. Say you have some data stored in a two-dimensional array:

```
In [9]: M = rng.integers(0, 10, (3, 4))
        print(M)
Out[9]: [[0 3 1 2]
```

```
[1 9 7 0]
[4 8 3 7]]
```

NumPy aggregations will apply across all elements of a multidimensional array:

```
In [10]: M.sum()
Out[10]: 45
```

Aggregation functions take an additional argument specifying the *axis* along which the aggregate is computed. For example, we can find the minimum value within each column by specifying axis=0:

```
In [11]: M.min(axis=0)
Out[11]: array([0, 3, 1, 0])
```

The function returns four values, corresponding to the four columns of numbers.

Similarly, we can find the maximum value within each row:

```
In [12]: M.max(axis=1)
Out[12]: array([3, 9, 8])
```

The way the axis is specified here can be confusing to users coming from other languages. The axis keyword specifies the dimension of the array that will be *collapsed*, rather than the dimension that will be returned. So, specifying axis=0 means that axis 0 will be collapsed: for two-dimensional arrays, values within each column will be aggregated.

Other Aggregation Functions

NumPy provides several other aggregation functions with a similar API, and additionally most have a NaN-safe counterpart that computes the result while ignoring missing values, which are marked by the special IEEE floating-point NaN value (see Chapter 16).

Table 7-1 provides a list of useful aggregation functions available in NumPy.

Table 7-1. Aggregation functions available in NumPy

Function name	NaN-safe version	Description
np.sum	np.nansum	Compute sum of elements
np.prod	np.nanprod	Compute product of elements
np.mean	np.nanmean	Compute mean of elements
np.std	np.nanstd	Compute standard deviation
np.var	np.nanvar	Compute variance
np.min	np.nanmin	Find minimum value
np.max	np.nanmax	Find maximum value
np.argmin	np.nanargmin	Find index of minimum value

Function name	NaN-safe version	Description
np.argmax	np.nanargmax	Find index of maximum value
np.median	np.nanmedian	Compute median of elements
np.percentile	np.nanpercentile	Compute rank-based statistics of elements
np.any	N/A	Evaluate whether any elements are true
np.all	N/A	Evaluate whether all elements are true

You will see these aggregates often throughout the rest of the book.

Example: What Is the Average Height of US Presidents?

Aggregates available in NumPy can act as summary statistics for a set of values. As a small example, let's consider the heights of all US presidents. This data is available in the file *president_heights.csv*, which is a comma-separated list of labels and values:

```
In [13]: !head -4 data/president_heights.csv
Out[13]: order,name,height(cm)
         1,George Washington,189
         2,John Adams,170
         3,Thomas Jefferson,189
```

We'll use the Pandas package, which we'll explore more fully in Part III, to read the file and extract this information (note that the heights are measured in centimeters):

```
In [14]: import pandas as pd
         data = pd.read_csv('data/president_heights.csv')
         heights = np.array(data['height(cm)'])
         print(heights)
Out[14]: [189 170 189 163 183 171 185 168 173 183 173 173 175 178 183 193 178 173
          174 183 183 168 170 178 182 180 183 178 182 188 175 179 183 193 182 183
          177 185 188 188 182 185 191 182]
```

Now that we have this data array, we can compute a variety of summary statistics:

```
In [15]: print("Mean height:       ", heights.mean())
         print("Standard deviation:", heights.std())
         print("Minimum height:    ", heights.min())
         print("Maximum height:    ", heights.max())
Out[15]: Mean height:        180.04545454545453
         Standard deviation: 6.983599441335736
         Minimum height:     163
         Maximum height:     193
```

Note that in each case, the aggregation operation reduced the entire array to a single summarizing value, which gives us information about the distribution of values. We may also wish to compute quantiles:

```
In [16]: print("25th percentile:   ", np.percentile(heights, 25))
         print("Median:            ", np.median(heights))
         print("75th percentile:   ", np.percentile(heights, 75))
```

```
Out[16]: 25th percentile:     174.75
         Median:              182.0
         75th percentile:     183.5
```

We see that the median height of US presidents is 182 cm, or just shy of six feet.

Of course, sometimes it's more useful to see a visual representation of this data, which we can accomplish using tools in Matplotlib (we'll discuss Matplotlib more fully in Part IV). For example, this code generates Figure 7-1:

```
In [17]: %matplotlib inline
         import matplotlib.pyplot as plt
         plt.style.use('seaborn-whitegrid')
```

```
In [18]: plt.hist(heights)
         plt.title('Height Distribution of US Presidents')
         plt.xlabel('height (cm)')
         plt.ylabel('number');
```

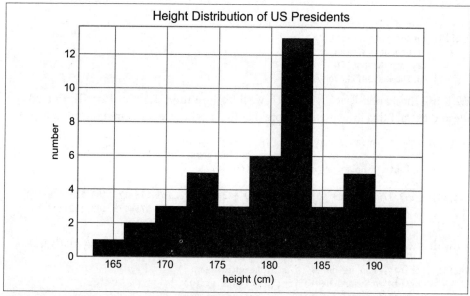

Figure 7-1. Histogram of presidential heights

## CHAPTER 8

Computation on Arrays: Broadcasting

We saw in Chapter 6 how NumPy's universal functions can be used to *vectorize* operations and thereby remove slow Python loops. This chapter discusses *broadcasting*: a set of rules by which NumPy lets you apply binary operations (e.g., addition, subtraction, multiplication, etc.) between arrays of different sizes and shapes.

Introducing Broadcasting

Recall that for arrays of the same size, binary operations are performed on an element-by-element basis:

```
In [1]: import numpy as np
In [2]: a = np.array([0, 1, 2])
        b = np.array([5, 5, 5])
        a + b
Out[2]: array([5, 6, 7])
```

Broadcasting allows these types of binary operations to be performed on arrays of different sizes—for example, we can just as easily add a scalar (think of it as a zero-dimensional array) to an array:

```
In [3]: a + 5
Out[3]: array([5, 6, 7])
```

We can think of this as an operation that stretches or duplicates the value 5 into the array [5, 5, 5], and adds the results.

We can similarly extend this idea to arrays of higher dimension. Observe the result when we add a one-dimensional array to a two-dimensional array:

```
In [4]: M = np.ones((3, 3))
        M
Out[4]: array([[1., 1., 1.],
               [1., 1., 1.],
               [1., 1., 1.]])

In [5]: M + a
Out[5]: array([[1., 2., 3.],
               [1., 2., 3.],
               [1., 2., 3.]])
```

Here the one-dimensional array a is stretched, or broadcasted, across the second dimension in order to match the shape of M.

While these examples are relatively easy to understand, more complicated cases can involve broadcasting of both arrays. Consider the following example:

```
In [6]: a = np.arange(3)
        b = np.arange(3)[:, np.newaxis]

        print(a)
        print(b)
Out[6]: [0 1 2]
        [[0]
         [1]
         [2]]

In [7]: a + b
Out[7]: array([[0, 1, 2],
               [1, 2, 3],
               [2, 3, 4]])
```

Just as before we stretched or broadcasted one value to match the shape of the other, here we've stretched *both* a and b to match a common shape, and the result is a two-dimensional array! The geometry of these examples is visualized in Figure 8-1.

The light boxes represent the broadcasted values. This way of thinking about broadcasting may raise questions about its efficiency in terms of memory use, but worry not: NumPy broadcasting does not actually copy the broadcasted values in memory. Still, this can be a useful mental model as we think about broadcasting.

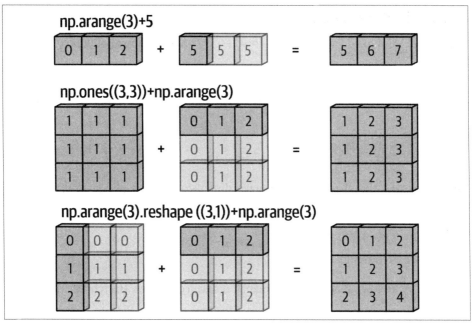

Figure 8-1. Visualization of NumPy broadcasting (adapted from a source published in the astroML documentation (http://astroml.org) and used with permission)[1]

Rules of Broadcasting

Broadcasting in NumPy follows a strict set of rules to determine the interaction between the two arrays:

Rule 1
> If the two arrays differ in their number of dimensions, the shape of the one with fewer dimensions is *padded* with ones on its leading (left) side.

Rule 2
> If the shape of the two arrays does not match in any dimension, the array with shape equal to 1 in that dimension is stretched to match the other shape.

Rule 3
> If in any dimension the sizes disagree and neither is equal to 1, an error is raised.

To make these rules clear, let's consider a few examples in detail.

---

1 Code to produce this plot can be found in the online appendix (*https://oreil.ly/gtOaU*).

Broadcasting Example 1

Suppose we want to add a two-dimensional array to a one-dimensional array:

```
In [8]: M = np.ones((2, 3))
        a = np.arange(3)
```

Let's consider an operation on these two arrays, which have the following shapes:

- `M.shape` is `(2, 3)`
- `a.shape` is `(3,)`

We see by rule 1 that the array a has fewer dimensions, so we pad it on the left with ones:

- `M.shape` remains `(2, 3)`
- `a.shape` becomes `(1, 3)`

By rule 2, we now see that the first dimension disagrees, so we stretch this dimension to match:

- `M.shape` remains `(2, 3)`
- `a.shape` becomes `(2, 3)`

The shapes now match, and we see that the final shape will be `(2, 3)`:

```
In [9]: M + a
Out[9]: array([[1., 2., 3.],
               [1., 2., 3.]])
```

Broadcasting Example 2

Now let's take a look at an example where both arrays need to be broadcast:

```
In [10]: a = np.arange(3).reshape((3, 1))
         b = np.arange(3)
```

Again, we'll start by determining the shapes of the arrays:

- `a.shape` is `(3, 1)`
- `b.shape` is `(3,)`

Rule 1 says we must pad the shape of b with ones:

- `a.shape` remains `(3, 1)`
- `b.shape` becomes `(1, 3)`

And rule 2 tells us that we must upgrade each of these 1s to match the corresponding size of the other array:

- a.shape becomes (3, 3)

- b.shape becomes (3, 3)

Because the results match, these shapes are compatible. We can see this here:

```
In [11]: a + b
Out[11]: array([[0, 1, 2],
                [1, 2, 3],
                [2, 3, 4]])
```

Broadcasting Example 3

Next, let's take a look at an example in which the two arrays are not compatible:

```
In [12]: M = np.ones((3, 2))
         a = np.arange(3)
```

This is just a slightly different situation than in the first example: the matrix M is transposed. How does this affect the calculation? The shapes of the arrays are as follows:

- M.shape is (3, 2)

- a.shape is (3,)

Again, rule 1 tells us that we must pad the shape of a with ones:

- M.shape remains (3, 2)

- a.shape becomes (1, 3)

By rule 2, the first dimension of a is then stretched to match that of M:

- M.shape remains (3, 2)

- a.shape becomes (3, 3)

Now we hit rule 3—the final shapes do not match, so these two arrays are incompatible, as we can observe by attempting this operation:

```
In [13]: M + a
ValueError: operands could not be broadcast together with shapes (3,2) (3,)
```

Note the potential confusion here: you could imagine making a and M compatible by, say, padding a's shape with ones on the right rather than the left. But this is not how the broadcasting rules work! That sort of flexibility might be useful in some cases, but it would lead to potential areas of ambiguity. If right-side padding is what you'd like, you can do this explicitly by reshaping the array (we'll use the `np.newaxis` keyword introduced in Chapter 5 for this):

```
In [14]: a[:, np.newaxis].shape
Out[14]: (3, 1)
```

```
In [15]: M + a[:, np.newaxis]
Out[15]: array([[1., 1.],
                [2., 2.],
                [3., 3.]])
```

While we've been focusing on the + operator here, these broadcasting rules apply to *any* binary ufunc. For example, here is the `logaddexp(a, b)` function, which computes `log(exp(a) + exp(b))` with more precision than the naive approach:

```
In [16]: np.logaddexp(M, a[:, np.newaxis])
Out[16]: array([[1.31326169, 1.31326169],
                [1.69314718, 1.69314718],
                [2.31326169, 2.31326169]])
```

For more information on the many available universal functions, refer to Chapter 6.

Broadcasting in Practice

Broadcasting operations form the core of many examples you'll see throughout this book. We'll now take a look at some instances of where they can be useful.

Centering an Array

In Chapter 6, we saw that ufuncs allow a NumPy user to remove the need to explicitly write slow Python loops. Broadcasting extends this ability. One commonly seen example in data science is subtracting the row-wise mean from an array of data. Imagine we have an array of 10 observations, each of which consists of 3 values. Using the standard convention (see Chapter 38), we'll store this in a 10 × 3 array:

```
In [17]: rng = np.random.default_rng(seed=1701)
         X = rng.random((10, 3))
```

We can compute the mean of each column using the `mean` aggregate across the first dimension:

```
In [18]: Xmean = X.mean(0)
         Xmean
Out[18]: array([0.38503638, 0.36991443, 0.63896043])
```

And now we can center the X array by subtracting the mean (this is a broadcasting operation):

```
In [19]: X_centered = X - Xmean
```

To double-check that we've done this correctly, we can check that the centered array has a mean near zero:

```
In [20]: X_centered.mean(0)
Out[20]: array([ 4.99600361e-17, -4.44089210e-17,  0.00000000e+00])
```

To within machine precision, the mean is now zero.

Plotting a Two-Dimensional Function

One place that broadcasting often comes in handy is in displaying images based on two-dimensional functions. If we want to define a function $z = f(x, y)$, broadcasting can be used to compute the function across the grid:

```
In [21]: # x and y have 50 steps from 0 to 5
         x = np.linspace(0, 5, 50)
         y = np.linspace(0, 5, 50)[:, np.newaxis]

         z = np.sin(x) ** 10 + np.cos(10 + y * x) * np.cos(x)
```

We'll use Matplotlib to plot this two-dimensional array, shown in Figure 8-2 (these tools will be discussed in full in Chapter 28):

```
In [22]: %matplotlib inline
         import matplotlib.pyplot as plt
```

```
In [23]: plt.imshow(z, origin='lower', extent=[0, 5, 0, 5])
         plt.colorbar();
```

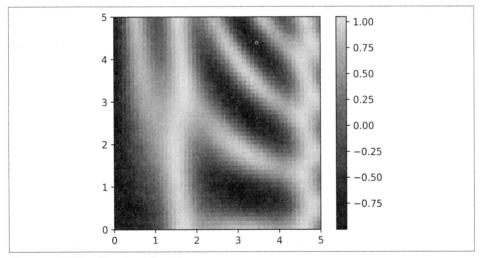

Figure 8-2. Visualization of a 2D array

The result is a compelling visualization of the two-dimensional function.

## CHAPTER 9

Comparisons, Masks, and Boolean Logic

This chapter covers the use of Boolean masks to examine and manipulate values within NumPy arrays. Masking comes up when you want to extract, modify, count, or otherwise manipulate values in an array based on some criterion: for example, you might wish to count all values greater than a certain value, or remove all outliers that are above some threshold. In NumPy, Boolean masking is often the most efficient way to accomplish these types of tasks.

Example: Counting Rainy Days

Imagine you have a series of data that represents the amount of precipitation each day for a year in a given city. For example, here we'll load the daily rainfall statistics for the city of Seattle in 2015, using Pandas (see Part III):

```
In [1]: import numpy as np
        from vega_datasets import data

        # Use DataFrame operations to extract rainfall as a NumPy array
        rainfall_mm = np.array(
            data.seattle_weather().set_index('date')['precipitation']['2015'])
        len(rainfall_mm)
Out[1]: 365
```

The array contains 365 values, giving daily rainfall in millimeters from January 1 to December 31, 2015.

As a first quick visualization, let's look at the histogram of rainy days in Figure 9-1, which was generated using Matplotlib (we will explore this tool more fully in Part IV):

```
In [2]: %matplotlib inline
        import matplotlib.pyplot as plt
        plt.style.use('seaborn-whitegrid')
```

```
In [3]: plt.hist(rainfall_mm, 40);
```

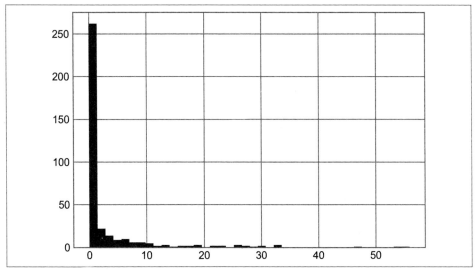

Figure 9-1. Histogram of 2015 rainfall in Seattle

This histogram gives us a general idea of what the data looks like: despite the city's rainy reputation, the vast majority of days in Seattle saw near zero measured rainfall in 2015. But this doesn't do a good job of conveying some information we'd like to see: for example, how many rainy days were there in the year? What was the average precipitation on those rainy days? How many days were there with more than 10 mm of rainfall?

One approach to this would be to answer these questions by hand: we could loop through the data, incrementing a counter each time we see values in some desired range. But for reasons discussed throughout this chapter, such an approach is very inefficient from the standpoint of both time writing code and time computing the result. We saw in Chapter 6 that NumPy's ufuncs can be used in place of loops to do fast element-wise arithmetic operations on arrays; in the same way, we can use other ufuncs to do element-wise *comparisons* over arrays, and we can then manipulate the results to answer the questions we have. We'll leave the data aside for now, and discuss some general tools in NumPy to use *masking* to quickly answer these types of questions.

Comparison Operators as Ufuncs

Chapter 6 introduced ufuncs, and focused in particular on arithmetic operators. We saw that using +, -, *, /, and other operators on arrays leads to element-wise operations. NumPy also implements comparison operators such as < (less than) and > (greater than) as element-wise ufuncs. The result of these comparison operators is always an array with a Boolean data type. All six of the standard comparison operations are available:

```
In [4]: x = np.array([1, 2, 3, 4, 5])

In [5]: x < 3  # less than
Out[5]: array([ True,  True, False, False, False])

In [6]: x > 3  # greater than
Out[6]: array([False, False, False,  True,  True])

In [7]: x <= 3  # less than or equal
Out[7]: array([ True,  True,  True, False, False])

In [8]: x >= 3  # greater than or equal
Out[8]: array([False, False,  True,  True,  True])

In [9]: x != 3  # not equal
Out[9]: array([ True,  True, False,  True,  True])

In [10]: x == 3  # equal
Out[10]: array([False, False,  True, False, False])
```

It is also possible to do an element-wise comparison of two arrays, and to include compound expressions:

```
In [11]: (2 * x) == (x ** 2)
Out[11]: array([False,  True, False, False, False])
```

As in the case of arithmetic operators, the comparison operators are implemented as ufuncs in NumPy; for example, when you write x < 3, internally NumPy uses np.less(x, 3). A summary of the comparison operators and their equivalent ufuncs is shown here:

Operator	Equivalent ufunc	Operator	Equivalent ufunc
==	np.equal	!=	np.not_equal
<	np.less	<=	np.less_equal
>	np.greater	>=	np.greater_equal

Just as in the case of arithmetic ufuncs, these will work on arrays of any size and shape. Here is a two-dimensional example:

```
In [12]: rng = np.random.default_rng(seed=1701)
         x = rng.integers(10, size=(3, 4))
         x
Out[12]: array([[9, 4, 0, 3],
                [8, 6, 3, 1],
                [3, 7, 4, 0]])

In [13]: x < 6
Out[13]: array([[False,  True,  True,  True],
                [False, False,  True,  True],
                [ True, False,  True,  True]])
```

In each case, the result is a Boolean array, and NumPy provides a number of straight-forward patterns for working with these Boolean results.

Working with Boolean Arrays

Given a Boolean array, there are a host of useful operations you can do. We'll work with x, the two-dimensional array we created earlier:

```
In [14]: print(x)
Out[14]: [[9 4 0 3]
          [8 6 3 1]
          [3 7 4 0]]
```

Counting Entries

To count the number of True entries in a Boolean array, np.count_nonzero is useful:

```
In [15]: # how many values less than 6?
         np.count_nonzero(x < 6)
Out[15]: 8
```

We see that there are eight array entries that are less than 6. Another way to get at this information is to use np.sum; in this case, False is interpreted as 0, and True is interpreted as 1:

```
In [16]: np.sum(x < 6)
Out[16]: 8
```

The benefit of np.sum is that, like with other NumPy aggregation functions, this summation can be done along rows or columns as well:

```
In [17]: # how many values less than 6 in each row?
         np.sum(x < 6, axis=1)
Out[17]: array([3, 2, 3])
```

This counts the number of values less than 6 in each row of the matrix.

If we're interested in quickly checking whether any or all the values are True, we can use (you guessed it) np.any or np.all:

```
In [18]: # are there any values greater than 8?
         np.any(x > 8)
Out[18]: True
```

```
In [19]: # are there any values less than zero?
         np.any(x < 0)
Out[19]: False
```

```
In [20]: # are all values less than 10?
         np.all(x < 10)
Out[20]: True
```

```
In [21]: # are all values equal to 6?
         np.all(x == 6)
Out[21]: False
```

`np.all` and `np.any` can be used along particular axes as well. For example:

```
In [22]: # are all values in each row less than 8?
         np.all(x < 8, axis=1)
Out[22]: array([False, False,  True])
```

Here all the elements in the third row are less than 8, while this is not the case for others.

Finally, a quick warning: as mentioned in Chapter 7, Python has built-in sum, any, and all functions. These have a different syntax than the NumPy versions, and in particular will fail or produce unintended results when used on multidimensional arrays. Be sure that you are using np.sum, np.any, and np.all for these examples!

Boolean Operators

We've already seen how we might count, say, all days with less than 20 mm of rain, or all days with more than 10 mm of rain. But what if we want to know how many days there were with more than 10 mm and less than 20 mm of rain? We can accomplish this with Python's *bitwise logic operators*, &, |, ^, and ~. Like with the standard arithmetic operators, NumPy overloads these as ufuncs that work element-wise on (usually Boolean) arrays.

For example, we can address this sort of compound question as follows:

```
In [23]: np.sum((rainfall_mm > 10) & (rainfall_mm < 20))
Out[23]: 16
```

This tells us that there were 16 days with rainfall of between 10 and 20 millimeters.

The parentheses here are important. Because of operator precedence rules, with the parentheses removed this expression would be evaluated as follows, which results in an error:

```
    rainfall_mm > (10 & rainfall_mm) < 20
```

Let's demonstrate a more complicated expression. Using De Morgan's laws, we can compute the same result in a different manner:

```
In [24]: np.sum(~( (rainfall_mm <= 10) | (rainfall_mm >= 20) ))
Out[24]: 16
```

Combining comparison operators and Boolean operators on arrays can lead to a wide range of efficient logical operations.

The following table summarizes the bitwise Boolean operators and their equivalent ufuncs:

Operator	Equivalent ufunc	Operator	Equivalent ufunc
&	np.bitwise_and		np.bitwise_or
^	np.bitwise_xor	~	np.bitwise_not

Using these tools, we can start to answer many of the questions we might have about our weather data. Here are some examples of results we can compute when combining Boolean operations with aggregations:

```
In [25]: print("Number days without rain: ", np.sum(rainfall_mm == 0))
         print("Number days with rain:    ", np.sum(rainfall_mm != 0))
         print("Days with more than 10 mm: ", np.sum(rainfall_mm > 10))
         print("Rainy days with < 5 mm:    ", np.sum((rainfall_mm > 0) &
                                                      (rainfall_mm < 5)))

Out[25]: Number days without rain:  221
         Number days with rain:     144
         Days with more than 10 mm: 34
         Rainy days with < 5 mm:    83
```

Boolean Arrays as Masks

In the preceding section we looked at aggregates computed directly on Boolean arrays. A more powerful pattern is to use Boolean arrays as masks, to select particular subsets of the data themselves. Let's return to our x array from before:

```
In [26]: x
Out[26]: array([[9, 4, 0, 3],
                [8, 6, 3, 1],
                [3, 7, 4, 0]])
```

Suppose we want an array of all values in the array that are less than, say, 5. We can obtain a Boolean array for this condition easily, as we've already seen:

```
In [27]: x < 5
Out[27]: array([[False,  True,  True,  True],
                [False, False,  True,  True],
                [ True, False,  True,  True]])
```

Now, to *select* these values from the array, we can simply index on this Boolean array; this is known as a *masking* operation:

```
In [28]: x[x < 5]
Out[28]: array([4, 0, 3, 3, 1, 3, 4, 0])
```

What is returned is a one-dimensional array filled with all the values that meet this condition; in other words, all the values in positions at which the mask array is True.

We are then free to operate on these values as we wish. For example, we can compute some relevant statistics on our Seattle rain data:

```
In [29]: # construct a mask of all rainy days
         rainy = (rainfall_mm > 0)
```

```
# construct a mask of all summer days (June 21st is the 172nd day)
days = np.arange(365)
summer = (days > 172) & (days < 262)

print("Median precip on rainy days in 2015 (mm):   ",
      np.median(rainfall_mm[rainy]))
print("Median precip on summer days in 2015 (mm):  ",
      np.median(rainfall_mm[summer]))
print("Maximum precip on summer days in 2015 (mm): ",
      np.max(rainfall_mm[summer]))
print("Median precip on non-summer rainy days (mm):",
      np.median(rainfall_mm[rainy & ~summer]))
```
```
Out[29]: Median precip on rainy days in 2015 (mm):    3.8
         Median precip on summer days in 2015 (mm):   0.0
         Maximum precip on summer days in 2015 (mm): 32.5
         Median precip on non-summer rainy days (mm): 4.1
```

By combining Boolean operations, masking operations, and aggregates, we can very quickly answer these sorts of questions about our dataset.

Using the Keywords and/or Versus the Operators &/|

One common point of confusion is the difference between the keywords and and or on the one hand, and the operators & and | on the other. When would you use one versus the other?

The difference is this: and and or operate on the object as a whole, while & and | operate on the elements within the object.

When you use and or or, it is equivalent to asking Python to treat the object as a single Boolean entity. In Python, all nonzero integers will evaluate as True. Thus:

```
In [30]: bool(42), bool(0)
Out[30]: (True, False)
```
```
In [31]: bool(42 and 0)
Out[31]: False
```
```
In [32]: bool(42 or 0)
Out[32]: True
```

When you use & and | on integers, the expression operates on the bitwise representation of the element, applying the *and* or the *or* to the individual bits making up the number:

```
In [33]: bin(42)
Out[33]: '0b101010'
```
```
In [34]: bin(59)
Out[34]: '0b111011'
```

```
In [35]: bin(42 & 59)
Out[35]: '0b101010'
```

```
In [36]: bin(42 | 59)
Out[36]: '0b111011'
```

Notice that the corresponding bits of the binary representation are compared in order to yield the result.

When you have an array of Boolean values in NumPy, this can be thought of as a string of bits where 1 = True and 0 = False, and & and | will operate similarly to in the preceding examples:

```
In [37]: A = np.array([1, 0, 1, 0, 1, 0], dtype=bool)
         B = np.array([1, 1, 1, 0, 1, 1], dtype=bool)
         A | B
Out[37]: array([ True,   True,   True, False,   True,   True])
```

But if you use or on these arrays it will try to evaluate the truth or falsehood of the entire array object, which is not a well-defined value:

```
In [38]: A or B
ValueError: The truth value of an array with more than one element is
        > ambiguous.
        a.any() or a.all()
```

Similarly, when evaluating a Boolean expression on a given array, you should use | or & rather than or or and:

```
In [39]: x = np.arange(10)
         (x > 4) & (x < 8)
Out[39]: array([False, False, False, False, False,  True,  True,  True, False,
                False])
```

Trying to evaluate the truth or falsehood of the entire array will give the same ValueError we saw previously:

```
In [40]: (x > 4) and (x < 8)
ValueError: The truth value of an array with more than one element is
        > ambiguous.
        a.any() or a.all()
```

So, remember this: and and or perform a single Boolean evaluation on an entire object, while & and | perform multiple Boolean evaluations on the content (the individual bits or bytes) of an object. For Boolean NumPy arrays, the latter is nearly always the desired operation.

## CHAPTER 10

Fancy Indexing

The previous chapters discussed how to access and modify portions of arrays using simple indices (e.g., arr[0]), slices (e.g., arr[:5]), and Boolean masks (e.g., arr[arr > 0]). In this chapter, we'll look at another style of array indexing, known as *fancy* or *vectorized* indexing, in which we pass arrays of indices in place of single scalars. This allows us to very quickly access and modify complicated subsets of an array's values.

Exploring Fancy Indexing

Fancy indexing is conceptually simple: it means passing an array of indices to access multiple array elements at once. For example, consider the following array:

```
In [1]: import numpy as np
        rng = np.random.default_rng(seed=1701)

        x = rng.integers(100, size=10)
        print(x)
Out[1]: [90 40  9 30 80 67 39 15 33 79]
```

Suppose we want to access three different elements. We could do it like this:

```
In [2]: [x[3], x[7], x[2]]
Out[2]: [30, 15, 9]
```

Alternatively, we can pass a single list or array of indices to obtain the same result:

```
In [3]: ind = [3, 7, 4]
        x[ind]
Out[3]: array([30, 15, 80])
```

When using arrays of indices, the shape of the result reflects the shape of the *index arrays* rather than the shape of the *array being indexed*:

```
In [4]: ind = np.array([[3, 7],
                        [4, 5]])
        x[ind]
Out[4]: array([[30, 15],
               [80, 67]])
```

Fancy indexing also works in multiple dimensions. Consider the following array:

```
In [5]: X = np.arange(12).reshape((3, 4))
        X
Out[5]: array([[ 0,  1,  2,  3],
               [ 4,  5,  6,  7],
               [ 8,  9, 10, 11]])
```

Like with standard indexing, the first index refers to the row, and the second to the column:

```
In [6]: row = np.array([0, 1, 2])
        col = np.array([2, 1, 3])
        X[row, col]
Out[6]: array([ 2,  5, 11])
```

Notice that the first value in the result is X[0, 2], the second is X[1, 1], and the third is X[2, 3]. The pairing of indices in fancy indexing follows all the broadcasting rules that were mentioned in Chapter 8. So, for example, if we combine a column vector and a row vector within the indices, we get a two-dimensional result:

```
In [7]: X[row[:, np.newaxis], col]
Out[7]: array([[ 2,  1,  3],
               [ 6,  5,  7],
               [10,  9, 11]])
```

Here, each row value is matched with each column vector, exactly as we saw in broadcasting of arithmetic operations. For example:

```
In [8]: row[:, np.newaxis] * col
Out[8]: array([[0, 0, 0],
               [2, 1, 3],
               [4, 2, 6]])
```

It is always important to remember with fancy indexing that the return value reflects the *broadcasted shape of the indices*, rather than the shape of the array being indexed.

Combined Indexing

For even more powerful operations, fancy indexing can be combined with the other indexing schemes we've seen. For example, given the array X:

```
In [9]: print(X)
Out[9]: [[ 0  1  2  3]
         [ 4  5  6  7]
         [ 8  9 10 11]]
```

We can combine fancy and simple indices:

```
In [10]: X[2, [2, 0, 1]]
Out[10]: array([10,  8,  9])
```

We can also combine fancy indexing with slicing:

```
In [11]: X[1:, [2, 0, 1]]
Out[11]: array([[ 6,  4,  5],
                [10,  8,  9]])
```

And we can combine fancy indexing with masking:

```
In [12]: mask = np.array([True, False, True, False])
         X[row[:, np.newaxis], mask]
Out[12]: array([[ 0,  2],
                [ 4,  6],
                [ 8, 10]])
```

All of these indexing options combined lead to a very flexible set of operations for efficiently accessing and modifying array values.

Example: Selecting Random Points

One common use of fancy indexing is the selection of subsets of rows from a matrix. For example, we might have an $N \times D$ matrix representing N points in D dimensions, such as the following points drawn from a two-dimensional normal distribution:

```
In [13]: mean = [0, 0]
         cov = [[1, 2],
                [2, 5]]
         X = rng.multivariate_normal(mean, cov, 100)
         X.shape
Out[13]: (100, 2)
```

Using the plotting tools we will discuss in Part IV, we can visualize these points as a scatter plot (Figure 10-1).

```
In [14]: %matplotlib inline
         import matplotlib.pyplot as plt
         plt.style.use('seaborn-whitegrid')

         plt.scatter(X[:, 0], X[:, 1]);
```

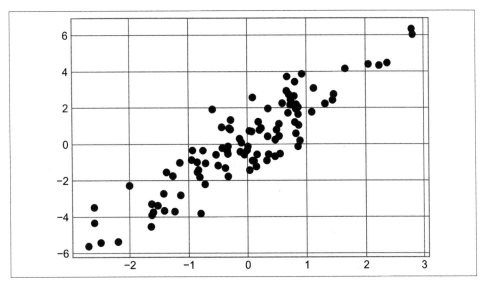

Figure 10-1. Normally distributed points

Let's use fancy indexing to select 20 random points. We'll do this by first choosing 20 random indices with no repeats, and using these indices to select a portion of the original array:

```
In [15]: indices = np.random.choice(X.shape[0], 20, replace=False)
         indices
Out[15]: array([82, 84, 10, 55, 14, 33,  4, 16, 34, 92, 99, 64,  8, 76, 68, 18, 59,
                80, 87, 90])
In [16]: selection = X[indices]  # fancy indexing here
         selection.shape
Out[16]: (20, 2)
```

Now to see which points were selected, let's overplot large circles at the locations of the selected points (see Figure 10-2).

```
In [17]: plt.scatter(X[:, 0], X[:, 1], alpha=0.3)
         plt.scatter(selection[:, 0], selection[:, 1],
                     facecolor='none', edgecolor='black', s=200);
```

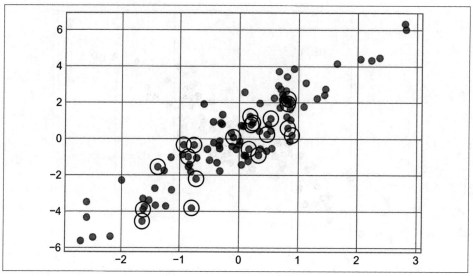

Figure 10-2. Random selection among points

This sort of strategy is often used to quickly partition datasets, as is often needed in train/test splitting for validation of statistical models (see Chapter 39), and in sampling approaches to answering statistical questions.

Modifying Values with Fancy Indexing

Just as fancy indexing can be used to access parts of an array, it can also be used to modify parts of an array. For example, imagine we have an array of indices and we'd like to set the corresponding items in an array to some value:

```
In [18]: x = np.arange(10)
         i = np.array([2, 1, 8, 4])
         x[i] = 99
         print(x)
Out[18]: [ 0 99 99  3 99  5  6  7 99  9]
```

We can use any assignment-type operator for this. For example:

```
In [19]: x[i] -= 10
         print(x)
Out[19]: [ 0 89 89  3 89  5  6  7 89  9]
```

Notice, though, that repeated indices with these operations can cause some potentially unexpected results. Consider the following:

```
In [20]: x = np.zeros(10)
         x[[0, 0]] = [4, 6]
         print(x)
Out[20]: [6. 0. 0. 0. 0. 0. 0. 0. 0. 0.]
```

Where did the 4 go? This operation first assigns x[0] = 4, followed by x[0] = 6. The result, of course, is that x[0] contains the value 6.

Fair enough, but consider this operation:

```
In [21]: i = [2, 3, 3, 4, 4, 4]
         x[i] += 1
         x
Out[21]: array([6., 0., 1., 1., 1., 0., 0., 0., 0., 0.])
```

You might expect that x[3] would contain the value 2 and x[4] would contain the value 3, as this is how many times each index is repeated. Why is this not the case? Conceptually, this is because x[i] += 1 is meant as a shorthand of x[i] = x[i] + 1. x[i] + 1 is evaluated, and then the result is assigned to the indices in x. With this in mind, it is not the augmentation that happens multiple times, but the assignment, which leads to the rather nonintuitive results.

So what if you want the other behavior where the operation is repeated? For this, you can use the at method of ufuncs and do the following:

```
In [22]: x = np.zeros(10)
         np.add.at(x, i, 1)
         print(x)
Out[22]: [0. 0. 1. 2. 3. 0. 0. 0. 0. 0.]
```

The at method does an in-place application of the given operator at the specified indices (here, i) with the specified value (here, 1). Another method that is similar in spirit is the reduceat method of ufuncs, which you can read about in the NumPy documentation (*https://oreil.ly/7ys9D*).

Example: Binning Data

You could use these ideas to efficiently do custom binned computations on data. For example, imagine we have 100 values and would like to quickly find where they fall within an array of bins. We could compute this using ufunc.at like this:

```
In [23]: rng = np.random.default_rng(seed=1701)
         x = rng.normal(size=100)

         # compute a histogram by hand
         bins = np.linspace(-5, 5, 20)
         counts = np.zeros_like(bins)

         # find the appropriate bin for each x
         i = np.searchsorted(bins, x)

         # add 1 to each of these bins
         np.add.at(counts, i, 1)
```

The counts now reflect the number of points within each bin—in other words, a histogram (see Figure 10-3).

```
In [24]: # plot the results
         plt.plot(bins, counts, drawstyle='steps');
```

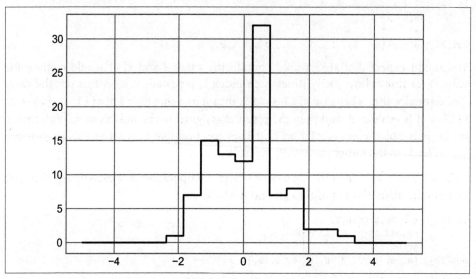

Figure 10-3. A histogram computed by hand

Of course, it would be inconvenient to have to do this each time you want to plot a histogram. This is why Matplotlib provides the `plt.hist` routine, which does the same in a single line:

```
    plt.hist(x, bins, histtype='step');
```

This function will create a nearly identical plot to the one just shown. To compute the binning, Matplotlib uses the `np.histogram` function, which does a very similar computation to what we did before. Let's compare the two here:

```
In [25]: print(f"NumPy histogram ({len(x)} points):")
         %timeit counts, edges = np.histogram(x, bins)

         print(f"Custom histogram ({len(x)} points):")
         %timeit np.add.at(counts, np.searchsorted(bins, x), 1)
Out[25]: NumPy histogram (100 points):
         33.8 µs ± 311 ns per loop (mean ± std. dev. of 7 runs, 10000 loops each)
         Custom histogram (100 points):
         17.6 µs ± 113 ns per loop (mean ± std. dev. of 7 runs, 100000 loops each)
```

Our own one-line algorithm is twice as fast as the optimized algorithm in NumPy! How can this be? If you dig into the `np.histogram` source code (you can do this in IPython by typing `np.histogram??`), you'll see that it's quite a bit more involved than the simple search-and-count that we've done; this is because NumPy's algorithm is more flexible, and particularly is designed for better performance when the number of data points becomes large:

```
In [26]: x = rng.normal(size=1000000)
         print(f"NumPy histogram ({len(x)} points):")
         %timeit counts, edges = np.histogram(x, bins)

         print(f"Custom histogram ({len(x)} points):")
         %timeit np.add.at(counts, np.searchsorted(bins, x), 1)
Out[26]: NumPy histogram (1000000 points):
         84.4 ms ± 2.82 ms per loop (mean ± std. dev. of 7 runs, 10 loops each)
         Custom histogram (1000000 points):
         128 ms ± 2.04 ms per loop (mean ± std. dev. of 7 runs, 10 loops each)
```

What this comparison shows is that algorithmic efficiency is almost never a simple question. An algorithm efficient for large datasets will not always be the best choice for small datasets, and vice versa (see Chapter 11). But the advantage of coding this algorithm yourself is that with an understanding of these basic methods, the sky is the limit: you're no longer constrained to built-in routines, but can create your own approaches to exploring the data. Key to efficiently using Python in data-intensive applications is not only knowing about general convenience routines like np.histo gram and when they're appropriate, but also knowing how to make use of lower-level functionality when you need more pointed behavior.

## CHAPTER 11

Sorting Arrays

Up to this point we have been concerned mainly with tools to access and operate on array data with NumPy. This chapter covers algorithms related to sorting values in NumPy arrays. These algorithms are a favorite topic in introductory computer science courses: if you've ever taken one, you probably have had dreams (or, depending on your temperament, nightmares) about *insertion sorts*, *selection sorts*, *merge sorts*, *quick sorts*, *bubble sorts*, and many, many more. All are means of accomplishing a similar task: sorting the values in a list or array.

Python has a couple of built-in functions and methods for sorting lists and other iterable objects. The sorted function accepts a list and returns a sorted version of it:

```
In [1]: L = [3, 1, 4, 1, 5, 9, 2, 6]
        sorted(L)  # returns a sorted copy
Out[1]: [1, 1, 2, 3, 4, 5, 6, 9]
```

By contrast, the sort method of lists will sort the list in-place:

```
In [2]: L.sort()  # acts in-place and returns None
        print(L)
Out[2]: [1, 1, 2, 3, 4, 5, 6, 9]
```

Python's sorting methods are quite flexible, and can handle any iterable object. For example, here we sort a string:

```
In [3]: sorted('python')
Out[3]: ['h', 'n', 'o', 'p', 't', 'y']
```

These built-in sorting methods are convenient, but as previously discussed, the dynamism of Python values means that they are less performant than routines designed specifically for uniform arrays of numbers. This is where NumPy's sorting routines come in.

Fast Sorting in NumPy: np.sort and np.argsort

The np.sort function is analogous to Python's built-in sorted function, and will efficiently return a sorted copy of an array:

```
In [4]: import numpy as np

        x = np.array([2, 1, 4, 3, 5])
        np.sort(x)
Out[4]: array([1, 2, 3, 4, 5])
```

Similarly to the sort method of Python lists, you can also sort an array in-place using
the array sort method:

```
In [5]: x.sort()
        print(x)
Out[5]: [1 2 3 4 5]
```

A related function is argsort, which instead returns the *indices* of the sorted elements:

```
In [6]: x = np.array([2, 1, 4, 3, 5])
        i = np.argsort(x)
        print(i)
Out[6]: [1 0 3 2 4]
```

The first element of this result gives the index of the smallest element, the second
value gives the index of the second smallest, and so on. These indices can then be
used (via fancy indexing) to construct the sorted array if desired:

```
In [7]: x[i]
Out[7]: array([1, 2, 3, 4, 5])
```

You'll see an application of argsort later in this chapter.

Sorting Along Rows or Columns

A useful feature of NumPy's sorting algorithms is the ability to sort along specific
rows or columns of a multidimensional array using the axis argument. For example:

```
In [8]: rng = np.random.default_rng(seed=42)
        X = rng.integers(0, 10, (4, 6))
        print(X)
Out[8]: [[0 7 6 4 4 8]
         [0 6 2 0 5 9]
         [7 7 7 7 5 1]
         [8 4 5 3 1 9]]

In [9]: # sort each column of X
        np.sort(X, axis=0)
Out[9]: array([[0, 4, 2, 0, 1, 1],
               [0, 6, 5, 3, 4, 8],
```

```
          [7, 7, 6, 4, 5, 9],
          [8, 7, 7, 7, 5, 9]])

In [10]: # sort each row of X
         np.sort(X, axis=1)
Out[10]: array([[0, 4, 4, 6, 7, 8],
          [0, 0, 2, 5, 6, 9],
          [1, 5, 7, 7, 7, 7],
          [1, 3, 4, 5, 8, 9]])
```

Keep in mind that this treats each row or column as an independent array, and any relationships between the row or column values will be lost!

Partial Sorts: Partitioning

Sometimes we're not interested in sorting the entire array, but simply want to find the *k* smallest values in the array. NumPy enables this with the `np.partition` function. `np.partition` takes an array and a number *k*; the result is a new array with the smallest *k* values to the left of the partition and the remaining values to the right:

```
In [11]: x = np.array([7, 2, 3, 1, 6, 5, 4])
         np.partition(x, 3)
Out[11]: array([2, 1, 3, 4, 6, 5, 7])
```

Notice that the first three values in the resulting array are the three smallest in the array, and the remaining array positions contain the remaining values. Within the two partitions, the elements have arbitrary order.

Similarly to sorting, we can partition along an arbitrary axis of a multidimensional array:

```
In [12]: np.partition(X, 2, axis=1)
Out[12]: array([[0, 4, 4, 7, 6, 8],
          [0, 0, 2, 6, 5, 9],
          [1, 5, 7, 7, 7, 7],
          [1, 3, 4, 5, 8, 9]])
```

The result is an array where the first two slots in each row contain the smallest values from that row, with the remaining values filling the remaining slots.

Finally, just as there is an `np.argsort` function that computes indices of the sort, there is an `np.argpartition` function that computes indices of the partition. We'll see both of these in action in the following section.

Example: k-Nearest Neighbors

Let's quickly see how we might use the `argsort` function along multiple axes to find the nearest neighbors of each point in a set. We'll start by creating a random set of 10 points on a two-dimensional plane. Using the standard convention, we'll arrange these in a 10 × 2 array:

```
In [13]: X = rng.random((10, 2))
```

To get an idea of how these points look, let's generate a quick scatter plot (see Figure 11-1).

```
In [14]: %matplotlib inline
         import matplotlib.pyplot as plt
         plt.style.use('seaborn-whitegrid')
         plt.scatter(X[:, 0], X[:, 1], s=100);
```

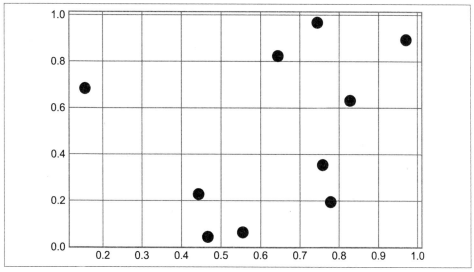

Figure 11-1. Visualization of points in the k-*neighbors example*

Now we'll compute the distance between each pair of points. Recall that the squared distance between two points is the sum of the squared differences in each dimension; using the efficient broadcasting (Chapter 8) and aggregation (Chapter 7) routines provided by NumPy we can compute the matrix of square distances in a single line of code:

```
In [15]: dist_sq = np.sum((X[:, np.newaxis] - X[np.newaxis, :]) ** 2, axis=-1)
```

This operation has a lot packed into it, and it might be a bit confusing if you're unfamiliar with NumPy's broadcasting rules. When you come across code like this, it can be useful to break it down into its component steps:

```
In [16]: # for each pair of points, compute differences in their coordinates
         differences = X[:, np.newaxis] - X[np.newaxis, :]
         differences.shape
Out[16]: (10, 10, 2)
```

```
In [17]: # square the coordinate differences
         sq_differences = differences ** 2
         sq_differences.shape
Out[17]: (10, 10, 2)
```

```
In [18]: # sum the coordinate differences to get the squared distance
         dist_sq = sq_differences.sum(-1)
         dist_sq.shape
Out[18]: (10, 10)
```

As a quick check of our logic, we should see that the diagonal of this matrix (i.e., the set of distances between each point and itself) is all zeros:

```
In [19]: dist_sq.diagonal()
Out[19]: array([0., 0., 0., 0., 0., 0., 0., 0., 0., 0.])
```

With the pairwise square distances converted, we can now use `np.argsort` to sort along each row. The leftmost columns will then give the indices of the nearest neighbors:

```
In [20]: nearest = np.argsort(dist_sq, axis=1)
         print(nearest)
Out[20]: [[0 9 3 5 4 8 1 6 2 7]
          [1 7 2 6 4 8 3 0 9 5]
          [2 7 1 6 4 3 8 0 9 5]
          [3 0 4 5 9 6 1 2 8 7]
          [4 6 3 1 2 7 0 5 9 8]
          [5 9 3 0 4 6 8 1 2 7]
          [6 4 2 1 7 3 0 5 9 8]
          [7 2 1 6 4 3 8 0 9 5]
          [8 0 1 9 3 4 7 2 6 5]
          [9 0 5 3 4 8 6 1 2 7]]
```

Notice that the first column gives the numbers 0 through 9 in order: this is due to the fact that each point's closest neighbor is itself, as we would expect.

By using a full sort here, we've actually done more work than we need to in this case. If we're simply interested in the nearest k neighbors, all we need to do is partition each row so that the smallest $k + 1$ squared distances come first, with larger distances filling the remaining positions of the array. We can do this with the `np.argpartition` function:

```
In [21]: K = 2
         nearest_partition = np.argpartition(dist_sq, K + 1, axis=1)
```

In order to visualize this network of neighbors, let's quickly plot the points along with lines representing the connections from each point to its two nearest neighbors (see Figure 11-2).

```
In [22]: plt.scatter(X[:, 0], X[:, 1], s=100)

         # draw lines from each point to its two nearest neighbors
         K = 2
```

```
for i in range(X.shape[0]):
    for j in nearest_partition[i, :K+1]:
        # plot a line from X[i] to X[j]
        # use some zip magic to make it happen:
        plt.plot(*zip(X[j], X[i]), color='black')
```

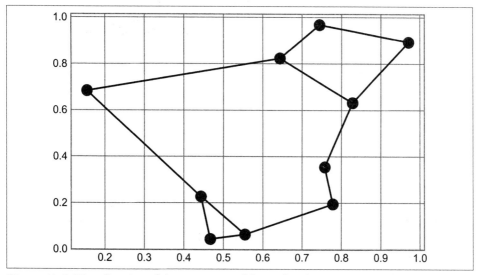

Figure 11-2. Visualization of the nearest neighbors of each point

Each point in the plot has lines drawn to its two nearest neighbors. At first glance, it might seem strange that some of the points have more than two lines coming out of them: this is due to the fact that if point A is one of the two nearest neighbors of point B, this does not necessarily imply that point B is one of the two nearest neighbors of point A.

Although the broadcasting and row-wise sorting of this approach might seem less straightforward than writing a loop, it turns out to be a very efficient way of operating on this data in Python. You might be tempted to do the same type of operation by manually looping through the data and sorting each set of neighbors individually, but this would almost certainly lead to a slower algorithm than the vectorized version we used. The beauty of this approach is that it's written in a way that's agnostic to the size of the input data: we could just as easily compute the neighbors among 100 or 1,000,000 points in any number of dimensions, and the code would look the same.

Finally, I'll note that when doing very large nearest neighbor searches, there are tree-based and/or approximate algorithms that can scale as $\mathcal{O}[N \log N]$ or better rather than the $\mathcal{O}[N^2]$ of the brute-force algorithm. One example of this is the KD-Tree, implemented in Scikit-Learn (*https://oreil.ly/lUFb8*).

## CHAPTER 12

Structured Data: NumPy's Structured Arrays

While often our data can be well represented by a homogeneous array of values, sometimes this is not the case. This chapter demonstrates the use of NumPy's *structured arrays* and *record arrays*, which provide efficient storage for compound, heterogeneous data. While the patterns shown here are useful for simple operations, scenarios like this often lend themselves to the use of Pandas `DataFrame`s, which we'll explore in Part III.

```
In [1]: import numpy as np
```

Imagine that we have several categories of data on a number of people (say, name, age, and weight), and we'd like to store these values for use in a Python program. It would be possible to store these in three separate arrays:

```
In [2]: name = ['Alice', 'Bob', 'Cathy', 'Doug']
        age = [25, 45, 37, 19]
        weight = [55.0, 85.5, 68.0, 61.5]
```

But this is a bit clumsy. There's nothing here that tells us that the three arrays are related; NumPy's structured arrays allow us to do this more naturally by using a single structure to store all of this data.

Recall that previously we created a simple array using an expression like this:

```
In [3]: x = np.zeros(4, dtype=int)
```

We can similarly create a structured array using a compound data type specification:

```
In [4]: # Use a compound data type for structured arrays
        data = np.zeros(4, dtype={'names':('name', 'age', 'weight'),
                                  'formats':('U10', 'i4', 'f8')})
```

```
        print(data.dtype)
Out[4]: [('name', '<U10'), ('age', '<i4'), ('weight', '<f8')]
```

Here `'U10'` translates to "Unicode string of maximum length 10," `'i4'` translates to "4-byte (i.e., 32-bit) integer," and `'f8'` translates to "8-byte (i.e., 64-bit) float." We'll discuss other options for these type codes in the following section.

Now that we've created an empty container array, we can fill the array with our lists of values:

```
In [5]: data['name'] = name
        data['age'] = age
        data['weight'] = weight
        print(data)
Out[5]: [('Alice', 25, 55. ) ('Bob', 45, 85.5) ('Cathy', 37, 68. )
         ('Doug', 19, 61.5)]
```

As we had hoped, the data is now conveniently arranged in one structured array.

The handy thing with structured arrays is that we can now refer to values either by index or by name:

```
In [6]: # Get all names
        data['name']
Out[6]: array(['Alice', 'Bob', 'Cathy', 'Doug'], dtype='<U10')

In [7]: # Get first row of data
        data[0]
Out[7]: ('Alice', 25, 55.)

In [8]: # Get the name from the last row
        data[-1]['name']
Out[8]: 'Doug'
```

Using Boolean masking, we can even do some more sophisticated operations, such as filtering on age:

```
In [9]: # Get names where age is under 30
        data[data['age'] < 30]['name']
Out[9]: array(['Alice', 'Doug'], dtype='<U10')
```

If you'd like to do any operations that are any more complicated than these, you should probably consider the Pandas package, covered in Part IV. As you'll see, Pandas provides a `DataFrame` object, which is a structure built on NumPy arrays that offers a variety of useful data manipulation functionality similar to what you've seen here, as well as much, much more.

Exploring Structured Array Creation

Structured array data types can be specified in a number of ways. Earlier, we saw the dictionary method:

```
In [10]: np.dtype({'names':('name', 'age', 'weight'),
                    'formats':('U10', 'i4', 'f8')})
Out[10]: dtype([('name', '<U10'), ('age', '<i4'), ('weight', '<f8')])
```

For clarity, numerical types can be specified using Python types or NumPy dtypes instead:

```
In [11]: np.dtype({'names':('name', 'age', 'weight'),
                    'formats':((np.str_, 10), int, np.float32)})
Out[11]: dtype([('name', '<U10'), ('age', '<i8'), ('weight', '<f4')])
```

A compound type can also be specified as a list of tuples:

```
In [12]: np.dtype([('name', 'S10'), ('age', 'i4'), ('weight', 'f8')])
Out[12]: dtype([('name', 'S10'), ('age', '<i4'), ('weight', '<f8')])
```

If the names of the types do not matter to you, you can specify the types alone in a comma-separated string:

```
In [13]: np.dtype('S10,i4,f8')
Out[13]: dtype([('f0', 'S10'), ('f1', '<i4'), ('f2', '<f8')])
```

The shortened string format codes may not be immediately intuitive, but they are built on simple principles. The first (optional) character < or >, means "little endian" or "big endian," respectively, and specifies the ordering convention for significant bits. The next character specifies the type of data: characters, bytes, ints, floating points, and so on (see Table 12-1). The last character or characters represent the size of the object in bytes.

Table 12-1. NumPy data types

Character	Description	Example
'b'	Byte	np.dtype('b')
'i'	Signed integer	np.dtype('i4') == np.int32
'u'	Unsigned integer	np.dtype('u1') == np.uint8
'f'	Floating point	np.dtype('f8') == np.int64
'c'	Complex floating point	np.dtype('c16') == np.complex128
'S', 'a'	String	np.dtype('S5')
'U'	Unicode string	np.dtype('U') == np.str_
'V'	Raw data (void)	np.dtype('V') == np.void

More Advanced Compound Types

It is possible to define even more advanced compound types. For example, you can create a type where each element contains an array or matrix of values. Here, we'll create a data type with a `mat` component consisting of a 3×3 floating-point matrix:

```
In [14]: tp = np.dtype([('id', 'i8'), ('mat', 'f8', (3, 3))])
         X = np.zeros(1, dtype=tp)
         print(X[0])
         print(X['mat'][0])
Out[14]: (0, [[0., 0., 0.], [0., 0., 0.], [0., 0., 0.]])
         [[0. 0. 0.]
          [0. 0. 0.]
          [0. 0. 0.]]
```

Now each element in the X array consists of an `id` and a 3×3 matrix. Why would you use this rather than a simple multidimensional array, or perhaps a Python dictionary? One reason is that this NumPy `dtype` directly maps onto a C structure definition, so the buffer containing the array content can be accessed directly within an appropriately written C program. If you find yourself writing a Python interface to a legacy C or Fortran library that manipulates structured data, structured arrays can provide a powerful interface.

Record Arrays: Structured Arrays with a Twist

NumPy also provides record arrays (instances of the `np.recarray` class), which are almost identical to the structured arrays just described, but with one additional feature: fields can be accessed as attributes rather than as dictionary keys. Recall that we previously accessed the ages in our sample dataset by writing:

```
In [15]: data['age']
Out[15]: array([25, 45, 37, 19], dtype=int32)
```

If we view our data as a record array instead, we can access this with slightly fewer keystrokes:

```
In [16]: data_rec = data.view(np.recarray)
         data_rec.age
Out[16]: array([25, 45, 37, 19], dtype=int32)
```

The downside is that for record arrays, there is some extra overhead involved in accessing the fields, even when using the same syntax:

```
In [17]: %timeit data['age']
         %timeit data_rec['age']
         %timeit data_rec.age
Out[17]: 121 ns ± 1.4 ns per loop (mean ± std. dev. of 7 runs, 1000000 loops each)
         2.41 µs ± 15.7 ns per loop (mean ± std. dev. of 7 runs, 100000 loops each)
         3.98 µs ± 20.5 ns per loop (mean ± std. dev. of 7 runs, 100000 loops each)
```

Whether the more convenient notation is worth the (slight) overhead will depend on your own application.

On to Pandas

This chapter on structured and record arrays is purposely located at the end of this part of the book, because it leads so well into the next package we will cover: Pandas. Structured arrays can come in handy in certain situations, like when you're using NumPy arrays to map onto binary data formats in C, Fortran, or another language. But for day-to-day use of structured data, the Pandas package is a much better choice; we'll explore it in depth in the chapters that follow.

## PART III

Data Manipulation with Pandas

In Part II, we dove into detail on NumPy and its `ndarray` object, which enables efficient storage and manipulation of dense typed arrays in Python. Here we'll build on this knowledge by looking in depth at the data structures provided by the Pandas library. Pandas is a newer package built on top of NumPy that provides an efficient implementation of a `DataFrame`. `DataFrames` are essentially multidimensional arrays with attached row and column labels, often with heterogeneous types and/or missing data. As well as offering a convenient storage interface for labeled data, Pandas implements a number of powerful data operations familiar to users of both database frameworks and spreadsheet programs.

As we've seen, NumPy's `ndarray` data structure provides essential features for the type of clean, well-organized data typically seen in numerical computing tasks. While it serves this purpose very well, its limitations become clear when we need more flexibility (e.g., attaching labels to data, working with missing data, etc.) and when attempting operations that do not map well to element-wise broadcasting (e.g., groupings, pivots, etc.), each of which is an important piece of analyzing the less structured data available in many forms in the world around us. Pandas, and in particular its `Series` and `DataFrame` objects, builds on the NumPy array structure and provides efficient access to these sorts of "data munging" tasks that occupy much of a data scientist's time.

In this part of the book, we will focus on the mechanics of using `Series`, `DataFrame`, and related structures effectively. We will use examples drawn from real datasets where appropriate, but these examples are not necessarily the focus.

Installing Pandas on your system requires NumPy, and if you're building the library from source, you will need the appropriate tools to compile the C and Cython sources on which Pandas is built. Details on the installation process can be found in the Pandas documentation (*http://pandas.pydata.org*). If you followed the advice outlined in the Preface and used the Anaconda stack, you already have Pandas installed.

Once Pandas is installed, you can import it and check the version; here is the version used by this book:

```
In [1]: import pandas
        pandas.__version__
Out[1]: '1.3.5'
```

Just as we generally import NumPy under the alias np, we will import Pandas under the alias pd:

```
In [2]: import pandas as pd
```

This import convention will be used throughout the remainder of this book.

Reminder About Built-in Documentation

As you read through this part of the book, don't forget that IPython gives you the ability to quickly explore the contents of a package (by using the tab completion feature) as well as the documentation of various functions (using the ? character). Refer back to Chapter 1 if you need a refresher on this.

For example, to display all the contents of the Pandas namespace, you can type:

```
In [3]: pd.<TAB>
```

And to display the built-in Pandas documentation, you can use this:

```
In [4]: pd?
```

See the Pandas website for more detailed documentation (*http://pandas.pydata.org*), along with tutorials and other resources.

## CHAPTER 13

Introducing Pandas Objects

At a very basic level, Pandas objects can be thought of as enhanced versions of NumPy structured arrays in which the rows and columns are identified with labels rather than simple integer indices. As we will see during the course of this chapter, Pandas provides a host of useful tools, methods, and functionality on top of the basic data structures, but nearly everything that follows will require an understanding of what these structures are. Thus, before we go any further, let's take a look at these three fundamental Pandas data structures: the Series, DataFrame, and Index.

We will start our code sessions with the standard NumPy and Pandas imports:

```
In [1]: import numpy as np
        import pandas as pd
```

The Pandas Series Object

A Pandas Series is a one-dimensional array of indexed data. It can be created from a list or array as follows:

```
In [2]: data = pd.Series([0.25, 0.5, 0.75, 1.0])
        data
Out[2]: 0    0.25
        1    0.50
        2    0.75
        3    1.00
        dtype: float64
```

The Series combines a sequence of values with an explicit sequence of indices, which we can access with the values and index attributes. The values are simply a familiar NumPy array:

```
In [3]: data.values
Out[3]: array([0.25, 0.5 , 0.75, 1.  ])
```

The index is an array-like object of type pd.Index, which we'll discuss in more detail momentarily:

```
In [4]: data.index
Out[4]: RangeIndex(start=0, stop=4, step=1)
```

Like with a NumPy array, data can be accessed by the associated index via the familiar Python square-bracket notation:

```
In [5]: data[1]
Out[5]: 0.5

In [6]: data[1:3]
Out[6]: 1    0.50
        2    0.75
        dtype: float64
```

As we will see, though, the Pandas Series is much more general and flexible than the one-dimensional NumPy array that it emulates.

Series as Generalized NumPy Array

From what we've seen so far, the Series object may appear to be basically inter-changeable with a one-dimensional NumPy array. The essential difference is that while the NumPy array has an *implicitly defined* integer index used to access the val-ues, the Pandas Series has an *explicitly defined* index associated with the values.

This explicit index definition gives the Series object additional capabilities. For example, the index need not be an integer, but can consist of values of any desired type. So, if we wish, we can use strings as an index:

```
In [7]: data = pd.Series([0.25, 0.5, 0.75, 1.0],
                         index=['a', 'b', 'c', 'd'])
        data
Out[7]: a    0.25
        b    0.50
        c    0.75
        d    1.00
        dtype: float64
```

And the item access works as expected:

```
In [8]: data['b']
Out[8]: 0.5
```

We can even use noncontiguous or nonsequential indices:

```
In [9]: data = pd.Series([0.25, 0.5, 0.75, 1.0],
                         index=[2, 5, 3, 7])
        data
Out[9]: 2    0.25
        5    0.50
        3    0.75
```

```
7    1.00
dtype: float64

In [10]: data[5]
Out[10]: 0.5
```

Series as Specialized Dictionary

In this way, you can think of a Pandas Series a bit like a specialization of a Python dictionary. A dictionary is a structure that maps arbitrary keys to a set of arbitrary values, and a Series is a structure that maps typed keys to a set of typed values. This typing is important: just as the type-specific compiled code behind a NumPy array makes it more efficient than a Python list for certain operations, the type information of a Pandas Series makes it more efficient than Python dictionaries for certain operations.

The Series-as-dictionary analogy can be made even more clear by constructing a Series object directly from a Python dictionary, here the five most populous US states according to the 2020 census:

```
In [11]: population_dict = {'California': 39538223, 'Texas': 29145505,
                            'Florida': 21538187, 'New York': 20201249,
                            'Pennsylvania': 13002700}
         population = pd.Series(population_dict)
         population
Out[11]: California      39538223
         Texas          29145505
         Florida        21538187
         New York       20201249
         Pennsylvania   13002700
         dtype: int64
```

From here, typical dictionary-style item access can be performed:

```
In [12]: population['California']
Out[12]: 39538223
```

Unlike a dictionary, though, the Series also supports array-style operations such as slicing:

```
In [13]: population['California':'Florida']
Out[13]: California      39538223
         Texas          29145505
         Florida        21538187
         dtype: int64
```

We'll discuss some of the quirks of Pandas indexing and slicing in Chapter 14.

Constructing Series Objects

We've already seen a few ways of constructing a Pandas Series from scratch. All of them are some version of the following:

```
pd.Series(data, index=index)
```

where index is an optional argument, and data can be one of many entities.

For example, data can be a list or NumPy array, in which case index defaults to an integer sequence:

```
In [14]: pd.Series([2, 4, 6])
Out[14]: 0    2
         1    4
         2    6
         dtype: int64
```

Or data can be a scalar, which is repeated to fill the specified index:

```
In [15]: pd.Series(5, index=[100, 200, 300])
Out[15]: 100    5
         200    5
         300    5
         dtype: int64
```

Or it can be a dictionary, in which case index defaults to the dictionary keys:

```
In [16]: pd.Series({2:'a', 1:'b', 3:'c'})
Out[16]: 2    a
         1    b
         3    c
         dtype: object
```

In each case, the index can be explicitly set to control the order or the subset of keys used:

```
In [17]: pd.Series({2:'a', 1:'b', 3:'c'}, index=[1, 2])
Out[17]: 1    b
         2    a
         dtype: object
```

The Pandas DataFrame Object

The next fundamental structure in Pandas is the DataFrame. Like the Series object discussed in the previous section, the DataFrame can be thought of either as a generalization of a NumPy array, or as a specialization of a Python dictionary. We'll now take a look at each of these perspectives.

DataFrame as Generalized NumPy Array

If a `Series` is an analog of a one-dimensional array with explicit indices, a `DataFrame` is an analog of a two-dimensional array with explicit row and column indices. Just as you might think of a two-dimensional array as an ordered sequence of aligned one-dimensional columns, you can think of a `DataFrame` as a sequence of aligned `Series` objects. Here, by "aligned" we mean that they share the same index.

To demonstrate this, let's first construct a new `Series` listing the area of each of the five states discussed in the previous section (in square kilometers):

```
In [18]: area_dict = {'California': 423967, 'Texas': 695662, 'Florida': 170312,
                      'New York': 141297, 'Pennsylvania': 119280}
         area = pd.Series(area_dict)
         area
Out[18]: California     423967
         Texas         695662
         Florida       170312
         New York      141297
         Pennsylvania  119280
         dtype: int64
```

Now that we have this along with the `population` Series from before, we can use a dictionary to construct a single two-dimensional object containing this information:

```
In [19]: states = pd.DataFrame({'population': population,
                               'area': area})
         states
Out[19]:               population    area
         California      39538223   423967
         Texas           29145505   695662
         Florida         21538187   170312
         New York        20201249   141297
         Pennsylvania    13002700   119280
```

Like the `Series` object, the `DataFrame` has an `index` attribute that gives access to the index labels:

```
In [20]: states.index
Out[20]: Index(['California', 'Texas', 'Florida', 'New York', 'Pennsylvania'],
             > dtype='object')
```

Additionally, the `DataFrame` has a `columns` attribute, which is an `Index` object holding the column labels:

```
In [21]: states.columns
Out[21]: Index(['population', 'area'], dtype='object')
```

Thus the `DataFrame` can be thought of as a generalization of a two-dimensional NumPy array, where both the rows and columns have a generalized index for accessing the data.

DataFrame as Specialized Dictionary

Similarly, we can also think of a DataFrame as a specialization of a dictionary. Where a dictionary maps a key to a value, a DataFrame maps a column name to a Series of column data. For example, asking for the 'area' attribute returns the Series object containing the areas we saw earlier:

```
In [22]: states['area']
Out[22]: California      423967
         Texas          695662
         Florida        170312
         New York       141297
         Pennsylvania   119280
         Name: area, dtype: int64
```

Notice the potential point of confusion here: in a two-dimensional NumPy array, data[0] will return the first *row*. For a DataFrame, data['col0'] will return the first *column*. Because of this, it is probably better to think about DataFrames as generalized dictionaries rather than generalized arrays, though both ways of looking at the situation can be useful. We'll explore more flexible means of indexing DataFrames in Chapter 14.

Constructing DataFrame Objects

A Pandas DataFrame can be constructed in a variety of ways. Here we'll explore several examples.

From a single Series object

A DataFrame is a collection of Series objects, and a single-column DataFrame can be constructed from a single Series:

```
In [23]: pd.DataFrame(population, columns=['population'])
Out[23]:              population
         California     39538223
         Texas          29145505
         Florida        21538187
         New York       20201249
         Pennsylvania   13002700
```

From a list of dicts

Any list of dictionaries can be made into a DataFrame. We'll use a simple list comprehension to create some data:

```
In [24]: data = [{'a': i, 'b': 2 * i}
                  for i in range(3)]
         pd.DataFrame(data)
Out[24]:    a  b
         0  0  0
```

```
        1  1  2
        2  2  4
```

Even if some keys in the dictionary are missing, Pandas will fill them in with NaN values (i.e., "Not a Number"; see Chapter 16):

```
In [25]: pd.DataFrame([{'a': 1, 'b': 2}, {'b': 3, 'c': 4}])
Out[25]:     a  b    c
         0  1.0  2  NaN
         1  NaN  3  4.0
```

From a dictionary of Series objects

As we saw before, a DataFrame can be constructed from a dictionary of Series objects as well:

```
In [26]: pd.DataFrame({'population': population,
                       'area': area})
Out[26]:              population     area
         California     39538223   423967
         Texas          29145505   695662
         Florida        21538187   170312
         New York       20201249   141297
         Pennsylvania   13002700   119280
```

From a two-dimensional NumPy array

Given a two-dimensional array of data, we can create a DataFrame with any specified column and index names. If omitted, an integer index will be used for each:

```
In [27]: pd.DataFrame(np.random.rand(3, 2),
                      columns=['foo', 'bar'],
                      index=['a', 'b', 'c'])
Out[27]:         foo       bar
         a  0.471098  0.317396
         b  0.614766  0.305971
         c  0.533596  0.512377
```

From a NumPy structured array

We covered structured arrays in Chapter 12. A Pandas DataFrame operates much like a structured array, and can be created directly from one:

```
In [28]: A = np.zeros(3, dtype=[('A', 'i8'), ('B', 'f8')])
         A
Out[28]: array([(0, 0.), (0, 0.), (0, 0.)], dtype=[('A', '<i8'), ('B', '<f8')])

In [29]: pd.DataFrame(A)
Out[29]:    A    B
         0  0  0.0
         1  0  0.0
         2  0  0.0
```

The Pandas Index Object

As you've seen, the Series and DataFrame objects both contain an explicit *index* that lets you reference and modify data. This Index object is an interesting structure in itself, and it can be thought of either as an *immutable array* or as an *ordered set* (technically a multiset, as Index objects may contain repeated values). Those views have some interesting consequences in terms of the operations available on Index objects. As a simple example, let's construct an Index from a list of integers:

```
In [30]: ind = pd.Index([2, 3, 5, 7, 11])
         ind
Out[30]: Int64Index([2, 3, 5, 7, 11], dtype='int64')
```

Index as Immutable Array

The Index in many ways operates like an array. For example, we can use standard Python indexing notation to retrieve values or slices:

```
In [31]: ind[1]
Out[31]: 3
```

```
In [32]: ind[::2]
Out[32]: Int64Index([2, 5, 11], dtype='int64')
```

Index objects also have many of the attributes familiar from NumPy arrays:

```
In [33]: print(ind.size, ind.shape, ind.ndim, ind.dtype)
Out[33]: 5 (5,) 1 int64
```

One difference between Index objects and NumPy arrays is that the indices are immutable—that is, they cannot be modified via the normal means:

```
In [34]: ind[1] = 0
TypeError: Index does not support mutable operations
```

This immutability makes it safer to share indices between multiple DataFrames and arrays, without the potential for side effects from inadvertent index modification.

Index as Ordered Set

Pandas objects are designed to facilitate operations such as joins across datasets, which depend on many aspects of set arithmetic. The Index object follows many of the conventions used by Python's built-in set data structure, so that unions, intersections, differences, and other combinations can be computed in a familiar way:

```
In [35]: indA = pd.Index([1, 3, 5, 7, 9])
         indB = pd.Index([2, 3, 5, 7, 11])

In [36]: indA.intersection(indB)
Out[36]: Int64Index([3, 5, 7], dtype='int64')

In [37]: indA.union(indB)
Out[37]: Int64Index([1, 2, 3, 5, 7, 9, 11], dtype='int64')

In [38]: indA.symmetric_difference(indB)
Out[38]: Int64Index([1, 2, 9, 11], dtype='int64')
```

## CHAPTER 14

Data Indexing and Selection

In Part II, we looked in detail at methods and tools to access, set, and modify values in NumPy arrays. These included indexing (e.g., `arr[2, 1]`), slicing (e.g., `arr[:, 1:5]`), masking (e.g., `arr[arr > 0]`), fancy indexing (e.g., `arr[0, [1, 5]]`), and combinations thereof (e.g., `arr[:, [1, 5]]`). Here we'll look at similar means of accessing and modifying values in Pandas `Series` and `DataFrame` objects. If you have used the NumPy patterns, the corresponding patterns in Pandas will feel very familiar, though there are a few quirks to be aware of.

We'll start with the simple case of the one-dimensional `Series` object, and then move on to the more complicated two-dimensional `DataFrame` object.

Data Selection in Series

As you saw in the previous chapter, a `Series` object acts in many ways like a one-dimensional NumPy array, and in many ways like a standard Python dictionary. If you keep these two overlapping analogies in mind, it will help you understand the patterns of data indexing and selection in these arrays.

Series as Dictionary

Like a dictionary, the `Series` object provides a mapping from a collection of keys to a collection of values:

```
In [1]: import pandas as pd
        data = pd.Series([0.25, 0.5, 0.75, 1.0],
                         index=['a', 'b', 'c', 'd'])
        data
Out[1]: a    0.25
        b    0.50
        c    0.75
```

```
        d    1.00
        dtype: float64

In [2]: data['b']
Out[2]: 0.5
```

We can also use dictionary-like Python expressions and methods to examine the keys/indices and values:

```
In [3]: 'a' in data
Out[3]: True

In [4]: data.keys()
Out[4]: Index(['a', 'b', 'c', 'd'], dtype='object')

In [5]: list(data.items())
Out[5]: [('a', 0.25), ('b', 0.5), ('c', 0.75), ('d', 1.0)]
```

Series objects can also be modified with a dictionary-like syntax. Just as you can extend a dictionary by assigning to a new key, you can extend a Series by assigning to a new index value:

```
In [6]: data['e'] = 1.25
        data
Out[6]: a    0.25
        b    0.50
        c    0.75
        d    1.00
        e    1.25
        dtype: float64
```

This easy mutability of the objects is a convenient feature: under the hood, Pandas is making decisions about memory layout and data copying that might need to take place, and the user generally does not need to worry about these issues.

Series as One-Dimensional Array

A Series builds on this dictionary-like interface and provides array-style item selection via the same basic mechanisms as NumPy arrays—that is, slices, masking, and fancy indexing. Examples of these are as follows:

```
In [7]: # slicing by explicit index
        data['a':'c']
Out[7]: a    0.25
        b    0.50
        c    0.75
        dtype: float64

In [8]: # slicing by implicit integer index
        data[0:2]
Out[8]: a    0.25
        b    0.50
        dtype: float64
```

```
In [9]: # masking
        data[(data > 0.3) & (data < 0.8)]
Out[9]: b    0.50
        c    0.75
        dtype: float64

In [10]: # fancy indexing
         data[['a', 'e']]
Out[10]: a    0.25
         e    1.25
         dtype: float64
```

Of these, slicing may be the source of the most confusion. Notice that when slicing with an explicit index (e.g., data['a':'c']), the final index is *included* in the slice, while when slicing with an implicit index (e.g., data[0:2]), the final index is *excluded* from the slice.

Indexers: loc and iloc

If your Series has an explicit integer index, an indexing operation such as data[1] will use the explicit indices, while a slicing operation like data[1:3] will use the implicit Python-style indices:

```
In [11]: data = pd.Series(['a', 'b', 'c'], index=[1, 3, 5])
         data
Out[11]: 1    a
         3    b
         5    c
         dtype: object

In [12]: # explicit index when indexing
         data[1]
Out[12]: 'a'

In [13]: # implicit index when slicing
         data[1:3]
Out[13]: 3    b
         5    c
         dtype: object
```

Because of this potential confusion in the case of integer indexes, Pandas provides some special *indexer* attributes that explicitly expose certain indexing schemes. These are not functional methods, but attributes that expose a particular slicing interface to the data in the Series.

First, the loc attribute allows indexing and slicing that always references the explicit index:

```
In [14]: data.loc[1]
Out[14]: 'a'
```

```
In [15]: data.loc[1:3]
Out[15]: 1    a
         3    b
         dtype: object
```

The `iloc` attribute allows indexing and slicing that always references the implicit Python-style index:

```
In [16]: data.iloc[1]
Out[16]: 'b'

In [17]: data.iloc[1:3]
Out[17]: 3    b
         5    c
         dtype: object
```

One guiding principle of Python code is that "explicit is better than implicit." The explicit nature of `loc` and `iloc` makes them helpful in maintaining clean and readable code; especially in the case of integer indexes, using them consistently can prevent subtle bugs due to the mixed indexing/slicing convention.

Data Selection in DataFrames

Recall that a `DataFrame` acts in many ways like a two-dimensional or structured array, and in other ways like a dictionary of `Series` structures sharing the same index. These analogies can be helpful to keep in mind as we explore data selection within this structure.

DataFrame as Dictionary

The first analogy we will consider is the `DataFrame` as a dictionary of related `Series` objects. Let's return to our example of areas and populations of states:

```
In [18]: area = pd.Series({'California': 423967, 'Texas': 695662,
                           'Florida': 170312, 'New York': 141297,
                           'Pennsylvania': 119280})
         pop = pd.Series({'California': 39538223, 'Texas': 29145505,
                         'Florida': 21538187, 'New York': 20201249,
                         'Pennsylvania': 13002700})
         data = pd.DataFrame({'area':area, 'pop':pop})
         data
Out[18]:               area       pop
         California   423967  39538223
         Texas        695662  29145505
         Florida      170312  21538187
         New York     141297  20201249
         Pennsylvania 119280  13002700
```

The individual Series that make up the columns of the DataFrame can be accessed via dictionary-style indexing of the column name:

```
In [19]: data['area']
Out[19]: California      423967
         Texas          695662
         Florida        170312
         New York       141297
         Pennsylvania   119280
         Name: area, dtype: int64
```

Equivalently, we can use attribute-style access with column names that are strings:

```
In [20]: data.area
Out[20]: California      423967
         Texas          695662
         Florida        170312
         New York       141297
         Pennsylvania   119280
         Name: area, dtype: int64
```

Though this is a useful shorthand, keep in mind that it does not work for all cases! For example, if the column names are not strings, or if the column names conflict with methods of the DataFrame, this attribute-style access is not possible. For example, the DataFrame has a pop method, so data.pop will point to this rather than the pop column:

```
In [21]: data.pop is data["pop"]
Out[21]: False
```

In particular, you should avoid the temptation to try column assignment via attributes (i.e., use data['pop'] = z rather than data.pop = z).

Like with the Series objects discussed earlier, this dictionary-style syntax can also be used to modify the object, in this case adding a new column:

```
In [22]: data['density'] = data['pop'] / data['area']
         data
Out[22]:                 area       pop     density
         California     423967  39538223   93.257784
         Texas          695662  29145505   41.896072
         Florida        170312  21538187  126.463121
         New York       141297  20201249  142.970120
         Pennsylvania   119280  13002700  109.009893
```

This shows a preview of the straightforward syntax of element-by-element arithmetic between Series objects; we'll dig into this further in Chapter 15.

DataFrame as Two-Dimensional Array

As mentioned previously, we can also view the DataFrame as an enhanced two-dimensional array. We can examine the raw underlying data array using the values attribute:

```
In [23]: data.values
Out[23]: array([[4.23967000e+05, 3.95382230e+07, 9.32577842e+01],
                [6.95662000e+05, 2.91455050e+07, 4.18960717e+01],
                [1.70312000e+05, 2.15381870e+07, 1.26463121e+02],
                [1.41297000e+05, 2.02012490e+07, 1.42970120e+02],
                [1.19280000e+05, 1.30027000e+07, 1.09009893e+02]])
```

With this picture in mind, many familiar array-like operations can be done on the DataFrame itself. For example, we can transpose the full DataFrame to swap rows and columns:

```
In [24]: data.T
Out[24]:            California        Texas       Florida      New York  Pennsylvania
         area     4.239670e+05  6.956620e+05  1.703120e+05  1.412970e+05  1.192800e+05
         pop      3.953822e+07  2.914550e+07  2.153819e+07  2.020125e+07  1.300270e+07
         density  9.325778e+01  4.189607e+01  1.264631e+02  1.429701e+02  1.090099e+02
```

When it comes to indexing of a DataFrame object, however, it is clear that the dictionary-style indexing of columns precludes our ability to simply treat it as a NumPy array. In particular, passing a single index to an array accesses a row:

```
In [25]: data.values[0]
Out[25]: array([4.23967000e+05, 3.95382230e+07, 9.32577842e+01])
```

and passing a single "index" to a DataFrame accesses a column:

```
In [26]: data['area']
Out[26]: California      423967
         Texas          695662
         Florida        170312
         New York       141297
         Pennsylvania   119280
         Name: area, dtype: int64
```

Thus, for array-style indexing, we need another convention. Here Pandas again uses the loc and iloc indexers mentioned earlier. Using the iloc indexer, we can index the underlying array as if it were a simple NumPy array (using the implicit Python-style index), but the DataFrame index and column labels are maintained in the result:

```
In [27]: data.iloc[:3, :2]
Out[27]:               area       pop
         California   423967  39538223
         Texas        695662  29145505
         Florida      170312  21538187
```

Similarly, using the loc indexer we can index the underlying data in an array-like style but using the explicit index and column names:

```
In [28]: data.loc[:'Florida', :'pop']
Out[28]:             area       pop
         California  423967  39538223
         Texas       695662  29145505
         Florida     170312  21538187
```

Any of the familiar NumPy-style data access patterns can be used within these indexers. For example, in the loc indexer we can combine masking and fancy indexing as follows:

```
In [29]: data.loc[data.density > 120, ['pop', 'density']]
Out[29]:          pop       density
         Florida  21538187  126.463121
         New York 20201249  142.970120
```

Any of these indexing conventions may also be used to set or modify values; this is done in the standard way that you might be accustomed to from working with NumPy:

```
In [30]: data.iloc[0, 2] = 90
         data
Out[30]:              area       pop      density
         California   423967  39538223   90.000000
         Texas        695662  29145505   41.896072
         Florida      170312  21538187  126.463121
         New York     141297  20201249  142.970120
         Pennsylvania 119280  13002700  109.009893
```

To build up your fluency in Pandas data manipulation, I suggest spending some time with a simple DataFrame and exploring the types of indexing, slicing, masking, and fancy indexing that are allowed by these various indexing approaches.

Additional Indexing Conventions

There are a couple of extra indexing conventions that might seem at odds with the preceding discussion, but nevertheless can be useful in practice. First, while *indexing* refers to columns, *slicing* refers to rows:

```
In [31]: data['Florida':'New York']
Out[31]:          area       pop      density
         Florida  170312  21538187  126.463121
         New York 141297  20201249  142.970120
```

Such slices can also refer to rows by number rather than by index:

```
In [32]: data[1:3]
Out[32]:         area       pop      density
         Texas   695662  29145505   41.896072
         Florida 170312  21538187  126.463121
```

Similarly, direct masking operations are interpreted row-wise rather than column-wise:

```
In [33]: data[data.density > 120]
Out[33]:             area       pop    density
         Florida   170312  21538187  126.463121
         New York  141297  20201249  142.970120
```

These two conventions are syntactically similar to those on a NumPy array, and while they may not precisely fit the mold of the Pandas conventions, they are included due to their practical utility.

## CHAPTER 15

Operating on Data in Pandas

One of the strengths of NumPy is that it allows us to perform quick element-wise operations, both with basic arithmetic (addition, subtraction, multiplication, etc.) and with more complicated operations (trigonometric functions, exponential and logarithmic functions, etc.). Pandas inherits much of this functionality from NumPy, and the ufuncs introduced in Chapter 6 are key to this.

Pandas includes a couple of useful twists, however: for unary operations like negation and trigonometric functions, these ufuncs will *preserve index and column labels* in the output, and for binary operations such as addition and multiplication, Pandas will automatically *align indices* when passing the objects to the ufunc. This means that keeping the context of data and combining data from different sources—both potentially error-prone tasks with raw NumPy arrays—become essentially foolproof with Pandas. We will additionally see that there are well-defined operations between one-dimensional `Series` structures and two-dimensional `DataFrame` structures.

Ufuncs: Index Preservation

Because Pandas is designed to work with NumPy, any NumPy ufunc will work on Pandas `Series` and `DataFrame` objects. Let's start by defining a simple `Series` and `DataFrame` on which to demonstrate this:

```
In [1]: import pandas as pd
        import numpy as np

In [2]: rng = np.random.default_rng(42)
        ser = pd.Series(rng.integers(0, 10, 4))
        ser
Out[2]: 0    0
        1    7
        2    6
```

```
        3    4
        dtype: int64
In [3]: df = pd.DataFrame(rng.integers(0, 10, (3, 4)),
                          columns=['A', 'B', 'C', 'D'])
        df
Out[3]:    A  B  C  D
        0  4  8  0  6
        1  2  0  5  9
        2  7  7  7  7
```

If we apply a NumPy ufunc on either of these objects, the result will be another Pandas object *with the indices preserved*:

```
In [4]: np.exp(ser)
Out[4]: 0       1.000000
        1    1096.633158
        2     403.428793
        3      54.598150
        dtype: float64
```

This is true also for more involved sequences of operations:

```
In [5]: np.sin(df * np.pi / 4)
Out[5]:              A             B          C          D
        0  1.224647e-16 -2.449294e-16   0.000000  -1.000000
        1  1.000000e+00  0.000000e+00  -0.707107   0.707107
        2 -7.071068e-01 -7.071068e-01  -0.707107  -0.707107
```

Any of the ufuncs discussed in Chapter 6 can be used in a similar manner.

Ufuncs: Index Alignment

For binary operations on two `Series` or `DataFrame` objects, Pandas will align indices in the process of performing the operation. This is very convenient when working with incomplete data, as we'll see in some of the examples that follow.

Index Alignment in Series

As an example, suppose we are combining two different data sources and wish to find only the top three US states by *area* and the top three US states by *population*:

```
In [6]: area = pd.Series({'Alaska': 1723337, 'Texas': 695662,
                          'California': 423967}, name='area')
        population = pd.Series({'California': 39538223, 'Texas': 29145505,
                               'Florida': 21538187}, name='population')
```

Let's see what happens when we divide these to compute the population density:

```
In [7]: population / area
Out[7]: Alaska              NaN
        California    93.257784
        Florida             NaN
```

```
       Texas        41.896072
       dtype: float64
```

The resulting array contains the *union* of indices of the two input arrays, which could be determined directly from these indices:

```
In [8]: area.index.union(population.index)
Out[8]: Index(['Alaska', 'California', 'Florida', 'Texas'], dtype='object')
```

Any item for which one or the other does not have an entry is marked with NaN, or "Not a Number," which is how Pandas marks missing data (see further discussion of missing data in Chapter 16). This index matching is implemented this way for any of Python's built-in arithmetic expressions; any missing values are marked by NaN:

```
In [9]: A = pd.Series([2, 4, 6], index=[0, 1, 2])
        B = pd.Series([1, 3, 5], index=[1, 2, 3])
        A + B
Out[9]: 0    NaN
        1    5.0
        2    9.0
        3    NaN
        dtype: float64
```

If using NaN values is not the desired behavior, the fill value can be modified using appropriate object methods in place of the operators. For example, calling A.add(B) is equivalent to calling A + B, but allows optional explicit specification of the fill value for any elements in A or B that might be missing:

```
In [10]: A.add(B, fill_value=0)
Out[10]: 0    2.0
         1    5.0
         2    9.0
         3    5.0
         dtype: float64
```

Index Alignment in DataFrames

A similar type of alignment takes place for *both* columns and indices when performing operations on DataFrame objects:

```
In [11]: A = pd.DataFrame(rng.integers(0, 20, (2, 2)),
                          columns=['a', 'b'])
         A
Out[11]:     a   b
         0  10   2
         1  16   9

In [12]: B = pd.DataFrame(rng.integers(0, 10, (3, 3)),
                          columns=['b', 'a', 'c'])
         B
Out[12]:     b   a   c
         0   5   3   1
```

```
          1  9  7  6
          2  4  8  5
In [13]: A + B
Out[12]:       a     b   c
          0  13.0   7.0 NaN
          1  23.0  18.0 NaN
          2   NaN   NaN NaN
```

Notice that indices are aligned correctly irrespective of their order in the two objects, and indices in the result are sorted. As was the case with `Series`, we can use the associated object's arithmetic methods and pass any desired `fill_value` to be used in place of missing entries. Here we'll fill with the mean of all values in A:

```
In [14]: A.add(B, fill_value=A.values.mean())
Out[14]:       a      b      c
          0  13.00   7.00  10.25
          1  23.00  18.00  15.25
          2  17.25  13.25  14.25
```

Table 15-1 lists Python operators and their equivalent Pandas object methods.

Table 15-1. Mapping between Python operators and Pandas methods

Python operator	Pandas method(s)
+	add
-	sub, subtract
*	mul, multiply
/	truediv, div, divide
//	floordiv
%	mod
**	pow

Ufuncs: Operations Between DataFrames and Series

When performing operations between a `DataFrame` and a `Series`, the index and column alignment is similarly maintained, and the result is similar to operations between a two-dimensional and one-dimensional NumPy array. Consider one common operation, where we find the difference of a two-dimensional array and one of its rows:

```
In [15]: A = rng.integers(10, size=(3, 4))
         A
Out[15]: array([[4, 4, 2, 0],
                [5, 8, 0, 8],
                [8, 2, 6, 1]])
```

```
In [16]: A - A[0]
Out[16]: array([[ 0,  0,  0,  0],
               [ 1,  4, -2,  8],
               [ 4, -2,  4,  1]])
```

According to NumPy's broadcasting rules (see Chapter 8), subtraction between a two-dimensional array and one of its rows is applied row-wise.

In Pandas, the convention similarly operates row-wise by default:

```
In [17]: df = pd.DataFrame(A, columns=['Q', 'R', 'S', 'T'])
         df - df.iloc[0]
Out[17]:    Q  R  S  T
         0  0  0  0  0
         1  1  4 -2  8
         2  4 -2  4  1
```

If you would instead like to operate column-wise, you can use the object methods mentioned earlier, while specifying the `axis` keyword:

```
In [18]: df.subtract(df['R'], axis=0)
Out[18]:    Q  R  S  T
         0  0  0 -2 -4
         1 -3  0 -8  0
         2  6  0  4 -1
```

Note that these `DataFrame`/`Series` operations, like the operations discussed previously, will automatically align indices between the two elements:

```
In [19]: halfrow = df.iloc[0, ::2]
         halfrow
Out[19]: Q    4
         S    2
         Name: 0, dtype: int64

In [20]: df - halfrow
Out[20]:      Q   R    S   T
         0  0.0 NaN  0.0 NaN
         1  1.0 NaN -2.0 NaN
         2  4.0 NaN  4.0 NaN
```

This preservation and alignment of indices and columns means that operations on data in Pandas will always maintain the data context, which prevents the common errors that might arise when working with heterogeneous and/or misaligned data in raw NumPy arrays.

## CHAPTER 16

Handling Missing Data

The difference between data found in many tutorials and data in the real world is that real-world data is rarely clean and homogeneous. In particular, many interesting datasets will have some amount of data missing. To make matters even more complicated, different data sources may indicate missing data in different ways.

In this chapter, we will discuss some general considerations for missing data, look at how Pandas chooses to represent it, and explore some built-in Pandas tools for handling missing data in Python. Here and throughout the book, I will refer to missing data in general as *null*, *NaN*, or *NA* values.

Trade-offs in Missing Data Conventions

A number of approaches have been developed to track the presence of missing data in a table or `DataFrame`. Generally, they revolve around one of two strategies: using a *mask* that globally indicates missing values, or choosing a *sentinel value* that indicates a missing entry.

In the masking approach, the mask might be an entirely separate Boolean array, or it might involve appropriation of one bit in the data representation to locally indicate the null status of a value.

In the sentinel approach, the sentinel value could be some data-specific convention, such as indicating a missing integer value with –9999 or some rare bit pattern, or it could be a more global convention, such as indicating a missing floating-point value with `NaN` (Not a Number), a special value that is part of the IEEE floating-point specification.

Neither of these approaches is without trade-offs. Use of a separate mask array requires allocation of an additional Boolean array, which adds overhead in both storage and computation. A sentinel value reduces the range of valid values that can be represented, and may require extra (often nonoptimized) logic in CPU and GPU arithmetic, because common special values like NaN are not available for all data types.

As in most cases where no universally optimal choice exists, different languages and systems use different conventions. For example, the R language uses reserved bit patterns within each data type as sentinel values indicating missing data, while the SciDB system uses an extra byte attached to every cell to indicate an NA state.

Missing Data in Pandas

The way in which Pandas handles missing values is constrained by its reliance on the NumPy package, which does not have a built-in notion of NA values for non-floating-point data types.

Perhaps Pandas could have followed R's lead in specifying bit patterns for each individual data type to indicate nullness, but this approach turns out to be rather unwieldy. While R has just 4 main data types, NumPy supports *far* more than this: for example, while R has a single integer type, NumPy supports 14 basic integer types once you account for available bit widths, signedness, and endianness of the encoding. Reserving a specific bit pattern in all available NumPy types would lead to an unwieldy amount of overhead in special-casing various operations for various types, likely even requiring a new fork of the NumPy package. Further, for the smaller data types (such as 8-bit integers), sacrificing a bit to use as a mask would significantly reduce the range of values it can represent.

Because of these constraints and trade-offs, Pandas has two "modes" of storing and manipulating null values:

- The default mode is to use a sentinel-based missing data scheme, with sentinel values NaN or None depending on the type of the data.
- Alternatively, you can opt in to using the nullable data types (dtypes) Pandas provides (discussed later in this chapter), which results in the creation an accompanying mask array to track missing entries. These missing entries are then presented to the user as the special pd.NA value.

In either case, the data operations and manipulations provided by the Pandas API will handle and propagate those missing entries in a predictable manner. But to develop some intuition into *why* these choices are made, let's dive quickly into the trade-offs inherent in None, NaN, and NA. As usual, we'll start by importing NumPy and Pandas:

```
In [1]: import numpy as np
        import pandas as pd
```

None as a Sentinel Value

For some data types, Pandas uses None as a sentinel value. None is a Python object, which means that any array containing None must have dtype=object—that is, it must be a sequence of Python objects.

For example, observe what happens if you pass None to a NumPy array:

```
In [2]: vals1 = np.array([1, None, 2, 3])
        vals1
Out[2]: array([1, None, 2, 3], dtype=object)
```

This dtype=object means that the best common type representation NumPy could infer for the contents of the array is that they are Python objects. The downside of using None in this way is that operations on the data will be done at the Python level, with much more overhead than the typically fast operations seen for arrays with native types:

```
In [3]: %timeit np.arange(1E6, dtype=int).sum()
Out[3]: 2.73 ms ± 288 µs per loop (mean ± std. dev. of 7 runs, 100 loops each)

In [4]: %timeit np.arange(1E6, dtype=object).sum()
Out[4]: 92.1 ms ± 3.42 ms per loop (mean ± std. dev. of 7 runs, 10 loops each)
```

Further, because Python does not support arithmetic operations with None, aggregations like sum or min will generally lead to an error:

```
In [5]: vals1.sum()
TypeError: unsupported operand type(s) for +: 'int' and 'NoneType'
```

For this reason, Pandas does not use None as a sentinel in its numerical arrays.

NaN: Missing Numerical Data

The other missing data sentinel, NaN is different; it is a special floating-point value recognized by all systems that use the standard IEEE floating-point representation:

```
In [6]: vals2 = np.array([1, np.nan, 3, 4])
        vals2
Out[6]: array([ 1., nan,  3.,  4.])
```

Notice that NumPy chose a native floating-point type for this array: this means that unlike the object array from before, this array supports fast operations pushed into compiled code. Keep in mind that NaN is a bit like a data virus—it infects any other object it touches.

Regardless of the operation, the result of arithmetic with NaN will be another NaN:

```
In [7]: 1 + np.nan
Out[7]: nan

In [8]: 0 * np.nan
Out[8]: nan
```

This means that aggregates over the values are well defined (i.e., they don't result in an error) but not always useful:

```
In [9]: vals2.sum(), vals2.min(), vals2.max()
Out[9]: (nan, nan, nan)
```

That said, NumPy does provide NaN-aware versions of aggregations that will ignore these missing values:

```
In [10]: np.nansum(vals2), np.nanmin(vals2), np.nanmax(vals2)
Out[10]: (8.0, 1.0, 4.0)
```

The main downside of NaN is that it is specifically a floating-point value; there is no equivalent NaN value for integers, strings, or other types.

NaN and None in Pandas

NaN and None both have their place, and Pandas is built to handle the two of them nearly interchangeably, converting between them where appropriate:

```
In [11]: pd.Series([1, np.nan, 2, None])
Out[11]: 0    1.0
         1    NaN
         2    2.0
         3    NaN
         dtype: float64
```

For types that don't have an available sentinel value, Pandas automatically typecasts when NA values are present. For example, if we set a value in an integer array to np.nan, it will automatically be upcast to a floating-point type to accommodate the NA:

```
In [12]: x = pd.Series(range(2), dtype=int)
         x
Out[12]: 0    0
         1    1
         dtype: int64

In [13]: x[0] = None
         x
Out[13]: 0    NaN
         1    1.0
         dtype: float64
```

Notice that in addition to casting the integer array to floating point, Pandas automatically converts the None to a NaN value.

While this type of magic may feel a bit hackish compared to the more unified approach to NA values in domain-specific languages like R, the Pandas sentinel/casting approach works quite well in practice and in my experience only rarely causes issues.

Table 16-1 lists the upcasting conventions in Pandas when NA values are introduced.

Table 16-1. Pandas handling of NAs by type

Typeclass	Conversion when storing NAs	NA sentinel value
floating	No change	np.nan
object	No change	None or np.nan
integer	Cast to float64	np.nan
boolean	Cast to object	None or np.nan

Keep in mind that in Pandas, string data is always stored with an object dtype.

Pandas Nullable Dtypes

In early versions of Pandas, NaN and None as sentinel values were the only missing data representations available. The primary difficulty this introduced was with regard to the implicit type casting: for example, there was no way to represent a true integer array with missing data.

To address this difficulty, Pandas later added *nullable dtypes*, which are distinguished from regular dtypes by capitalization of their names (e.g., pd.Int32 versus np.int32). For backward compatibility, these nullable dtypes are only used if specifically requested.

For example, here is a Series of integers with missing data, created from a list containing all three available markers of missing data:

```
In [14]: pd.Series([1, np.nan, 2, None, pd.NA], dtype='Int32')
Out[14]: 0       1
         1    <NA>
         2       2
         3    <NA>
         4    <NA>
         dtype: Int32
```

This representation can be used interchangeably with the others in all the operations explored through the rest of this chapter.

Operating on Null Values

As we have seen, Pandas treats None, NaN, and NA as essentially interchangeable for indicating missing or null values. To facilitate this convention, Pandas provides several methods for detecting, removing, and replacing null values in Pandas data structures. They are:

isnull
: Generates a Boolean mask indicating missing values

notnull
: Opposite of isnull

dropna
: Returns a filtered version of the data

fillna
: Returns a copy of the data with missing values filled or imputed

We will conclude this chapter with a brief exploration and demonstration of these routines.

Detecting Null Values

Pandas data structures have two useful methods for detecting null data: isnull and notnull. Either one will return a Boolean mask over the data. For example:

```
In [15]: data = pd.Series([1, np.nan, 'hello', None])

In [16]: data.isnull()
Out[16]: 0    False
         1     True
         2    False
         3     True
         dtype: bool
```

As mentioned in Chapter 14, Boolean masks can be used directly as a Series or DataFrame index:

```
In [17]: data[data.notnull()]
Out[17]: 0        1
         2    hello
         dtype: object
```

The isnull() and notnull() methods produce similar Boolean results for DataFrame objects.

Dropping Null Values

In addition to these masking methods, there are the convenience methods `dropna` (which removes NA values) and `fillna` (which fills in NA values). For a `Series`, the result is straightforward:

```
In [18]: data.dropna()
Out[18]: 0        1
         2    hello
         dtype: object
```

For a `DataFrame`, there are more options. Consider the following `DataFrame`:

```
In [19]: df = pd.DataFrame([[1,      np.nan, 2],
                            [2,      3,      5],
                            [np.nan, 4,      6]])
         df
Out[19]:     0    1  2
         0  1.0  NaN  2
         1  2.0  3.0  5
         2  NaN  4.0  6
```

We cannot drop single values from a `DataFrame`; we can only drop entire rows or columns. Depending on the application, you might want one or the other, so `dropna` includes a number of options for a `DataFrame`.

By default, `dropna` will drop all rows in which *any* null value is present:

```
In [20]: df.dropna()
Out[20]:     0    1  2
         1  2.0  3.0  5
```

Alternatively, you can drop NA values along a different axis. Using `axis=1` or `axis='columns'` drops all columns containing a null value:

```
In [21]: df.dropna(axis='columns')
Out[21]:    2
         0  2
         1  5
         2  6
```

But this drops some good data as well; you might rather be interested in dropping rows or columns with *all* NA values, or a majority of NA values. This can be specified through the `how` or `thresh` parameters, which allow fine control of the number of nulls to allow through.

The default is how='any', such that any row or column containing a null value will be dropped. You can also specify how='all', which will only drop rows/columns that contain *all* null values:

```
In [22]: df[3] = np.nan
         df
Out[22]:     0    1  2    3
         0  1.0  NaN  2  NaN
         1  2.0  3.0  5  NaN
         2  NaN  4.0  6  NaN

In [23]: df.dropna(axis='columns', how='all')
Out[23]:     0    1  2
         0  1.0  NaN  2
         1  2.0  3.0  5
         2  NaN  4.0  6
```

For finer-grained control, the thresh parameter lets you specify a minimum number of non-null values for the row/column to be kept:

```
In [24]: df.dropna(axis='rows', thresh=3)
Out[24]:     0    1  2    3
         1  2.0  3.0  5  NaN
```

Here, the first and last rows have been dropped because they each contain only two non-null values.

Filling Null Values

Sometimes rather than dropping NA values, you'd like to replace them with a valid value. This value might be a single number like zero, or it might be some sort of imputation or interpolation from the good values. You could do this in-place using the isnull method as a mask, but because it is such a common operation Pandas provides the fillna method, which returns a copy of the array with the null values replaced.

Consider the following Series:

```
In [25]: data = pd.Series([1, np.nan, 2, None, 3], index=list('abcde'),
                          dtype='Int32')
         data
Out[25]: a       1
         b    <NA>
         c       2
         d    <NA>
         e       3
         dtype: Int32
```

We can fill NA entries with a single value, such as zero:

```
In [26]: data.fillna(0)
Out[26]: a    1
         b    0
         c    2
         d    0
         e    3
         dtype: Int32
```

We can specify a forward fill to propagate the previous value forward:

```
In [27]: # forward fill
         data.fillna(method='ffill')
Out[27]: a    1
         b    1
         c    2
         d    2
         e    3
         dtype: Int32
```

Or we can specify a backward fill to propagate the next values backward:

```
In [28]: # back fill
         data.fillna(method='bfill')
Out[28]: a    1
         b    2
         c    2
         d    3
         e    3
         dtype: Int32
```

In the case of a DataFrame, the options are similar, but we can also specify an axis along which the fills should take place:

```
In [29]: df
Out[29]:      0    1  2   3
         0  1.0  NaN  2 NaN
         1  2.0  3.0  5 NaN
         2  NaN  4.0  6 NaN

In [30]: df.fillna(method='ffill', axis=1)
Out[30]:      0    1    2    3
         0  1.0  1.0  2.0  2.0
         1  2.0  3.0  5.0  5.0
         2  NaN  4.0  6.0  6.0
```

Notice that if a previous value is not available during a forward fill, the NA value remains.

## CHAPTER 17

Hierarchical Indexing

Up to this point we've been focused primarily on one-dimensional and two-dimensional data, stored in Pandas `Series` and `DataFrame` objects, respectively. Often it is useful to go beyond this and store higher-dimensional data—that is, data indexed by more than one or two keys. Early Pandas versions provided `Panel` and `Panel4D` objects that could be thought of as 3D or 4D analogs to the 2D `DataFrame`, but they were somewhat clunky to use in practice. A far more common pattern for handling higher-dimensional data is to make use of *hierarchical indexing* (also known as *multi-indexing*) to incorporate multiple index *levels* within a single index. In this way, higher-dimensional data can be compactly represented within the familiar one-dimensional `Series` and two-dimensional `DataFrame` objects. (If you're interested in true *N*-dimensional arrays with Pandas-style flexible indices, you can look into the excellent Xarray package (*https://xarray.pydata.org*).)

In this chapter, we'll explore the direct creation of `MultiIndex` objects; considerations when indexing, slicing, and computing statistics across multiply indexed data; and useful routines for converting between simple and hierarchically indexed representations of data.

We begin with the standard imports:

```
In [1]: import pandas as pd
        import numpy as np
```

A Multiply Indexed Series

Let's start by considering how we might represent two-dimensional data within a one-dimensional `Series`. For concreteness, we will consider a series of data where each point has a character and numerical key.

The Bad Way

Suppose you would like to track data about states from two different years. Using the Pandas tools we've already covered, you might be tempted to simply use Python tuples as keys:

```
In [2]: index = [('California', 2010), ('California', 2020),
                 ('New York', 2010), ('New York', 2020),
                 ('Texas', 2010), ('Texas', 2020)]
        populations = [37253956, 39538223,
                       19378102, 20201249,
                       25145561, 29145505]
        pop = pd.Series(populations, index=index)
        pop
Out[2]: (California, 2010)    37253956
        (California, 2020)    39538223
        (New York, 2010)      19378102
        (New York, 2020)      20201249
        (Texas, 2010)         25145561
        (Texas, 2020)         29145505
        dtype: int64
```

With this indexing scheme, you can straightforwardly index or slice the series based on this tuple index:

```
In [3]: pop[('California', 2020):('Texas', 2010)]
Out[3]: (California, 2020)    39538223
        (New York, 2010)      19378102
        (New York, 2020)      20201249
        (Texas, 2010)         25145561
        dtype: int64
```

But the convenience ends there. For example, if you need to select all values from 2010, you'll need to do some messy (and potentially slow) munging to make it happen:

```
In [4]: pop[[i for i in pop.index if i[1] == 2010]]
Out[4]: (California, 2010)    37253956
        (New York, 2010)      19378102
        (Texas, 2010)         25145561
        dtype: int64
```

This produces the desired result, but is not as clean (or as efficient for large datasets) as the slicing syntax we've grown to love in Pandas.

The Better Way: The Pandas MultiIndex

Fortunately, Pandas provides a better way. Our tuple-based indexing is essentially a rudimentary multi-index, and the Pandas MultiIndex type gives us the types of operations we wish to have. We can create a multi-index from the tuples as follows:

```
In [5]: index = pd.MultiIndex.from_tuples(index)
```

The `MultiIndex` represents multiple *levels* of indexing—in this case, the state names and the years—as well as multiple *labels* for each data point which encode these levels.

If we reindex our series with this `MultiIndex`, we see the hierarchical representation of the data:

```
In [6]: pop = pop.reindex(index)
        pop
Out[6]: California  2010    37253956
                    2020    39538223
        New York    2010    19378102
                    2020    20201249
        Texas       2010    25145561
                    2020    29145505
        dtype: int64
```

Here the first two columns of the Series representation show the multiple index values, while the third column shows the data. Notice that some entries are missing in the first column: in this multi-index representation, any blank entry indicates the same value as the line above it.

Now to access all data for which the second index is 2020, we can use the Pandas slicing notation:

```
In [7]: pop[:, 2020]
Out[7]: California   39538223
        New York     20201249
        Texas        29145505
        dtype: int64
```

The result is a singly indexed Series with just the keys we're interested in. This syntax is much more convenient (and the operation is much more efficient!) than the home-spun tuple-based multi-indexing solution that we started with. We'll now further discuss this sort of indexing operation on hierarchically indexed data.

MultiIndex as Extra Dimension

You might notice something else here: we could easily have stored the same data using a simple `DataFrame` with index and column labels. In fact, Pandas is built with this equivalence in mind. The `unstack` method will quickly convert a multiply indexed `Series` into a conventionally indexed `DataFrame`:

```
In [8]: pop_df = pop.unstack()
        pop_df
Out[8]:                 2010        2020
        California   37253956    39538223
        New York     19378102    20201249
        Texas        25145561    29145505
```

Naturally, the stack method provides the opposite operation:

```
In [9]: pop_df.stack()
Out[9]: California  2010    37253956
                    2020    39538223
        New York    2010    19378102
                    2020    20201249
        Texas       2010    25145561
                    2020    29145505
        dtype: int64
```

Seeing this, you might wonder why would we would bother with hierarchical indexing at all. The reason is simple: just as we were able to use multi-indexing to manipulate two-dimensional data within a one-dimensional Series, we can also use it to manipulate data of three or more dimensions in a Series or DataFrame. Each extra level in a multi-index represents an extra dimension of data; taking advantage of this property gives us much more flexibility in the types of data we can represent. Concretely, we might want to add another column of demographic data for each state at each year (say, population under 18); with a MultiIndex this is as easy as adding another column to the DataFrame:

```
In [10]: pop_df = pd.DataFrame({'total': pop,
                                'under18': [9284094, 8898092,
                                            4318033, 4181528,
                                            6879014, 7432474]})
         pop_df
Out[10]:                      total   under18
         California 2010    37253956  9284094
                    2020    39538223  8898092
         New York   2010    19378102  4318033
                    2020    20201249  4181528
         Texas      2010    25145561  6879014
                    2020    29145505  7432474
```

In addition, all the ufuncs and other functionality discussed in Chapter 15 work with hierarchical indices as well. Here we compute the fraction of people under 18 by year, given the above data:

```
In [11]: f_u18 = pop_df['under18'] / pop_df['total']
         f_u18.unstack()
Out[11]:                 2010      2020
         California  0.249211  0.225050
         New York    0.222831  0.206994
         Texas       0.273568  0.255013
```

This allows us to easily and quickly manipulate and explore even high-dimensional data.

Methods of MultiIndex Creation

The most straightforward way to construct a multiply indexed `Series` or `DataFrame` is to simply pass a list of two or more index arrays to the constructor. For example:

```
In [12]: df = pd.DataFrame(np.random.rand(4, 2),
                           index=[['a', 'a', 'b', 'b'], [1, 2, 1, 2]],
                           columns=['data1', 'data2'])
         df
Out[12]:        data1     data2
     a 1  0.748464  0.561409
       2  0.379199  0.622461
     b 1  0.701679  0.687932
       2  0.436200  0.950664
```

The work of creating the `MultiIndex` is done in the background.

Similarly, if you pass a dictionary with appropriate tuples as keys, Pandas will automatically recognize this and use a `MultiIndex` by default:

```
In [13]: data = {('California', 2010): 37253956,
                 ('California', 2020): 39538223,
                 ('New York', 2010): 19378102,
                 ('New York', 2020): 20201249,
                 ('Texas', 2010): 25145561,
                 ('Texas', 2020): 29145505}
         pd.Series(data)
Out[13]: California  2010     37253956
                     2020     39538223
         New York    2010     19378102
                     2020     20201249
         Texas       2010     25145561
                     2020     29145505
         dtype: int64
```

Nevertheless, it is sometimes useful to explicitly create a `MultiIndex`; we'll look at a couple of methods for doing this next.

Explicit MultiIndex Constructors

For more flexibility in how the index is constructed, you can instead use the constructor methods available in the `pd.MultiIndex` class. For example, as we did before, you can construct a `MultiIndex` from a simple list of arrays giving the index values within each level:

```
In [14]: pd.MultiIndex.from_arrays([['a', 'a', 'b', 'b'], [1, 2, 1, 2]])
Out[14]: MultiIndex([('a', 1),
                      ('a', 2),
                      ('b', 1),
                      ('b', 2)],
                    )
```

Or you can construct it from a list of tuples giving the multiple index values of each point:

```
In [15]: pd.MultiIndex.from_tuples([('a', 1), ('a', 2), ('b', 1), ('b', 2)])
Out[15]: MultiIndex([('a', 1),
                      ('a', 2),
                      ('b', 1),
                      ('b', 2)],
                     )
```

You can even construct it from a Cartesian product of single indices:

```
In [16]: pd.MultiIndex.from_product([['a', 'b'], [1, 2]])
Out[16]: MultiIndex([('a', 1),
                      ('a', 2),
                      ('b', 1),
                      ('b', 2)],
                     )
```

Similarly, you can construct a MultiIndex directly using its internal encoding by passing levels (a list of lists containing available index values for each level) and codes (a list of lists that reference these labels):

```
In [17]: pd.MultiIndex(levels=[['a', 'b'], [1, 2]],
                       codes=[[0, 0, 1, 1], [0, 1, 0, 1]])
Out[17]: MultiIndex([('a', 1),
                      ('a', 2),
                      ('b', 1),
                      ('b', 2)],
                     )
```

Any of these objects can be passed as the index argument when creating a Series or DataFrame, or be passed to the reindex method of an existing Series or DataFrame.

MultiIndex Level Names

Sometimes it is convenient to name the levels of the MultiIndex. This can be accomplished by passing the names argument to any of the previously discussed MultiIndex constructors, or by setting the names attribute of the index after the fact:

```
In [18]: pop.index.names = ['state', 'year']
         pop
Out[18]: state       year
         California  2010    37253956
                     2020    39538223
         New York    2010    19378102
                     2020    20201249
         Texas       2010    25145561
                     2020    29145505
         dtype: int64
```

With more involved datasets, this can be a useful way to keep track of the meaning of various index values.

MultiIndex for Columns

In a DataFrame, the rows and columns are completely symmetric, and just as the rows can have multiple levels of indices, the columns can have multiple levels as well. Consider the following, which is a mock-up of some (somewhat realistic) medical data:

```
In [19]: # hierarchical indices and columns
         index = pd.MultiIndex.from_product([[2013, 2014], [1, 2]],
                                            names=['year', 'visit'])
         columns = pd.MultiIndex.from_product([['Bob', 'Guido', 'Sue'],
                                               ['HR', 'Temp']],
                                              names=['subject', 'type'])

         # mock some data
         data = np.round(np.random.randn(4, 6), 1)
         data[:, ::2] *= 10
         data += 37

         # create the DataFrame
         health_data = pd.DataFrame(data, index=index, columns=columns)
         health_data
Out[19]: subject      Bob        Guido        Sue
         type          HR  Temp    HR  Temp    HR  Temp
         year visit
         2013 1       30.0 38.0   56.0 38.3   45.0 35.8
              2       47.0 37.1   27.0 36.0   37.0 36.4
         2014 1       51.0 35.9   24.0 36.7   32.0 36.2
              2       49.0 36.3   48.0 39.2   31.0 35.7
```

This is fundamentally four-dimensional data, where the dimensions are the subject, the measurement type, the year, and the visit number. With this in place we can, for example, index the top-level column by the person's name and get a full DataFrame containing just that person's information:

```
In [20]: health_data['Guido']
Out[20]: type          HR  Temp
         year visit
         2013 1       56.0 38.3
              2       27.0 36.0
         2014 1       24.0 36.7
              2       48.0 39.2
```

Indexing and Slicing a MultiIndex

Indexing and slicing on a MultiIndex is designed to be intuitive, and it helps if you think about the indices as added dimensions. We'll first look at indexing multiply indexed Series, and then multiply indexed DataFrame objects.

Multiply Indexed Series

Consider the multiply indexed `Series` of state populations we saw earlier:

```
In [21]: pop
Out[21]: state       year
         California  2010    37253956
                     2020    39538223
         New York    2010    19378102
                     2020    20201249
         Texas       2010    25145561
                     2020    29145505
         dtype: int64
```

We can access single elements by indexing with multiple terms:

```
In [22]: pop['California', 2010]
Out[22]: 37253956
```

The `MultiIndex` also supports *partial indexing*, or indexing just one of the levels in the index. The result is another `Series`, with the lower-level indices maintained:

```
In [23]: pop['California']
Out[23]: year
         2010    37253956
         2020    39538223
         dtype: int64
```

Partial slicing is available as well, as long as the `MultiIndex` is sorted (see the discussion in "Sorted and Unsorted Indices" on page 141):

```
In [24]: pop.loc['california':'new york']
Out[24]: state       year
         california  2010    37253956
                     2020    39538223
         new york    2010    19378102
                     2020    20201249
         dtype: int64
```

with sorted indices, partial indexing can be performed on lower levels by passing an empty slice in the first index:

```
In [25]: pop[:, 2010]
Out[25]: state
         california    37253956
         new york      19378102
         texas         25145561
         dtype: int64
```

Other types of indexing and selection (discussed in Chapter 14) work as well; for example, selection based on Boolean masks:

```
In [26]: pop[pop > 22000000]
Out[26]: state       year
```

```
     California   2010      37253956
                  2020      39538223
     Texas        2010      25145561
                  2020      29145505
     dtype: int64
```

Selection based on fancy indexing also works:

```
In [27]: pop[['California', 'Texas']]
Out[27]: state       year
         California   2010      37253956
                      2020      39538223
         Texas        2010      25145561
                      2020      29145505
         dtype: int64
```

Multiply Indexed DataFrames

A multiply indexed DataFrame behaves in a similar manner. Consider our toy medical DataFrame from before:

```
In [28]: health_data
Out[28]: subject      Bob           Guido         Sue
         type         HR   Temp     HR   Temp     HR   Temp
         year visit
         2013 1       30.0 38.0     56.0 38.3     45.0 35.8
              2       47.0 37.1     27.0 36.0     37.0 36.4
         2014 1       51.0 35.9     24.0 36.7     32.0 36.2
              2       49.0 36.3     48.0 39.2     31.0 35.7
```

Remember that columns are primary in a DataFrame, and the syntax used for multiply indexed Series applies to the columns. For example, we can recover Guido's heart rate data with a simple operation:

```
In [29]: health_data['Guido', 'HR']
Out[29]: year  visit
         2013  1        56.0
               2        27.0
         2014  1        24.0
               2        48.0
         Name: (Guido, HR), dtype: float64
```

Also, as with the single-index case, we can use the loc, iloc, and ix indexers introduced in Chapter 14. For example:

```
In [30]: health_data.iloc[:2, :2]
Out[30]: subject      Bob
         type         HR   Temp
         year visit
         2013 1       30.0 38.0
              2       47.0 37.1
```

These indexers provide an array-like view of the underlying two-dimensional data, but each individual index in `loc` or `iloc` can be passed a tuple of multiple indices. For example:

```
In [31]: health_data.loc[:, ('Bob', 'HR')]
Out[31]: year  visit
         2013  1          30.0
               2          47.0
         2014  1          51.0
               2          49.0
         Name: (Bob, HR), dtype: float64
```

Working with slices within these index tuples is not especially convenient; trying to create a slice within a tuple will lead to a syntax error:

```
In [32]: health_data.loc[(:, 1), (:, 'HR')]
SyntaxError: invalid syntax (3311942670.py, line 1)
```

You could get around this by building the desired slice explicitly using Python's built-in `slice` function, but a better way in this context is to use an `IndexSlice` object, which Pandas provides for precisely this situation. For example:

```
In [33]: idx = pd.IndexSlice
         health_data.loc[idx[:, 1], idx[:, 'HR']]
Out[33]: subject      Bob Guido  Sue
         type          HR   HR   HR
         year visit
         2013 1       30.0 56.0 45.0
         2014 1       51.0 24.0 32.0
```

As you can see, there are many ways to interact with data in multiply indexed `Series` and `DataFrames`, and as with many tools in this book the best way to become familiar with them is to try them out!

Rearranging Multi-Indexes

One of the keys to working with multiply indexed data is knowing how to effectively transform the data. There are a number of operations that will preserve all the information in the dataset, but rearrange it for the purposes of various computations. We saw a brief example of this in the `stack` and `unstack` methods, but there are many more ways to finely control the rearrangement of data between hierarchical indices and columns, and we'll explore them here.

Sorted and Unsorted Indices

Earlier I briefly mentioned a caveat, but I should emphasize it more here. *Many of the MultiIndex slicing operations will fail if the index is not sorted.* Let's take a closer look.

We'll start by creating some simple multiply indexed data where the indices are *not lexographically sorted*:

```
In [34]: index = pd.MultiIndex.from_product([['a', 'c', 'b'], [1, 2]])
         data = pd.Series(np.random.rand(6), index=index)
         data.index.names = ['char', 'int']
         data
Out[34]: char  int
         a     1      0.280341
               2      0.097290
         c     1      0.206217
               2      0.431771
         b     1      0.100183
               2      0.015851
         dtype: float64
```

If we try to take a partial slice of this index, it will result in an error:

```
In [35]: try:
             data['a':'b']
         except KeyError as e:
             print("KeyError", e)
KeyError 'Key length (1) was greater than MultiIndex lexsort depth (0)'
```

Although it is not entirely clear from the error message, this is the result of the Multi Index not being sorted. For various reasons, partial slices and other similar operations require the levels in the MultiIndex to be in sorted (i.e., lexographical) order. Pandas provides a number of convenience routines to perform this type of sorting, such as the sort_index and sortlevel methods of the DataFrame. We'll use the simplest, sort_index, here:

```
In [36]: data = data.sort_index()
         data
Out[36]: char  int
         a     1      0.280341
               2      0.097290
         b     1      0.100183
               2      0.015851
         c     1      0.206217
               2      0.431771
         dtype: float64
```

With the index sorted in this way, partial slicing will work as expected:

```
In [37]: data['a':'b']
Out[37]: char  int
         a     1      0.280341
               2      0.097290
         b     1      0.100183
               2      0.015851
         dtype: float64
```

Stacking and Unstacking Indices

As we saw briefly before, it is possible to convert a dataset from a stacked multi-index to a simple two-dimensional representation, optionally specifying the level to use:

```
In [38]: pop.unstack(level=0)
Out[38]: year            2010      2020
         state
         California  37253956  39538223
         New York    19378102  20201249
         Texas       25145561  29145505

In [39]: pop.unstack(level=1)
Out[39]: state       year
         California  2010      37253956
                     2020      39538223
         New York    2010      19378102
                     2020      20201249
         Texas       2010      25145561
                     2020      29145505
         dtype: int64
```

The opposite of `unstack` is `stack`, which here can be used to recover the original series:

```
In [40]: pop.unstack().stack()
Out[40]: state       year
         California  2010      37253956
                     2020      39538223
         New York    2010      19378102
                     2020      20201249
         Texas       2010      25145561
                     2020      29145505
         dtype: int64
```

Index Setting and Resetting

Another way to rearrange hierarchical data is to turn the index labels into columns; this can be accomplished with the `reset_index` method. Calling this on the population dictionary will result in a `DataFrame` with `state` and `year` columns holding the information that was formerly in the index. For clarity, we can optionally specify the name of the data for the column representation:

```
In [41]: pop_flat = pop.reset_index(name='population')
         pop_flat
Out[41]:        state  year  population
         0  California  2010    37253956
         1  California  2020    39538223
         2    New York  2010    19378102
         3    New York  2020    20201249
         4       Texas  2010    25145561
         5       Texas  2020    29145505
```

A common pattern is to build a MultiIndex from the column values. This can be done with the set_index method of the DataFrame, which returns a multiply indexed DataFrame:

```
In [42]: pop_flat.set_index(['state', 'year'])
Out[42]:                 population
         state      year
         California 2010 37253956
                    2020 39538223
         New York   2010 19378102
                    2020 20201249
         Texas      2010 25145561
                    2020 29145505
```

In practice, this type of reindexing is one of the more useful patterns when exploring real-world datasets.

## CHAPTER 18

Combining Datasets: concat and append

Some of the most interesting studies of data come from combining different data sources. These operations can involve anything from very straightforward concatenation of two different datasets to more complicated database-style joins and merges that correctly handle any overlaps between the datasets. Series and DataFrames are built with this type of operation in mind, and Pandas includes functions and methods that make this sort of data wrangling fast and straightforward.

Here we'll take a look at simple concatenation of Series and DataFrames with the pd.concat function; later we'll dive into more sophisticated in-memory merges and joins implemented in Pandas.

We begin with the standard imports:

```
In [1]: import pandas as pd
        import numpy as np
```

For convenience, we'll define this function, which creates a DataFrame of a particular form that will be useful in the following examples:

```
In [2]: def make_df(cols, ind):
            """Quickly make a DataFrame"""
            data = {c: [str(c) + str(i) for i in ind]
                    for c in cols}
            return pd.DataFrame(data, ind)

        # example DataFrame
        make_df('ABC', range(3))
Out[2]:    A   B   C
        0  A0  B0  C0
        1  A1  B1  C1
        2  A2  B2  C2
```

In addition, we'll create a quick class that allows us to display multiple `DataFrames` side by side. The code makes use of the special _repr_html_ method, which IPython/Jupyter uses to implement its rich object display:

```
In [3]: class display(object):
            """Display HTML representation of multiple objects"""
            template = """<div style="float: left; padding: 10px;">
            <p style='font-family:"Courier New", Courier, monospace'>{0}{1}
            """
            def __init__(self, *args):
                self.args = args

            def _repr_html_(self):
                return '\n'.join(self.template.format(a, eval(a)._repr_html_())
                                 for a in self.args)

            def __repr__(self):
                return '\n\n'.join(a + '\n' + repr(eval(a))
                                   for a in self.args)
```

The use of this will become clearer as we continue our discussion in the following section.

Recall: Concatenation of NumPy Arrays

Concatenation of `Series` and `DataFrame` objects behaves similarly to concatenation of NumPy arrays, which can be done via the `np.concatenate` function, as discussed in Chapter 5. Recall that with it, you can combine the contents of two or more arrays into a single array:

```
In [4]: x = [1, 2, 3]
        y = [4, 5, 6]
        z = [7, 8, 9]
        np.concatenate([x, y, z])
Out[4]: array([1, 2, 3, 4, 5, 6, 7, 8, 9])
```

The first argument is a list or tuple of arrays to concatenate. Additionally, in the case of multidimensional arrays, it takes an `axis` keyword that allows you to specify the axis along which the result will be concatenated:

```
In [5]: x = [[1, 2],
             [3, 4]]
        np.concatenate([x, x], axis=1)
Out[5]: array([[1, 2, 1, 2],
               [3, 4, 3, 4]])
```

Simple Concatenation with pd.concat

The pd.concat function provides a similar syntax to np.concatenate but contains a number of options that we'll discuss momentarily:

```
# Signature in Pandas v1.3.5
pd.concat(objs, axis=0, join='outer', ignore_index=False, keys=None,
          levels=None, names=None, verify_integrity=False,
          sort=False, copy=True)
```

pd.concat can be used for a simple concatenation of Series or DataFrame objects, just as np.concatenate can be used for simple concatenations of arrays:

```
In [6]: ser1 = pd.Series(['A', 'B', 'C'], index=[1, 2, 3])
        ser2 = pd.Series(['D', 'E', 'F'], index=[4, 5, 6])
        pd.concat([ser1, ser2])
Out[6]: 1    A
        2    B
        3    C
        4    D
        5    E
        6    F
        dtype: object
```

It also works to concatenate higher-dimensional objects, such as DataFrames:

```
In [7]: df1 = make_df('AB', [1, 2])
        df2 = make_df('AB', [3, 4])
        display('df1', 'df2', 'pd.concat([df1, df2])')
Out[7]: df1              df2            pd.concat([df1, df2])
           A   B            A   B           A   B
        1  A1  B1        3  A3  B3       1  A1  B1
        2  A2  B2        4  A4  B4       2  A2  B2
                                        3  A3  B3
                                        4  A4  B4
```

It's default behavior is to concatenate row-wise within the DataFrame (i.e., axis=0). Like np.concatenate, pd.concat allows specification of an axis along which concatenation will take place. Consider the following example:

```
In [8]: df3 = make_df('AB', [0, 1])
        df4 = make_df('CD', [0, 1])
        display('df3', 'df4', "pd.concat([df3, df4], axis='columns')")
Out[8]: df3              df4            pd.concat([df3, df4], axis='columns')
           A   B            C   D           A   B   C   D
        0  A0  B0        0  C0  D0       0  A0  B0  C0  D0
        1  A1  B1        1  C1  D1       1  A1  B1  C1  D1
```

We could have equivalently specified axis=1; here we've used the more intuitive axis='columns'.

Duplicate Indices

One important difference between `np.concatenate` and `pd.concat` is that Pandas concatenation *preserves indices*, even if the result will have duplicate indices! Consider this short example:

```
In [9]: x = make_df('AB', [0, 1])
        y = make_df('AB', [2, 3])
        y.index = x.index  # make indices match
        display('x', 'y', 'pd.concat([x, y])')
Out[9]: x              y              pd.concat([x, y])
           A   B          A   B          A   B
        0  A0  B0      0  A2  B2      0  A0  B0
        1  A1  B1      1  A3  B3      1  A1  B1
                                      0  A2  B2
                                      1  A3  B3
```

Notice the repeated indices in the result. While this is valid within `DataFrames`, the outcome is often undesirable. `pd.concat` gives us a few ways to handle it.

Treating repeated indices as an error

If you'd like to simply verify that the indices in the result of `pd.concat` do not overlap, you can include the `verify_integrity` flag. With this set to `True`, the concatenation will raise an exception if there are duplicate indices. Here is an example, where for clarity we'll catch and print the error message:

```
In [10]: try:
             pd.concat([x, y], verify_integrity=True)
         except ValueError as e:
             print("ValueError:", e)
ValueError: Indexes have overlapping values: Int64Index([0, 1], dtype='int64')
```

Ignoring the index

Sometimes the index itself does not matter, and you would prefer it to simply be ignored. This option can be specified using the `ignore_index` flag. With this set to `True`, the concatenation will create a new integer index for the resulting `DataFrame`:

```
In [11]: display('x', 'y', 'pd.concat([x, y], ignore_index=True)')
Out[11]: x              y              pd.concat([x, y], ignore_index=True)
            A   B          A   B          A   B
        0   A0  B0      0  A2  B2      0  A0  B0
        1   A1  B1      1  A3  B3      1  A1  B1
                                      2  A2  B2
                                      3  A3  B3
```

Adding MultiIndex keys

Another option is to use the keys option to specify a label for the data sources; the result will be a hierarchically indexed series containing the data:

```
In [12]: display('x', 'y', "pd.concat([x, y], keys=['x', 'y'])")
Out[12]: x                y                    pd.concat([x, y], keys=['x', 'y'])
            A   B              A   B                  A   B
         0  A0  B0         0  A2  B2        x  0  A0  B0
         1  A1  B1         1  A3  B3           1  A1  B1
                                           y  0  A2  B2
                                              1  A3  B3
```

We can use the tools discussed in Chapter 17 to transform this multiply indexed DataFrame into the representation we're interested in.

Concatenation with Joins

In the short examples we just looked at, we were mainly concatenating DataFrames with shared column names. In practice, data from different sources might have different sets of column names, and pd.concat offers several options in this case. Consider the concatenation of the following two DataFrames, which have some (but not all!) columns in common:

```
In [13]: df5 = make_df('ABC', [1, 2])
         df6 = make_df('BCD', [3, 4])
         display('df5', 'df6', 'pd.concat([df5, df6])')
Out[13]: df5                     df6                   pd.concat([df5, df6])
            A   B   C              B   C   D              A    B   C   D
         1  A1  B1  C1         3  B3  C3  D3         1   A1   B1  C1  NaN
         2  A2  B2  C2         4  B4  C4  D4         2   A2   B2  C2  NaN
                                                    3  NaN   B3  C3  D3
                                                    4  NaN   B4  C4  D4
```

The default behavior is to fill entries for which no data is available with NA values. To change this, we can adjust the join parameter of the concat function. By default, the join is a union of the input columns (join='outer'), but we can change this to an intersection of the columns using join='inner':

```
In [14]: display('df5', 'df6',
            "pd.concat([df5, df6], join='inner')")
Out[14]: df5                     df6
            A   B   C              B   C   D
         1  A1  B1  C1         3  B3  C3  D3
         2  A2  B2  C2         4  B4  C4  D4

         pd.concat([df5, df6], join='inner')
            B   C
         1  B1  C1
         2  B2  C2
```

```
            3   B3   C3
            4   B4   C4
```

Another useful pattern is to use the `reindex` method before concatenation for finer control over which columns are dropped:

```
In [15]: pd.concat([df5, df6.reindex(df5.columns, axis=1)])
Out[15]:      A    B   C
          1   A1   B1  C1
          2   A2   B2  C2
          3  NaN   B3  C3
          4  NaN   B4  C4
```

The append Method

Because direct array concatenation is so common, `Series` and `DataFrame` objects have an append method that can accomplish the same thing in fewer keystrokes. For example, in place of `pd.concat([df1, df2])`, you can use `df1.append(df2)`:

```
In [16]: display('df1', 'df2', 'df1.append(df2)')
Out[16]: df1              df2             df1.append(df2)
              A    B            A   B             A    B
          1   A1   B1       3   A3  B3        1   A1   B1
          2   A2   B2       4   A4  B4        2   A2   B2
                                             3   A3   B3
                                             4   A4   B4
```

Keep in mind that unlike the `append` and `extend` methods of Python lists, the `append` method in Pandas does not modify the original object; instead it creates a new object with the combined data. It also is not a very efficient method, because it involves creation of a new index *and* data buffer. Thus, if you plan to do multiple `append` operations, it is generally better to build a list of `DataFrame` objects and pass them all at once to the `concat` function.

In the next chapter, we'll look at a more powerful approach to combining data from multiple sources: the database-style merges/joins implemented in `pd.merge`. For more information on `concat`, `append`, and related functionality, see "Merge, Join, Concatenate and Compare" in the Pandas documentation (*https://oreil.ly/cY16c*).

## CHAPTER 19

Combining Datasets: merge and join

One important feature offered by Pandas is its high-performance, in-memory join and merge operations, which you may be familiar with if you have ever worked with databases. The main interface for this is the `pd.merge` function, and we'll see a few examples of how this can work in practice.

For convenience, we will again define the `display` function from the previous chapter after the usual imports:

```
In [1]: import pandas as pd
        import numpy as np

        class display(object):
            """Display HTML representation of multiple objects"""
            template = """<div style="float: left; padding: 10px;">
            <p style='font-family:"Courier New", Courier, monospace'>{0}{1}
            """
            def __init__(self, *args):
                self.args = args

            def _repr_html_(self):
                return '\n'.join(self.template.format(a, eval(a)._repr_html_())
                                 for a in self.args)

            def __repr__(self):
                return '\n\n'.join(a + '\n' + repr(eval(a))
                                   for a in self.args)
```

Relational Algebra

The behavior implemented in `pd.merge` is a subset of what is known as *relational algebra*, which is a formal set of rules for manipulating relational data that forms the conceptual foundation of operations available in most databases. The strength of the relational algebra approach is that it proposes several fundamental operations, which become the building blocks of more complicated operations on any dataset. With this lexicon of fundamental operations implemented efficiently in a database or other program, a wide range of fairly complicated composite operations can be performed.

Pandas implements several of these fundamental building blocks in the pd.merge function and the related join method of Series and DataFrame objects. As you will see, these let you efficiently link data from different sources.

Categories of Joins

The pd.merge function implements a number of types of joins: *one-to-one*, *many-to-one*, and *many-to-many*. All three types of joins are accessed via an identical call to the pd.merge interface; the type of join performed depends on the form of the input data. We'll start with some simple examples of the three types of merges, and discuss detailed options a bit later.

One-to-One Joins

Perhaps the simplest type of merge is the one-to-one join, which is in many ways similar to the column-wise concatenation you saw in Chapter 18. As a concrete example, consider the following two DataFrame objects, which contain information on several employees in a company:

```
In [2]: df1 = pd.DataFrame({'employee': ['Bob', 'Jake', 'Lisa', 'Sue'],
                            'group': ['Accounting', 'Engineering',
                                      'Engineering', 'HR']})
        df2 = pd.DataFrame({'employee': ['Lisa', 'Bob', 'Jake', 'Sue'],
                            'hire_date': [2004, 2008, 2012, 2014]})
        display('df1', 'df2')
Out[2]: df1                           df2
            employee        group         employee  hire_date
        0        Bob   Accounting     0       Lisa       2004
        1       Jake  Engineering     1        Bob       2008
        2       Lisa  Engineering     2       Jake       2012
        3        Sue           HR     3        Sue       2014
```

To combine this information into a single DataFrame, we can use the pd.merge function:

```
In [3]: df3 = pd.merge(df1, df2)
        df3
Out[3]:     employee        group  hire_date
        0        Bob   Accounting       2008
        1       Jake  Engineering       2012
        2       Lisa  Engineering       2004
        3        Sue           HR       2014
```

The pd.merge function recognizes that each DataFrame has an employee column, and automatically joins using this column as a key. The result of the merge is a new Data Frame that combines the information from the two inputs. Notice that the order of entries in each column is not necessarily maintained: in this case, the order of the employee column differs between df1 and df2, and the pd.merge function correctly accounts for this. Additionally, keep in mind that the merge in general discards the index, except in the special case of merges by index (see the left_index and right_index keywords, discussed momentarily).

Many-to-One Joins

Many-to-one joins are joins in which one of the two key columns contains duplicate entries. For the many-to-one case, the resulting DataFrame will preserve those duplicate entries as appropriate. Consider the following example of a many-to-one join:

```
In [4]: df4 = pd.DataFrame({'group': ['Accounting', 'Engineering', 'HR'],
                            'supervisor': ['Carly', 'Guido', 'Steve']})
        display('df3', 'df4', 'pd.merge(df3, df4)')
Out[4]: df3                                 df4
          employee         group  hire_date               group supervisor
        0      Bob    Accounting       2008     0    Accounting      Carly
        1     Jake   Engineering       2012     1   Engineering      Guido
        2     Lisa   Engineering       2004     2            HR      Steve
        3      Sue            HR       2014

        pd.merge(df3, df4)
          employee         group  hire_date supervisor
        0      Bob    Accounting       2008      Carly
        1     Jake   Engineering       2012      Guido
        2     Lisa   Engineering       2004      Guido
        3      Sue            HR       2014      Steve
```

The resulting DataFrame has an additional column with the "supervisor" information, where the information is repeated in one or more locations as required by the inputs.

Many-to-Many Joins

Many-to-many joins may be a bit confusing conceptually, but are nevertheless well defined. If the key column in both the left and right arrays contains duplicates, then the result is a many-to-many merge. This will be perhaps most clear with a concrete example. Consider the following, where we have a DataFrame showing one or more skills associated with a particular group.

By performing a many-to-many join, we can recover the skills associated with any individual person:

```
In [5]: df5 = pd.DataFrame({'group': ['Accounting', 'Accounting',
                            'Engineering', 'Engineering', 'HR', 'HR'],
```

```
                        'skills': ['math', 'spreadsheets', 'software', 'math',
                                   'spreadsheets', 'organization']})
         display('df1', 'df5', "pd.merge(df1, df5)")
Out[5]: df1                              df5
        employee        group                    group        skills
      0      Bob   Accounting      0   Accounting           math
      1     Jake  Engineering      1   Accounting   spreadsheets
      2     Lisa  Engineering      2  Engineering       software
      3      Sue           HR      3  Engineering           math
                                   4           HR   spreadsheets
                                   5           HR   organization


        pd.merge(df1, df5)
        employee        group        skills
      0      Bob   Accounting          math
      1      Bob   Accounting  spreadsheets
      2     Jake  Engineering      software
      3     Jake  Engineering          math
      4     Lisa  Engineering      software
      5     Lisa  Engineering          math
      6      Sue           HR  spreadsheets
      7      Sue           HR  organization
```

These three types of joins can be used with other Pandas tools to implement a wide array of functionality. But in practice, datasets are rarely as clean as the one we're working with here. In the following section we'll consider some of the options provided by `pd.merge` that enable you to tune how the join operations work.

Specification of the Merge Key

We've already seen the default behavior of `pd.merge`: it looks for one or more matching column names between the two inputs, and uses this as the key. However, often the column names will not match so nicely, and `pd.merge` provides a variety of options for handling this.

The on Keyword

Most simply, you can explicitly specify the name of the key column using the on keyword, which takes a column name or a list of column names:

```
In [6]: display('df1', 'df2', "pd.merge(df1, df2, on='employee')")
Out[6]: df1                          df2
        employee        group        employee  hire_date
      0      Bob   Accounting      0      Lisa       2004
      1     Jake  Engineering      1       Bob       2008
      2     Lisa  Engineering      2      Jake       2012
      3      Sue           HR      3       Sue       2014


        pd.merge(df1, df2, on='employee')
        employee        group  hire_date
```

```
0      Bob   Accounting   2008
1     Jake  Engineering   2012
2     Lisa  Engineering   2004
3      Sue           HR   2014
```

This option works only if both the left and right `DataFrame`s have the specified column name.

The left_on and right_on Keywords

At times you may wish to merge two datasets with different column names; for example, we may have a dataset in which the employee name is labeled as "name" rather than "employee". In this case, we can use the `left_on` and `right_on` keywords to specify the two column names:

```
In [7]: df3 = pd.DataFrame({'name': ['Bob', 'Jake', 'Lisa', 'Sue'],
                            'salary': [70000, 80000, 120000, 90000]})
        display('df1', 'df3', 'pd.merge(df1, df3, left_on="employee",
            right_on="name")')
Out[7]: df1                         df3
          employee        group          name   salary
        0      Bob   Accounting     0    Bob    70000
        1     Jake  Engineering     1   Jake    80000
        2     Lisa  Engineering     2   Lisa   120000
        3      Sue           HR     3    Sue    90000

        pd.merge(df1, df3, left_on="employee", right_on="name")
          employee        group  name    salary
        0      Bob   Accounting   Bob    70000
        1     Jake  Engineering  Jake    80000
        2     Lisa  Engineering  Lisa   120000
        3      Sue           HR   Sue    90000
```

The result has a redundant column that we can drop if desired—for example, by using the `DataFrame.drop()` method:

```
In [8]: pd.merge(df1, df3, left_on="employee", right_on="name").drop('name', axis=1)
Out[8]:   employee        group  salary
        0      Bob   Accounting   70000
        1     Jake  Engineering   80000
        2     Lisa  Engineering  120000
        3      Sue           HR   90000
```

The left_index and right_index Keywords

Sometimes, rather than merging on a column, you would instead like to merge on an index. For example, your data might look like this:

```
In [9]: df1a = df1.set_index('employee')
        df2a = df2.set_index('employee')
        display('df1a', 'df2a')
```

```
Out[9]: df1a                           df2a
                        group                   hire_date
        employee                        employee
        Bob         Accounting          Lisa        2004
        Jake        Engineering         Bob         2008
        Lisa        Engineering         Jake        2012
        Sue                 HR          Sue         2014
```

You can use the index as the key for merging by specifying the left_index and/or right_index flags in pd.merge():

```
In [10]: display('df1a', 'df2a',
                  "pd.merge(df1a, df2a, left_index=True, right_index=True)")
Out[10]: df1a                           df2a
                        group                   hire_date
        employee                        employee
        Bob         Accounting          Lisa        2004
        Jake        Engineering         Bob         2008
        Lisa        Engineering         Jake        2012
        Sue                 HR          Sue         2014

        pd.merge(df1a, df2a, left_index=True, right_index=True)
                        group  hire_date
        employee
        Bob         Accounting     2008
        Jake        Engineering    2012
        Lisa        Engineering    2004
        Sue                 HR     2014
```

For convenience, Pandas includes the DataFrame.join() method, which performs an index-based merge without extra keywords:

```
In [11]: df1a.join(df2a)
Out[11]:                 group  hire_date
        employee
        Bob         Accounting     2008
        Jake        Engineering    2012
        Lisa        Engineering    2004
        Sue                 HR     2014
```

If you'd like to mix indices and columns, you can combine left_index with right_on or left_on with right_index to get the desired behavior:

```
In [12]: display('df1a', 'df3', "pd.merge(df1a, df3, left_index=True,
                  right_on='name')")
Out[12]: df1a                           df3
                        group           name   salary
        employee                     0  Bob    70000
        Bob         Accounting       1  Jake   80000
        Jake        Engineering      2  Lisa   120000
        Lisa        Engineering      3  Sue    90000
        Sue                 HR
```

```
pd.merge(df1a, df3, left_index=True, right_on='name')
        group   name   salary
0   Accounting    Bob    70000
1  Engineering   Jake    80000
2  Engineering   Lisa   120000
3           HR    Sue    90000
```

All of these options also work with multiple indices and/or multiple columns; the interface for this behavior is very intuitive. For more information on this, see the "Merge, Join, and Concatenate" section (*https://oreil.ly/ffyAp*) of the Pandas documentation.

Specifying Set Arithmetic for Joins

In all the preceding examples we have glossed over one important consideration in performing a join: the type of set arithmetic used in the join. This comes up when a value appears in one key column but not the other. Consider this example:

```
In [13]: df6 = pd.DataFrame({'name': ['Peter', 'Paul', 'Mary'],
                             'food': ['fish', 'beans', 'bread']},
                            columns=['name', 'food'])
         df7 = pd.DataFrame({'name': ['Mary', 'Joseph'],
                             'drink': ['wine', 'beer']},
                            columns=['name', 'drink'])
         display('df6', 'df7', 'pd.merge(df6, df7)')
Out[13]: df6                      df7
             name   food              name drink
         0  Peter   fish          0    Mary  wine
         1   Paul  beans          1  Joseph  beer
         2   Mary  bread

         pd.merge(df6, df7)
             name   food drink
         0   Mary  bread  wine
```

Here we have merged two datasets that have only a single "name" entry in common: Mary. By default, the result contains the *intersection* of the two sets of inputs; this is what is known as an *inner join*. We can specify this explicitly using the how keyword, which defaults to "inner":

```
In [14]: pd.merge(df6, df7, how='inner')
Out[14]:     name   food drink
         0   Mary  bread  wine
```

Other options for the how keyword are 'outer', 'left', and 'right'. An *outer join* returns a join over the union of the input columns and fills in missing values with NAs:

```
In [15]: display('df6', 'df7', "pd.merge(df6, df7, how='outer')")
Out[15]: df6                      df7
             name   food              name drink
```

```
0   Peter   fish     0    Mary  wine
1   Paul    beans    1  Joseph  beer
2   Mary    bread

pd.merge(df6, df7, how='outer')
      name    food drink
0    Peter    fish   NaN
1     Paul   beans   NaN
2     Mary   bread  wine
3   Joseph     NaN  beer
```

The *left join* and *right join* return joins over the left entries and right entries, respectively. For example:

```
In [16]: display('df6', 'df7', "pd.merge(df6, df7, how='left')")
Out[16]: df6                     df7
             name    food              name drink
         0  Peter    fish     0     Mary  wine
         1   Paul   beans     1   Joseph  beer
         2   Mary   bread

         pd.merge(df6, df7, how='left')
               name    food drink
         0    Peter    fish   NaN
         1     Paul   beans   NaN
         2     Mary   bread  wine
```

The output rows now correspond to the entries in the left input. Using how='right' works in a similar manner.

All of these options can be applied straightforwardly to any of the preceding join types.

Overlapping Column Names: The suffixes Keyword

Last, you may end up in a case where your two input DataFrames have conflicting column names. Consider this example:

```
In [17]: df8 = pd.DataFrame({'name': ['Bob', 'Jake', 'Lisa', 'Sue'],
                             'rank': [1, 2, 3, 4]})
         df9 = pd.DataFrame({'name': ['Bob', 'Jake', 'Lisa', 'Sue'],
                             'rank': [3, 1, 4, 2]})
         display('df8', 'df9', 'pd.merge(df8, df9, on="name")')
Out[17]: df8                     df9
             name rank              name rank
         0   Bob     1     0    Bob     3
         1  Jake     2     1   Jake     1
         2  Lisa     3     2   Lisa     4
         3   Sue     4     3    Sue     2

         pd.merge(df8, df9, on="name")
             name  rank_x  rank_y
```

```
0    Bob    1    3
1    Jake   2    1
2    Lisa   3    4
3    Sue    4    2
```

Because the output would have two conflicting column names, the merge function automatically appends the suffixes _x and _y to make the output columns unique. If these defaults are inappropriate, it is possible to specify a custom suffix using the suf fixes keyword:

```
In [18]: pd.merge(df8, df9, on="name", suffixes=["_L", "_R"])
Out[18]:    name  rank_L  rank_R
         0    Bob       1       3
         1   Jake       2       1
         2   Lisa       3       4
         3    Sue       4       2
```

These suffixes work in any of the possible join patterns, and also work if there are multiple overlapping columns.

In Chapter 20, we'll dive a bit deeper into relational algebra. For further discussion, see "Merge, Join, Concatenate and Compare" (*https://oreil.ly/l8zZ1*) in the Pandas documentation.

Example: US States Data

Merge and join operations come up most often when combining data from different sources. Here we will consider an example of some data about US states and their populations (*https://oreil.ly/aq6Xb*):

```
In [19]: # Following are commands to download the data
         # repo = "https://raw.githubusercontent.com/jakevdp/data-USstates/master"
         # !cd data && curl -O {repo}/state-population.csv
         # !cd data && curl -O {repo}/state-areas.csv
         # !cd data && curl -O {repo}/state-abbrevs.csv
```

Let's take a look at the three datasets, using the Pandas read_csv function:

```
In [20]: pop = pd.read_csv('data/state-population.csv')
         areas = pd.read_csv('data/state-areas.csv')
         abbrevs = pd.read_csv('data/state-abbrevs.csv')

         display('pop.head()', 'areas.head()', 'abbrevs.head()')
Out[20]: pop.head()
            state/region     ages  year  population
         0            AL  under18  2012   1117489.0
         1            AL    total  2012   4817528.0
         2            AL  under18  2010   1130966.0
         3            AL    total  2010   4785570.0
         4            AL  under18  2011   1125763.0
```

```
areas.head()
        state   area (sq. mi)
0    Alabama           52423
1     Alaska          656425
2    Arizona          114006
3   Arkansas           53182
4 California          163707

abbrevs.head()
        state abbreviation
0    Alabama           AL
1     Alaska           AK
2    Arizona           AZ
3   Arkansas           AR
4 California           CA
```

Given this information, say we want to compute a relatively straightforward result: rank US states and territories by their 2010 population density. We clearly have the data here to find this result, but we'll have to combine the datasets to do so.

We'll start with a many-to-one merge that will give us the full state names within the population DataFrame. We want to merge based on the state/region column of pop and the abbreviation column of abbrevs. We'll use how='outer' to make sure no data is thrown away due to mismatched labels:

```
In [21]: merged = pd.merge(pop, abbrevs, how='outer',
                           left_on='state/region', right_on='abbreviation')
         merged = merged.drop('abbreviation', axis=1) # drop duplicate info
         merged.head()
Out[21]:   state/region     ages  year  population     state
         0           AL  under18  2012   1117489.0   Alabama
         1           AL    total  2012   4817528.0   Alabama
         2           AL  under18  2010   1130966.0   Alabama
         3           AL    total  2010   4785570.0   Alabama
         4           AL  under18  2011   1125763.0   Alabama
```

Let's double-check whether there were any mismatches here, which we can do by looking for rows with nulls:

```
In [22]: merged.isnull().any()
Out[22]: state/region    False
         ages            False
         year            False
         population       True
         state            True
         dtype: bool
```

Some of the population values are null; let's figure out which these are!

```
In [23]: merged[merged['population'].isnull()].head()
Out[23]:      state/region     ages  year  population state
         2448           PR  under18  1990         NaN   NaN
         2449           PR    total  1990         NaN   NaN
```

```
2450        PR    total  1991        NaN  NaN
2451        PR  under18  1991        NaN  NaN
2452        PR    total  1993        NaN  NaN
```

It appears that all the null population values are from Puerto Rico prior to the year 2000; this is likely due to this data not being available in the original source.

More importantly, we see that some of the new `state` entries are also null, which means that there was no corresponding entry in the `abbrevs` key! Let's figure out which regions lack this match:

```
In [24]: merged.loc[merged['state'].isnull(), 'state/region'].unique()
Out[24]: array(['PR', 'USA'], dtype=object)
```

We can quickly infer the issue: our population data includes entries for Puerto Rico (PR) and the United States as a whole (USA), while these entries do not appear in the state abbreviation key. We can fix these quickly by filling in appropriate entries:

```
In [25]: merged.loc[merged['state/region'] == 'PR', 'state'] = 'Puerto Rico'
         merged.loc[merged['state/region'] == 'USA', 'state'] = 'United States'
         merged.isnull().any()
Out[25]: state/region    False
         ages            False
         year            False
         population       True
         state           False
         dtype: bool
```

No more nulls in the `state` column: we're all set!

Now we can merge the result with the area data using a similar procedure. Examining our results, we will want to join on the `state` column in both:

```
In [26]: final = pd.merge(merged, areas, on='state', how='left')
         final.head()
Out[26]:    state/region     ages  year   population     state  area (sq. mi)
         0            AL  under18  2012  1117489.0  Alabama        52423.0
         1            AL    total  2012  4817528.0  Alabama        52423.0
         2            AL  under18  2010  1130966.0  Alabama        52423.0
         3            AL    total  2010  4785570.0  Alabama        52423.0
         4            AL  under18  2011  1125763.0  Alabama        52423.0
```

Again, let's check for nulls to see if there were any mismatches:

```
In [27]: final.isnull().any()
Out[27]: state/region    False
         ages            False
         year            False
         population       True
         state           False
         area (sq. mi)    True
         dtype: bool
```

There are nulls in the area column; we can take a look to see which regions were ignored here:

```
In [28]: final['state'][final['area (sq. mi)'].isnull()].unique()
Out[28]: array(['United States'], dtype=object)
```

We see that our areas DataFrame does not contain the area of the United States as a whole. We could insert the appropriate value (using the sum of all state areas, for instance), but in this case we'll just drop the null values because the population density of the entire United States is not relevant to our current discussion:

```
In [29]: final.dropna(inplace=True)
         final.head()
Out[29]:    state/region    ages  year  population    state  area (sq. mi)
         0            AL  under18  2012  1117489.0  Alabama        52423.0
         1            AL    total  2012  4817528.0  Alabama        52423.0
         2            AL  under18  2010  1130966.0  Alabama        52423.0
         3            AL    total  2010  4785570.0  Alabama        52423.0
         4            AL  under18  2011  1125763.0  Alabama        52423.0
```

Now we have all the data we need. To answer the question of interest, let's first select the portion of the data corresponding with the year 2010, and the total population. We'll use the query function to do this quickly (this requires the NumExpr package to be installed; see Chapter 24):

```
In [30]: data2010 = final.query("year == 2010 & ages == 'total'")
         data2010.head()
Out[30]:     state/region   ages  year  population       state  area (sq. mi)
         3             AL  total  2010   4785570.0     Alabama        52423.0
         91            AK  total  2010    713868.0      Alaska       656425.0
         101           AZ  total  2010   6408790.0     Arizona       114006.0
         189           AR  total  2010   2922280.0    Arkansas        53182.0
         197           CA  total  2010  37333601.0  California       163707.0
```

Now let's compute the population density and display it in order. We'll start by re-indexing our data on the state, and then compute the result:

```
In [31]: data2010.set_index('state', inplace=True)
         density = data2010['population'] / data2010['area (sq. mi)']
```

```
In [32]: density.sort_values(ascending=False, inplace=True)
         density.head()
Out[32]: state
         District of Columbia    8898.897059
         Puerto Rico             1058.665149
         New Jersey              1009.253268
         Rhode Island             681.339159
         Connecticut              645.600649
         dtype: float64
```

The result is a ranking of US states, plus Washington, DC, and Puerto Rico, in order of their 2010 population density, in residents per square mile. We can see that by far the densest region in this dataset is Washington, DC (i.e., the District of Columbia); among states, the densest is New Jersey.

We can also check the end of the list:

```
In [33]: density.tail()
Out[33]: state
         South Dakota    10.583512
         North Dakota     9.537565
         Montana          6.736171
         Wyoming          5.768079
         Alaska           1.087509
         dtype: float64
```

We see that the least dense state, by far, is Alaska, averaging slightly over one resident per square mile.

This type of data merging is a common task when trying to answer questions using real-world data sources. I hope that this example has given you an idea of some of the ways you can combine the tools we've covered in order to gain insight from your data!

## CHAPTER 20

Aggregation and Grouping

A fundamental piece of many data analysis tasks is efficient summarization: computing aggregations like sum, mean, median, min, and max, in which a single number summarizes aspects of a potentially large dataset. In this chapter, we'll explore aggregations in Pandas, from simple operations akin to what we've seen on NumPy arrays to more sophisticated operations based on the concept of a groupby.

For convenience, we'll use the same display magic function that we used in the previous chapters:

```
In [1]: import numpy as np
        import pandas as pd

        class display(object):
            """Display HTML representation of multiple objects"""
            template = """<div style="float: left; padding: 10px;">
            <p style='font-family:"Courier New", Courier, monospace'>{0}{1}
            """
            def __init__(self, *args):
                self.args = args

            def _repr_html_(self):
                return '\n'.join(self.template.format(a, eval(a)._repr_html_())
                                 for a in self.args)

            def __repr__(self):
                return '\n\n'.join(a + '\n' + repr(eval(a))
                                   for a in self.args)
```

Planets Data

Here we will use the Planets dataset, available via the Seaborn package (*http://seaborn.pydata.org*) (see Chapter 36). It gives information on planets that astronomers have discovered around other stars (known as *extrasolar planets*, or *exoplanets* for short). It can be downloaded with a simple Seaborn command:

```
In [2]: import seaborn as sns
        planets = sns.load_dataset('planets')
        planets.shape
Out[2]: (1035, 6)
```

```
In [3]: planets.head()
Out[3]:           method  number  orbital_period   mass  distance  year
        0  Radial Velocity       1         269.300   7.10     77.40  2006
        1  Radial Velocity       1         874.774   2.21     56.95  2008
        2  Radial Velocity       1         763.000   2.60     19.84  2011
        3  Radial Velocity       1         326.030  19.40    110.62  2007
        4  Radial Velocity       1         516.220  10.50    119.47  2009
```

This has some details on the more than one thousand extrasolar planets discovered up to 2014.

Simple Aggregation in Pandas

In Chapter 7, we explored some of the data aggregations available for NumPy arrays. As with a one-dimensional NumPy array, for a Pandas Series the aggregates return a single value:

```
In [4]: rng = np.random.RandomState(42)
        ser = pd.Series(rng.rand(5))
        ser
Out[4]: 0    0.374540
        1    0.950714
        2    0.731994
        3    0.598658
        4    0.156019
        dtype: float64
```

```
In [5]: ser.sum()
Out[5]: 2.811925491708157
```

```
In [6]: ser.mean()
Out[6]: 0.5623850983416314
```

For a `DataFrame`, by default the aggregates return results within each column:

```
In [7]: df = pd.DataFrame({'A': rng.rand(5),
                           'B': rng.rand(5)})
        df
Out[7]:          A         B
        0  0.155995  0.020584
        1  0.058084  0.969910
        2  0.866176  0.832443
        3  0.601115  0.212339
        4  0.708073  0.181825

In [8]: df.mean()
Out[8]: A    0.477888
        B    0.443420
        dtype: float64
```

By specifying the `axis` argument, you can instead aggregate within each row:

```
In [9]: df.mean(axis='columns')
Out[9]: 0    0.088290
        1    0.513997
        2    0.849309
        3    0.406727
        4    0.444949
        dtype: float64
```

Pandas `Series` and `DataFrame` objects include all of the common aggregates mentioned in Chapter 7; in addition, there is a convenience method, `describe`, that computes several common aggregates for each column and returns the result. Let's use this on the Planets data, for now dropping rows with missing values:

```
In [10]: planets.dropna().describe()
Out[10]:          number  orbital_period        mass    distance         year
        count  498.00000      498.000000  498.000000  498.000000   498.000000
        mean     1.73494      835.778671    2.509320   52.068213  2007.377510
        std      1.17572     1469.128259    3.636274   46.596041     4.167284
        min      1.00000        1.328300    0.003600    1.350000  1989.000000
        25%      1.00000       38.272250    0.212500   24.497500  2005.000000
        50%      1.00000      357.000000    1.245000   39.940000  2009.000000
        75%      2.00000      999.600000    2.867500   59.332500  2011.000000
        max      6.00000    17337.500000   25.000000  354.000000  2014.000000
```

This method helps us understand the overall properties of a dataset. For example, we see in the `year` column that although exoplanets were discovered as far back as 1989, half of all planets in the dataset were not discovered until 2010 or after. This is largely thanks to the *Kepler* mission, which aimed to find eclipsing planets around other stars using a specially designed space telescope.

Table 20-1 summarizes some other built-in Pandas aggregations.

Table 20-1. Listing of Pandas aggregation methods

Aggregation	Returns
count	Total number of items
first, last	First and last item
mean, median	Mean and median
min, max	Minimum and maximum
std, var	Standard deviation and variance
mad	Mean absolute deviation
prod	Product of all items
sum	Sum of all items

These are all methods of DataFrame and Series objects.

To go deeper into the data, however, simple aggregates are often not enough. The next level of data summarization is the groupby operation, which allows you to quickly and efficiently compute aggregates on subsets of data.

groupby: Split, Apply, Combine

Simple aggregations can give you a flavor of your dataset, but often we would prefer to aggregate conditionally on some label or index: this is implemented in the so-called groupby operation. The name "group by" comes from a command in the SQL database language, but it is perhaps more illuminative to think of it in the terms first coined by Hadley Wickham of Rstats fame: *split, apply, combine*.

Split, Apply, Combine

A canonical example of this split-apply-combine operation, where the "apply" is a summation aggregation, is illustrated Figure 20-1.

Figure 20-1 shows what the groupby operation accomplishes:

- The *split* step involves breaking up and grouping a DataFrame depending on the value of the specified key.
- The *apply* step involves computing some function, usually an aggregate, transformation, or filtering, within the individual groups.
- The *combine* step merges the results of these operations into an output array.

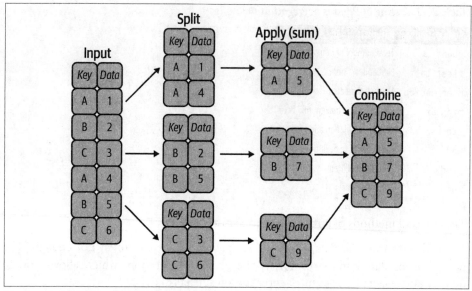

Figure 20-1. A visual representation of a groupby operation[1]

While this could certainly be done manually using some combination of the masking, aggregation, and merging commands covered earlier, an important realization is that *the intermediate splits do not need to be explicitly instantiated.* Rather, the groupby can (often) do this in a single pass over the data, updating the sum, mean, count, min, or other aggregate for each group along the way. The power of the groupby is that it abstracts away these steps: the user need not think about *how* the computation is done under the hood, but rather can think about the *operation as a whole.*

As a concrete example, let's take a look at using Pandas for the computation shown in the following table. We'll start by creating the input DataFrame:

```
In [11]: df = pd.DataFrame({'key': ['A', 'B', 'C', 'A', 'B', 'C'],
                            'data': range(6)}, columns=['key', 'data'])
         df
Out[11]: key  data
       0   A     0
       1   B     1
       2   C     2
       3   A     3
       4   B     4
       5   C     5
```

---

1 Code to produce this figure can be found in the online appendix (*https://oreil.ly/zHqzu*).

The most basic split-apply-combine operation can be computed with the `groupby` method of the `DataFrame`, passing the name of the desired key column:

```
In [12]: df.groupby('key')
Out[12]: <pandas.core.groupby.generic.DataFrameGroupBy object at 0x11d241e20>
```

Notice that what is returned is a `DataFrameGroupBy` object, not a set of `DataFrame` objects. This object is where the magic is: you can think of it as a special view of the `DataFrame`, which is poised to dig into the groups but does no actual computation until the aggregation is applied. This "lazy evaluation" approach means that common aggregates can be implemented efficiently in a way that is almost transparent to the user.

To produce a result, we can apply an aggregate to this `DataFrameGroupBy` object, which will perform the appropriate apply/combine steps to produce the desired result:

```
In [13]: df.groupby('key').sum()
Out[13]:      data
         key
         A     3
         B     5
         C     7
```

The `sum` method is just one possibility here; you can apply most Pandas or NumPy aggregation functions, as well as most `DataFrame` operations, as you will see in the following discussion.

The GroupBy Object

The `GroupBy` object is a flexible abstraction: in many ways, it can be treated as simply a collection of `DataFrame`s, though it is doing more sophisticated things under the hood. Let's see some examples using the Planets data.

Perhaps the most important operations made available by a `GroupBy` are *aggregate*, *filter*, *transform*, and *apply*. We'll discuss each of these more fully in the next section, but before that let's take a look at some of the other functionality that can be used with the basic `GroupBy` operation.

Column indexing

The `GroupBy` object supports column indexing in the same way as the `DataFrame`, and returns a modified `GroupBy` object. For example:

```
In [14]: planets.groupby('method')
Out[14]: <pandas.core.groupby.generic.DataFrameGroupBy object at 0x11d1bc820>

In [15]: planets.groupby('method')['orbital_period']
Out[15]: <pandas.core.groupby.generic.SeriesGroupBy object at 0x11d1bcd60>
```

Here we've selected a particular `Series` group from the original `DataFrame` group by reference to its column name. As with the `GroupBy` object, no computation is done until we call some aggregate on the object:

```
In [16]: planets.groupby('method')['orbital_period'].median()
Out[16]: method
         Astrometry                        631.180000
         Eclipse Timing Variations        4343.500000
         Imaging                         27500.000000
         Microlensing                     3300.000000
         Orbital Brightness Modulation       0.342887
         Pulsar Timing                      66.541900
         Pulsation Timing Variations      1170.000000
         Radial Velocity                   360.200000
         Transit                             5.714932
         Transit Timing Variations          57.011000
         Name: orbital_period, dtype: float64
```

This gives an idea of the general scale of orbital periods (in days) that each method is sensitive to.

Iteration over groups

The `GroupBy` object supports direct iteration over the groups, returning each group as a `Series` or `DataFrame`:

```
In [17]: for (method, group) in planets.groupby('method'):
             print("{0:30s} shape={1}".format(method, group.shape))
Out[17]: Astrometry                     shape=(2, 6)
         Eclipse Timing Variations      shape=(9, 6)
         Imaging                        shape=(38, 6)
         Microlensing                   shape=(23, 6)
         Orbital Brightness Modulation  shape=(3, 6)
         Pulsar Timing                  shape=(5, 6)
         Pulsation Timing Variations    shape=(1, 6)
         Radial Velocity                shape=(553, 6)
         Transit                        shape=(397, 6)
         Transit Timing Variations      shape=(4, 6)
```

This can be useful for manual inspection of groups for the sake of debugging, but it is often much faster to use the built-in `apply` functionality, which we will discuss momentarily.

Dispatch methods

Through some Python class magic, any method not explicitly implemented by the `GroupBy` object will be passed through and called on the groups, whether they are `DataFrame` or `Series` objects. For example, using the `describe` method is equivalent to calling `describe` on the `DataFrame` representing each group:

```
In [18]: planets.groupby('method')['year'].describe().unstack()
Out[18]:         method
         count   Astrometry                       2.0
                 Eclipse Timing Variations        9.0
                 Imaging                         38.0
                 Microlensing                    23.0
                 Orbital Brightness Modulation    3.0
                                                  ...
         max     Pulsar Timing                 2011.0
                 Pulsation Timing Variations   2007.0
                 Radial Velocity               2014.0
                 Transit                       2014.0
                 Transit Timing Variations     2014.0
         Length: 80, dtype: float64
```

Looking at this table helps us to better understand the data: for example, the vast majority of planets until 2014 were discovered by the Radial Velocity and Transit methods, though the latter method became common more recently. The newest methods seem to be Transit Timing Variation and Orbital Brightness Modulation, which were not used to discover a new planet until 2011.

Notice that these dispatch methods are applied *to each individual group*, and the results are then combined within GroupBy and returned. Again, any valid DataFrame/ Series method can be called in a similar manner on the corresponding GroupBy object.

Aggregate, Filter, Transform, Apply

The preceding discussion focused on aggregation for the combine operation, but there are more options available. In particular, GroupBy objects have aggregate, fil ter, transform, and apply methods that efficiently implement a variety of useful operations before combining the grouped data.

For the purpose of the following subsections, we'll use this DataFrame:

```
In [19]: rng = np.random.RandomState(0)
         df = pd.DataFrame({'key': ['A', 'B', 'C', 'A', 'B', 'C'],
                            'data1': range(6),
                            'data2': rng.randint(0, 10, 6)},
                           columns = ['key', 'data1', 'data2'])
         df
Out[19]:    key  data1  data2
         0    A      0      5
         1    B      1      0
         2    C      2      3
         3    A      3      3
         4    B      4      7
         5    C      5      9
```

Aggregation

You're now familiar with GroupBy aggregations with `sum`, `median`, and the like, but the `aggregate` method allows for even more flexibility. It can take a string, a function, or a list thereof, and compute all the aggregates at once. Here is a quick example combining all of these:

```
In [20]: df.groupby('key').aggregate(['min', np.median, max])
Out[20]:      data1              data2
         min median max    min median max
     key
     A     0    1.5   3      3    4.0   5
     B     1    2.5   4      0    3.5   7
     C     2    3.5   5      3    6.0   9
```

Another common pattern is to pass a dictionary mapping column names to operations to be applied on that column:

```
In [21]: df.groupby('key').aggregate({'data1': 'min',
                                       'data2': 'max'})
Out[21]:      data1  data2
     key
     A      0     5
     B      1     7
     C      2     9
```

Filtering

A filtering operation allows you to drop data based on the group properties. For example, we might want to keep all groups in which the standard deviation is larger than some critical value:

```
In [22]: def filter_func(x):
             return x['data2'].std() > 4

         display('df', "df.groupby('key').std()",
                 "df.groupby('key').filter(filter_func)")
Out[22]: df                         df.groupby('key').std()
         key  data1  data2              data1      data2
     0    A     0     5          key
     1    B     1     0          A     2.12132   1.414214
     2    C     2     3          B     2.12132   4.949747
     3    A     3     3          C     2.12132   4.242641
     4    B     4     7
     5    C     5     9


         df.groupby('key').filter(filter_func)
         key  data1  data2
     1    B     1     0
     2    C     2     3
     4    B     4     7
     5    C     5     9
```

The filter function should return a Boolean value specifying whether the group passes the filtering. Here, because group A does not have a standard deviation greater than 4, it is dropped from the result.

Transformation

While aggregation must return a reduced version of the data, transformation can return some transformed version of the full data to recombine. For such a transformation, the output is the same shape as the input. A common example is to center the data by subtracting the group-wise mean:

```
In [23]: def center(x):
             return x - x.mean()
         df.groupby('key').transform(center)
Out[23]:    data1  data2
         0   -1.5    1.0
         1   -1.5   -3.5
         2   -1.5   -3.0
         3    1.5   -1.0
         4    1.5    3.5
         5    1.5    3.0
```

The apply method

The apply method lets you apply an arbitrary function to the group results. The function should take a DataFrame and returns either a Pandas object (e.g., DataFrame, Series) or a scalar; the behavior of the combine step will be tailored to the type of output returned.

For example, here is an apply operation that normalizes the first column by the sum of the second:

```
In [24]: def norm_by_data2(x):
             # x is a DataFrame of group values
             x['data1'] /= x['data2'].sum()
             return x

         df.groupby('key').apply(norm_by_data2)
Out[24]:    key     data1  data2
         0   A   0.000000      5
         1   B   0.142857      0
         2   C   0.166667      3
         3   A   0.375000      3
         4   B   0.571429      7
         5   C   0.416667      9
```

apply within a GroupBy is flexible: the only criterion is that the function takes a DataFrame and returns a Pandas object or scalar. What you do in between is up to you!

Specifying the Split Key

In the simple examples presented before, we split the DataFrame on a single column name. This is just one of many options by which the groups can be defined, and we'll go through some other options for group specification here.

A list, array, series, or index providing the grouping keys

The key can be any series or list with a length matching that of the DataFrame. For example:

```
In [25]: L = [0, 1, 0, 1, 2, 0]
         df.groupby(L).sum()
Out[25]:    data1  data2
         0      7     17
         1      4      3
         2      4      7
```

Of course, this means there's another, more verbose way of accomplishing the df.groupby('key') from before:

```
In [26]: df.groupby(df['key']).sum()
Out[26]:      data1  data2
         key
         A        3      8
         B        5      7
         C        7     12
```

A dictionary or series mapping index to group

Another method is to provide a dictionary that maps index values to the group keys:

```
In [27]: df2 = df.set_index('key')
         mapping = {'A': 'vowel', 'B': 'consonant', 'C': 'consonant'}
         display('df2', 'df2.groupby(mapping).sum()')
Out[27]: df2                        df2.groupby(mapping).sum()
             data1  data2                        data1  data2
         key                        key
         A        0      5          consonant        12     19
         B        1      0          vowel             3      8
         C        2      3
         A        3      3
         B        4      7
         C        5      9
```

Any Python function

Similar to mapping, you can pass any Python function that will input the index value and output the group:

```
In [28]: df2.groupby(str.lower).mean()
Out[28]:     data1  data2
         key
         a    1.5    4.0
         b    2.5    3.5
         c    3.5    6.0
```

A list of valid keys

Further, any of the preceding key choices can be combined to group on a multi-index:

```
In [29]: df2.groupby([str.lower, mapping]).mean()
Out[29]:               data1  data2
         key key
         a   vowel       1.5    4.0
         b   consonant   2.5    3.5
         c   consonant   3.5    6.0
```

Grouping Example

As an example of this, in a few lines of Python code we can put all these together and count discovered planets by method and by decade:

```
In [30]: decade = 10 * (planets['year'] // 10)
         decade = decade.astype(str) + 's'
         decade.name = 'decade'
         planets.groupby(['method', decade])['number'].sum().unstack().fillna(0)
Out[30]: decade                        1980s  1990s  2000s  2010s
         method
         Astrometry                     0.0    0.0    0.0    2.0
         Eclipse Timing Variations      0.0    0.0    5.0   10.0
         Imaging                        0.0    0.0   29.0   21.0
         Microlensing                   0.0    0.0   12.0   15.0
         Orbital Brightness Modulation  0.0    0.0    0.0    5.0
         Pulsar Timing                  0.0    9.0    1.0    1.0
         Pulsation Timing Variations    0.0    0.0    1.0    0.0
         Radial Velocity                1.0   52.0  475.0  424.0
         Transit                        0.0    0.0   64.0  712.0
         Transit Timing Variations      0.0    0.0    0.0    9.0
```

This shows the power of combining many of the operations we've discussed up to this point when looking at realistic datasets: we quickly gain a coarse understanding of when and how extrasolar planets were detected in the years after the first discovery.

I would suggest digging into these few lines of code and evaluating the individual steps to make sure you understand exactly what they are doing to the result. It's certainly a somewhat complicated example, but understanding these pieces will give you the means to similarly explore your own data.

## CHAPTER 21

Pivot Tables

We have seen how the groupby abstraction lets us explore relationships within a dataset. A *pivot table* is a similar operation that is commonly seen in spreadsheets and other programs that operate on tabular data. The pivot table takes simple columnwise data as input, and groups the entries into a two-dimensional table that provides a multidimensional summarization of the data. The difference between pivot tables and groupby can sometimes cause confusion; it helps me to think of pivot tables as essentially a *multidimensional* version of groupby aggregation. That is, you split-apply-combine, but both the split and the combine happen across not a one-dimensional index, but across a two-dimensional grid.

Motivating Pivot Tables

For the examples in this section, we'll use the database of passengers on the *Titanic*, available through the Seaborn library (see Chapter 36):

```
In [1]: import numpy as np
        import pandas as pd
        import seaborn as sns
        titanic = sns.load_dataset('titanic')

In [2]: titanic.head()
Out[2]:    survived  pclass     sex   age  sibsp  parch     fare embarked  class \
        0         0       3    male  22.0      1      0   7.2500        S  Third
        1         1       1  female  38.0      1      0  71.2833        C  First
        2         1       3  female  26.0      0      0   7.9250        S  Third
        3         1       1  female  35.0      1      0  53.1000        S  First
        4         0       3    male  35.0      0      0   8.0500        S  Third
```

```
        who  adult_male  deck  embark_town  alive  alone
0       man        True   NaN  Southampton     no  False
1     woman       False     C    Cherbourg    yes  False
2     woman       False   NaN  Southampton    yes   True
3     woman       False     C  Southampton    yes  False
4       man        True   NaN  Southampton     no   True
```

As the output shows, this contains a number of data points on each passenger on that
ill-fated voyage, including sex, age, class, fare paid, and much more.

Pivot Tables by Hand

To start learning more about this data, we might begin by grouping according to sex,
survival status, or some combination thereof. If you read the previous chapter, you
might be tempted to apply a groupby operation—for example, let's look at survival
rate by sex:

```
In [3]: titanic.groupby('sex')[['survived']].mean()
Out[3]:        survived
        sex
        female  0.742038
        male    0.188908
```

This gives us some initial insight: overall, three of every four females on board survived, while only one in five males survived!

This is useful, but we might like to go one step deeper and look at survival rates by
both sex and, say, class. Using the vocabulary of groupby, we might proceed using a
process like this: we first *group by* class and sex, then *select* survival, *apply* a mean
aggregate, *combine* the resulting groups, and finally *unstack* the hierarchical index to
reveal the hidden multidimensionality. In code:

```
In [4]: titanic.groupby(['sex', 'class'])['survived'].aggregate('mean').unstack()
Out[4]: class      First    Second     Third
        sex
        female  0.968085  0.921053  0.500000
        male    0.368852  0.157407  0.135447
```

This gives us a better idea of how both sex and class affected survival, but the code is
starting to look a bit garbled. While each step of this pipeline makes sense in light of
the tools we've previously discussed, the long string of code is not particularly easy to
read or use. This two-dimensional groupby is common enough that Pandas includes
a convenience routine, pivot_table, which succinctly handles this type of multidimensional aggregation.

Pivot Table Syntax

Here is the equivalent to the preceding operation using the `DataFrame.pivot_table` method:

```
In [5]: titanic.pivot_table('survived', index='sex', columns='class', aggfunc='mean')
Out[5]: class      First     Second    Third
        sex
        female  0.968085  0.921053  0.500000
        male    0.368852  0.157407  0.135447
```

This is eminently more readable than the manual `groupby` approach, and produces the same result. As you might expect of an early 20th-century transatlantic cruise, the survival gradient favors both higher classes and people recorded as females in the data. First-class females survived with near certainty (hi, Rose!), while only one in eight or so third-class males survived (sorry, Jack!).

Multilevel Pivot Tables

Just as in a `groupby`, the grouping in pivot tables can be specified with multiple levels and via a number of options. For example, we might be interested in looking at age as a third dimension. We'll bin the age using the `pd.cut` function:

```
In [6]: age = pd.cut(titanic['age'], [0, 18, 80])
        titanic.pivot_table('survived', ['sex', age], 'class')
Out[6]: class             First     Second    Third
        sex    age
        female (0, 18]   0.909091  1.000000  0.511628
               (18, 80]  0.972973  0.900000  0.423729
        male   (0, 18]   0.800000  0.600000  0.215686
               (18, 80]  0.375000  0.071429  0.133663
```

We can apply the same strategy when working with the columns as well; let's add info on the fare paid, using `pd.qcut` to automatically compute quantiles:

```
In [7]: fare = pd.qcut(titanic['fare'], 2)
        titanic.pivot_table('survived', ['sex', age], [fare, 'class'])
Out[7]: fare             (-0.001, 14.454]                 (14.454, 512.329]  \
        class                First   Second    Third                 First
        sex    age
        female (0, 18]         NaN  1.000000  0.714286              0.909091
               (18, 80]        NaN  0.880000  0.444444              0.972973
        male   (0, 18]         NaN  0.000000  0.260870              0.800000
               (18, 80]        0.0  0.098039  0.125000              0.391304

        fare
        class              Second     Third
        sex    age
        female (0, 18]   1.000000  0.318182
               (18, 80]  0.914286  0.391304
```

```
male    (0, 18]   0.818182  0.178571
        (18, 80]  0.030303  0.192308
```

The result is a four-dimensional aggregation with hierarchical indices (see Chapter 17), shown in a grid demonstrating the relationship between the values.

Additional Pivot Table Options

The full call signature of the `DataFrame.pivot_table` method is as follows:

```
# call signature as of Pandas 1.3.5
DataFrame.pivot_table(data, values=None, index=None, columns=None,
                      aggfunc='mean', fill_value=None, margins=False,
                      dropna=True, margins_name='All', observed=False,
                      sort=True)
```

We've already seen examples of the first three arguments; here we'll look at some of the remaining ones. Two of the options, `fill_value` and `dropna`, have to do with missing data and are fairly straightforward; I will not show examples of them here.

The `aggfunc` keyword controls what type of aggregation is applied, which is a mean by default. As with `groupby`, the aggregation specification can be a string representing one of several common choices (`'sum'`, `'mean'`, `'count'`, `'min'`, `'max'`, etc.) or a function that implements an aggregation (e.g., `np.sum()`, `min()`, `sum()`, etc.). Additionally, it can be specified as a dictionary mapping a column to any of the desired options:

```
In [8]: titanic.pivot_table(index='sex', columns='class',
                            aggfunc={'survived':sum, 'fare':'mean'})
Out[8]:              fare                      survived
        class      First     Second     Third  First Second Third
        sex
        female  106.125798  21.970121  16.118810    91     70    72
        male     67.226127  19.741782  12.661633    45     17    47
```

Notice also here that we've omitted the `values` keyword; when specifying a mapping for `aggfunc`, this is determined automatically.

At times it's useful to compute totals along each grouping. This can be done via the `margins` keyword:

```
In [9]: titanic.pivot_table('survived', index='sex', columns='class', margins=True)
Out[9]: class      First     Second     Third       All
        sex
        female  0.968085  0.921053  0.500000  0.742038
        male    0.368852  0.157407  0.135447  0.188908
        All     0.629630  0.472826  0.242363  0.383838
```

Here, this automatically gives us information about the class-agnostic survival rate by sex, the sex-agnostic survival rate by class, and the overall survival rate of 38%. The margin label can be specified with the `margins_name` keyword; it defaults to `"All"`.

Example: Birthrate Data

As another example, let's take a look at the freely available data on births in the US (*https://oreil.ly/2NWnk*), provided by the Centers for Disease Control (CDC). (This dataset has been analyzed rather extensively by Andrew Gelman and his group; see, for example, the blog post on signal processing using Gaussian processes (*https://oreil.ly/5EqEp*)):[1]

```
In [10]: # shell command to download the data:
         # !cd data && curl -O \
         # https://raw.githubusercontent.com/jakevdp/data-CDCbirths/master/births.csv
In [11]: births = pd.read_csv('data/births.csv')
```

Taking a look at the data, we see that it's relatively simple—it contains the number of births grouped by date and gender:

```
In [12]: births.head()
Out[12]:    year  month  day gender  births
         0  1969      1  1.0      F    4046
         1  1969      1  1.0      M    4440
         2  1969      1  2.0      F    4454
         3  1969      1  2.0      M    4548
         4  1969      1  3.0      F    4548
```

We can start to understand this data a bit more by using a pivot table. Let's add a decade column, and take a look at male and female births as a function of decade:

```
In [13]: births['decade'] = 10 * (births['year'] // 10)
         births.pivot_table('births', index='decade', columns='gender',
                            aggfunc='sum')
Out[13]: gender          F         M
         decade
         1960      1753634   1846572
         1970     16263075  17121550
         1980     18310351  19243452
         1990     19479454  20420553
         2000     18229309  19106428
```

We see that male births outnumber female births in every decade. To see this trend a bit more clearly, we can use the built-in plotting tools in Pandas to visualize the total number of births by year, as shown in Figure 21-1 (see Part IV for a discussion of plotting with Matplotlib):

```
In [14]: %matplotlib inline
         import matplotlib.pyplot as plt
```

---

1 The CDC dataset used in this section uses the sex assigned at birth, which it calls "gender," and limits the data to male and female. While gender is a spectrum independent of biology, I will be using the same terminology while discussing this dataset for consistency and clarity.

```
plt.style.use('seaborn-whitegrid')
births.pivot_table(
    'births', index='year', columns='gender', aggfunc='sum').plot()
plt.ylabel('total births per year');
```

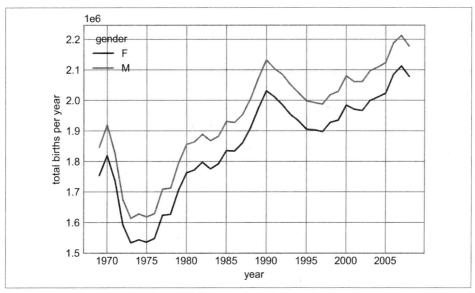

Figure 21-1. Total number of US births by year and gender[2]

With a simple pivot table and the `plot` method, we can immediately see the annual trend in births by gender. By eye, it appears that over the past 50 years male births have outnumbered female births by around 5%.

Though this doesn't necessarily relate to the pivot table, there are a few more interesting features we can pull out of this dataset using the Pandas tools covered up to this point. We must start by cleaning the data a bit, removing outliers caused by mistyped dates (e.g., June 31st) or missing values (e.g., June 99th). One easy way to remove these all at once is to cut outliers; we'll do this via a robust sigma-clipping operation:

```
In [15]: quartiles = np.percentile(births['births'], [25, 50, 75])
         mu = quartiles[1]
         sig = 0.74 * (quartiles[2] - quartiles[0])
```

This final line is a robust estimate of the sample standard deviation, where the 0.74 comes from the interquartile range of a Gaussian distribution (you can learn more about sigma-clipping operations in a book I coauthored with Željko Ivezić, Andrew J. Connolly, and Alexander Gray *Statistics, Data Mining, and Machine Learning in Astronomy* (Princeton University Press)).

---

2 A full-color version of this figure can be found on GitHub (*https://oreil.ly/PDSH_GitHub*).

With this, we can use the query method (discussed further in Chapter 24) to filter out rows with births outside these values:

```
In [16]: births = births.query('(births > @mu - 5 * @sig) &
                                (births < @mu + 5 * @sig)')
```

Next we set the day column to integers; previously it had been a string column because some columns in the dataset contained the value 'null':

```
In [17]: # set 'day' column to integer; it originally was a string due to nulls
         births['day'] = births['day'].astype(int)
```

Finally, we can combine the day, month, and year to create a date index (see Chapter 23). This allows us to quickly compute the weekday corresponding to each row:

```
In [18]: # create a datetime index from the year, month, day
         births.index = pd.to_datetime(10000 * births.year +
                                       100 * births.month +
                                       births.day, format='%Y%m%d')

         births['dayofweek'] = births.index.dayofweek
```

Using this, we can plot births by weekday for several decades (see Figure 21-2).

```
In [19]: import matplotlib.pyplot as plt
         import matplotlib as mpl

         births.pivot_table('births', index='dayofweek',
                            columns='decade', aggfunc='mean').plot()
         plt.gca().set(xticks=range(7),
                       xticklabels=['Mon', 'Tues', 'Wed', 'Thurs',
                                    'Fri', 'Sat', 'Sun'])
         plt.ylabel('mean births by day');
```

Apparently births are slightly less common on weekends than on weekdays! Note that the 1990s and 2000s are missing because starting in 1989, the CDC data contains only the month of birth.

Another interesting view is to plot the mean number of births by the day of the year. Let's first group the data by month and day separately:

```
In [20]: births_by_date = births.pivot_table('births',
                                              [births.index.month, births.index.day])
         births_by_date.head()
Out[20]:        births
         1 1    4009.225
           2    4247.400
           3    4500.900
           4    4571.350
           5    4603.625
```

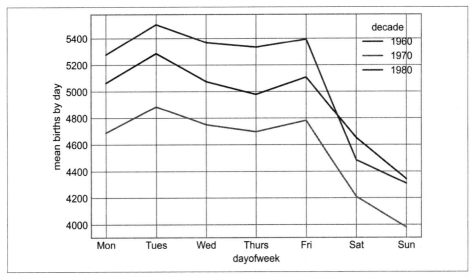

Figure 21-2. Average daily births by day of week and decade[3]

The result is a multi-index over months and days. To make this visualizable, let's turn these months and days into dates by associating them with a dummy year variable (making sure to choose a leap year so February 29th is correctly handled!):

```
In [21]: from datetime import datetime
         births_by_date.index = [datetime(2012, month, day)
                                 for (month, day) in births_by_date.index]
         births_by_date.head()
Out[21]:             births
         2012-01-01  4009.225
         2012-01-02  4247.400
         2012-01-03  4500.900
         2012-01-04  4571.350
         2012-01-05  4603.625
```

Focusing on the month and day only, we now have a time series reflecting the average number of births by date of the year. From this, we can use the plot method to plot the data. It reveals some interesting trends, as you can see in Figure 21-3.

```
In [22]: # Plot the results
         fig, ax = plt.subplots(figsize=(12, 4))
         births_by_date.plot(ax=ax);
```

---

3 A full-color version of this figure can be found on GitHub (*https://oreil.ly/PDSH_GitHub*).

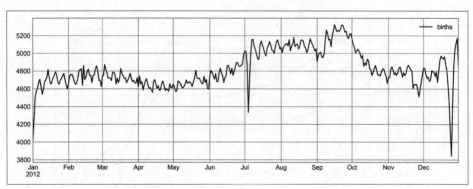

Figure 21-3. Average daily births by date[4]

In particular, the striking feature of this graph is the dip in birthrate on US holidays (e.g., Independence Day, Labor Day, Thanksgiving, Christmas, New Year's Day), although this likely reflects trends in scheduled/induced births rather than some deep psychosomatic effect on natural births. For more discussion of this trend, see the analysis and links in Andrew Gelman's blog post (*https://oreil.ly/ugVHI*) on the subject. We'll return to this figure in Chapter 32, where we will use Matplotlib's tools to annotate this plot.

Looking at this short example, you can see that many of the Python and Pandas tools we've seen to this point can be combined and used to gain insight from a variety of datasets. We will see some more sophisticated applications of these data manipulations in future chapters!

---

4 A full-size version of this figure can be found on GitHub (*https://oreil.ly/PDSH_GitHub*).

## CHAPTER 22

Vectorized String Operations

One strength of Python is its relative ease in handling and manipulating string data. Pandas builds on this and provides a comprehensive set of *vectorized string operations* that are an important part of the type of munging required when working with (read: cleaning up) real-world data. In this chapter, we'll walk through some of the Pandas string operations, and then take a look at using them to partially clean up a very messy dataset of recipes collected from the internet.

Introducing Pandas String Operations

We saw in previous chapters how tools like NumPy and Pandas generalize arithmetic operations so that we can easily and quickly perform the same operation on many array elements. For example:

```
In [1]: import numpy as np
        x = np.array([2, 3, 5, 7, 11, 13])
        x * 2
Out[1]: array([ 4,  6, 10, 14, 22, 26])
```

This *vectorization* of operations simplifies the syntax of operating on arrays of data: we no longer have to worry about the size or shape of the array, but just about what operation we want done. For arrays of strings, NumPy does not provide such simple access, and thus you're stuck using a more verbose loop syntax:

```
In [2]: data = ['peter', 'Paul', 'MARY', 'gUIDO']
        [s.capitalize() for s in data]
Out[2]: ['Peter', 'Paul', 'Mary', 'Guido']
```

This is perhaps sufficient to work with some data, but it will break if there are any missing values, so this approach requires putting in extra checks:

```
In [3]: data = ['peter', 'Paul', None, 'MARY', 'gUIDO']
        [s if s is None else s.capitalize() for s in data]
Out[3]: ['Peter', 'Paul', None, 'Mary', 'Guido']
```

This manual approach is not only verbose and inconvenient, it can be error-prone.

Pandas includes features to address both this need for vectorized string operations as well as the need for correctly handling missing data via the `str` attribute of Pandas `Series` and `Index` objects containing strings. So, for example, if we create a Pandas `Series` with this data we can directly call the `str.capitalize` method, which has missing value handling built in:

```
In [4]: import pandas as pd
        names = pd.Series(data)
        names.str.capitalize()
Out[4]: 0    Peter
        1    Paul
        2    None
        3    Mary
        4    Guido
        dtype: object
```

Tables of Pandas String Methods

If you have a good understanding of string manipulation in Python, most of the Pandas string syntax is intuitive enough that it's probably sufficient to just list the available methods. We'll start with that here, before diving deeper into a few of the subtleties. The examples in this section use the following `Series` object:

```
In [5]: monte = pd.Series(['Graham Chapman', 'John Cleese', 'Terry Gilliam',
                           'Eric Idle', 'Terry Jones', 'Michael Palin'])
```

Methods Similar to Python String Methods

Nearly all of Python's built-in string methods are mirrored by a Pandas vectorized string method. The following Pandas `str` methods mirror Python string methods:

len	lower	translate	islower	ljust
upper	startswith	isupper	rjust	find
endswith	isnumeric	center	rfind	isalnum
isdecimal	zfill	index	isalpha	split
strip	rindex	isdigit	rsplit	rstrip
capitalize	isspace	partition	lstrip	swapcase

Notice that these have various return values. Some, like `lower`, return a series of strings:

```
In [6]: monte.str.lower()
Out[6]: 0    graham chapman
        1       john cleese
        2     terry gilliam
        3         eric idle
        4       terry jones
        5     michael palin
        dtype: object
```

But some others return numbers:

```
In [7]: monte.str.len()
Out[7]: 0    14
        1    11
        2    13
        3     9
        4    11
        5    13
        dtype: int64
```

Or Boolean values:

```
In [8]: monte.str.startswith('T')
Out[8]: 0    False
        1    False
        2     True
        3    False
        4     True
        5    False
        dtype: bool
```

Still others return lists or other compound values for each element:

```
In [9]: monte.str.split()
Out[9]: 0    [Graham, Chapman]
        1       [John, Cleese]
        2     [Terry, Gilliam]
        3         [Eric, Idle]
        4       [Terry, Jones]
        5     [Michael, Palin]
        dtype: object
```

We'll see further manipulations of this kind of series-of-lists object as we continue our discussion.

Methods Using Regular Expressions

In addition, there are several methods that accept regular expressions (regexps) to examine the content of each string element, and follow some of the API conventions of Python's built-in `re` module (see Table 22-1).

Table 22-1. Mapping between Pandas methods and functions in Python's re module

Method	Description
match	Calls re.match on each element, returning a Boolean.
extract	Calls re.match on each element, returning matched groups as strings.
findall	Calls re.findall on each element
replace	Replaces occurrences of pattern with some other string
contains	Calls re.search on each element, returning a boolean
count	Counts occurrences of pattern
split	Equivalent to str.split, but accepts regexps
rsplit	Equivalent to str.rsplit, but accepts regexps

With these, we can do a wide range of operations. For example, we can extract the first name from each element by asking for a contiguous group of characters at the beginning of each element:

```
In [10]: monte.str.extract('([A-Za-z]+)', expand=False)
Out[10]: 0     Graham
         1       John
         2      Terry
         3       Eric
         4      Terry
         5    Michael
         dtype: object
```

Or we can do something more complicated, like finding all names that start and end with a consonant, making use of the start-of-string (^) and end-of-string ($) regular expression characters:

```
In [11]: monte.str.findall(r'^[^AEIOU].*[^aeiou]$')
Out[11]: 0    [Graham Chapman]
         1                  []
         2    [Terry Gilliam]
         3                  []
         4      [Terry Jones]
         5    [Michael Palin]
         dtype: object
```

The ability to concisely apply regular expressions across Series or DataFrame entries opens up many possibilities for analysis and cleaning of data.

Miscellaneous Methods

Finally, Table 22-2 lists miscellaneous methods that enable other convenient operations.

Table 22-2. Other Pandas string methods

Method	Description
get	Indexes each element
slice	Slices each element
slice_replace	Replaces slice in each element with the passed value
cat	Concatenates strings
repeat	Repeats values
normalize	Returns Unicode form of strings
pad	Adds whitespace to left, right, or both sides of strings
wrap	Splits long strings into lines with length less than a given width
join	Joins strings in each element of the Series with the passed separator
get_dummies	Extracts dummy variables as a DataFrame

Vectorized item access and slicing

The get and slice operations, in particular, enable vectorized element access from each array. For example, we can get a slice of the first three characters of each array using str.slice(0, 3). This behavior is also available through Python's normal indexing syntax; for example, df.str.slice(0, 3) is equivalent to df.str[0:3]:

```
In [12]: monte.str[0:3]
Out[12]: 0    Gra
         1    Joh
         2    Ter
         3    Eri
         4    Ter
         5    Mic
         dtype: object
```

Indexing via df.str.get(i) and df.str[i] are likewise similar.

These indexing methods also let you access elements of arrays returned by split. For example, to extract the last name of each entry, combine split with str indexing:

```
In [13]: monte.str.split().str[-1]
Out[13]: 0    Chapman
         1     Cleese
         2    Gilliam
         3       Idle
         4      Jones
         5      Palin
         dtype: object
```

Indicator variables

Another method that requires a bit of extra explanation is the get_dummies method. This is useful when your data has a column containing some sort of coded indicator.

For example, we might have a dataset that contains information in the form of codes, such as A = "born in America," B = "born in the United Kingdom," C = "likes cheese," D = "likes spam":

```
In [14]: full_monte = pd.DataFrame({'name': monte,
                                     'info': ['B|C|D', 'B|D', 'A|C',
                                              'B|D', 'B|C', 'B|C|D']})
         full_monte
Out[14]:             name    info
         0  Graham Chapman  B|C|D
         1     John Cleese    B|D
         2   Terry Gilliam    A|C
         3      Eric Idle     B|D
         4     Terry Jones    B|C
         5  Michael Palin   B|C|D
```

The `get_dummies` routine lets us split out these indicator variables into a `DataFrame`:

```
In [15]: full_monte['info'].str.get_dummies('|')
Out[15]:    A  B  C  D
         0  0  1  1  1
         1  0  1  0  1
         2  1  0  1  0
         3  0  1  0  1
         4  0  1  1  0
         5  0  1  1  1
```

With these operations as building blocks, you can construct an endless range of string processing procedures when cleaning your data.

We won't dive further into these methods here, but I encourage you to read through "Working with Text Data" (*https://oreil.ly/oYgWA*) in the Pandas online documentation, or to refer to the resources listed in "Further Resources" on page 221.

Example: Recipe Database

These vectorized string operations become most useful in the process of cleaning up messy, real-world data. Here I'll walk through an example of that, using an open recipe database compiled from various sources on the web. Our goal will be to parse the recipe data into ingredient lists, so we can quickly find a recipe based on some ingredients we have on hand. The scripts used to compile this can be found on GitHub (*https://oreil.ly/3S0Rg*), and the link to the most recent version of the database is found there as well.

This database is about 30 MB, and can be downloaded and unzipped with these commands:

```
In [16]: # repo = "https://raw.githubusercontent.com/jakevdp/open-recipe-data/master"
         # !cd data && curl -O {repo}/recipeitems.json.gz
         # !gunzip data/recipeitems.json.gz
```

The database is in JSON format, so we will use `pd.read_json` to read it (`lines=True` is required for this dataset because each line of the file is a JSON entry):

```
In [17]: recipes = pd.read_json('data/recipeitems.json', lines=True)
         recipes.shape
Out[17]: (173278, 17)
```

We see there are nearly 175,000 recipes, and 17 columns. Let's take a look at one row to see what we have:

```
In [18]: recipes.iloc[0]
Out[18]: _id                                {'$oid': '5160756b96cc62079cc2db15'}
         name                                        Drop Biscuits and Sausage Gravy
         ingredients                 Biscuits\n3 cups All-purpose Flour\n2 Tablespo...
         url                          http://thepioneerwoman.com/cooking/2013/03/dro...
         image                        http://static.thepioneerwoman.com/cooking/file...
         ts                                               {'$date': 1365276011104}
         cookTime                                                             PT30M
         source                                                    thepioneerwoman
         recipeYield                                                            12
         datePublished                                                 2013-03-11
         prepTime                                                             PT10M
         description                  Late Saturday afternoon, after Marlboro Man ha...
         totalTime                                                            NaN
         creator                                                              NaN
         recipeCategory                                                       NaN
         dateModified                                                         NaN
         recipeInstructions                                                   NaN
         Name: 0, dtype: object
```

There is a lot of information there, but much of it is in a very messy form, as is typical of data scraped from the web. In particular, the ingredient list is in string format; we're going to have to carefully extract the information we're interested in. Let's start by taking a closer look at the ingredients:

```
In [19]: recipes.ingredients.str.len().describe()
Out[19]: count    173278.000000
         mean        244.617926
         std         146.705285
         min           0.000000
         25%         147.000000
         50%         221.000000
         75%         314.000000
         max        9067.000000
         Name: ingredients, dtype: float64
```

The ingredient lists average 250 characters long, with a minimum of 0 and a maximum of nearly 10,000 characters!

Just out of curiosity, let's see which recipe has the longest ingredient list:

```
In [20]: recipes.name[np.argmax(recipes.ingredients.str.len())]
Out[20]: 'Carrot Pineapple Spice &amp; Brownie Layer Cake with Whipped Cream &amp;
         > Cream Cheese Frosting and Marzipan Carrots'
```

We can do other aggregate explorations; for example, we can see how many of the recipes are for breakfast foods (using regular expression syntax to match both lower-case and capital letters):

```
In [21]: recipes.description.str.contains('[Bb]reakfast').sum()
Out[21]: 3524
```

Or how many of the recipes list cinnamon as an ingredient:

```
In [22]: recipes.ingredients.str.contains('[Cc]innamon').sum()
Out[22]: 10526
```

We could even look to see whether any recipes misspell the ingredient as "cinamon":

```
In [23]: recipes.ingredients.str.contains('[Cc]inamon').sum()
Out[23]: 11
```

This is the type of data exploration that is possible with Pandas string tools. It is data munging like this that Python really excels at.

A Simple Recipe Recommender

Let's go a bit further, and start working on a simple recipe recommendation system: given a list of ingredients, we want to find any recipes that use all those ingredients. While conceptually straightforward, the task is complicated by the heterogeneity of the data: there is no easy operation, for example, to extract a clean list of ingredients from each row. So, we will cheat a bit: we'll start with a list of common ingredients, and simply search to see whether they are in each recipe's ingredient list. For simplicity, let's just stick with herbs and spices for the time being:

```
In [24]: spice_list = ['salt', 'pepper', 'oregano', 'sage', 'parsley',
                       'rosemary', 'tarragon', 'thyme', 'paprika', 'cumin']
```

We can then build a Boolean `DataFrame` consisting of `True` and `False` values, indicating whether each ingredient appears in the list:

```
In [25]: import re
         spice_df = pd.DataFrame({
             spice: recipes.ingredients.str.contains(spice, re.IGNORECASE)
             for spice in spice_list})
         spice_df.head()
Out[25]:     salt  pepper  oregano   sage  parsley  rosemary  tarragon  thyme  \
         0  False   False    False   True    False     False     False  False
         1  False   False    False  False    False     False     False  False
         2   True    True    False  False    False     False     False  False
         3  False   False    False  False    False     False     False  False
         4  False   False    False  False    False     False     False  False
```

```
     paprika   cumin
  0   False   False
  1   False   False
  2   False    True
  3   False   False
  4   False   False
```

Now, as an example, let's say we'd like to find a recipe that uses parsley, paprika, and tarragon. We can compute this very quickly using the query method of DataFrames, discussed further in Chapter 24:

```
In [26]: selection = spice_df.query('parsley & paprika & tarragon')
         len(selection)
Out[26]: 10
```

We find only 10 recipes with this combination. Let's use the index returned by this selection to discover the names of those recipes:

```
In [27]: recipes.name[selection.index]
Out[27]: 2069         All cremat with a Little Gem, dandelion and wa...
         74964                          Lobster with Thermidor butter
         93768      Burton's Southern Fried Chicken with White Gravy
         113926               Mijo's Slow Cooker Shredded Beef
         137686              Asparagus Soup with Poached Eggs
         140530                            Fried Oyster Po'boys
         158475          Lamb shank tagine with herb tabbouleh
         158486            Southern fried chicken in buttermilk
         163175      Fried Chicken Sliders with Pickles + Slaw
         165243                    Bar Tartine Cauliflower Salad
         Name: name, dtype: object
```

Now that we have narrowed down our recipe selection from 175,000 to 10, we are in a position to make a more informed decision about what we'd like to cook for dinner.

Going Further with Recipes

Hopefully this example has given you a bit of a flavor (heh) of the types of data cleaning operations that are efficiently enabled by Pandas string methods. Of course, building a robust recipe recommendation system would require a *lot* more work! Extracting full ingredient lists from each recipe would be an important piece of the task; unfortunately, the wide variety of formats used makes this a relatively time-consuming process. This points to the truism that in data science, cleaning and munging of real-world data often comprises the majority of the work—and Pandas provides the tools that can help you do this efficiently.

CHAPTER 23
Working with Time Series

Pandas was originally developed in the context of financial modeling, so as you might expect, it contains an extensive set of tools for working with dates, times, and time-indexed data. Date and time data comes in a few flavors, which we will discuss here:

Timestamps
> Particular moments in time (e.g., July 4, 2021 at 7:00 a.m.).

Time intervals and periods
> A length of time between a particular beginning and end point; for example, the month of June 2021. Periods usually reference a special case of time intervals in which each interval is of uniform length and does not overlap (e.g., 24-hour-long periods comprising days).

Time deltas or durations
> An exact length of time (e.g., a duration of 22.56 seconds).

This chapter will introduce how to work with each of these types of date/time data in Pandas. This is by no means a complete guide to the time series tools available in Python or Pandas, but instead is intended as a broad overview of how you as a user should approach working with time series. We will start with a brief discussion of tools for dealing with dates and times in Python, before moving more specifically to a discussion of the tools provided by Pandas. Finally, we will review some short examples of working with time series data in Pandas.

Dates and Times in Python

The Python world has a number of available representations of dates, times, deltas, and time spans. While the time series tools provided by Pandas tend to be the most useful for data science applications, it is helpful to see their relationship to other tools used in Python.

Native Python Dates and Times: datetime and dateutil

Python's basic objects for working with dates and times reside in the built-in datetime module. Along with the third-party `dateutil` module, you can use this to quickly perform a host of useful functionalities on dates and times. For example, you can manually build a date using the `datetime` type:

```
In [1]: from datetime import datetime
        datetime(year=2021, month=7, day=4)
Out[1]: datetime.datetime(2021, 7, 4, 0, 0)
```

Or, using the `dateutil` module, you can parse dates from a variety of string formats:

```
In [2]: from dateutil import parser
        date = parser.parse("4th of July, 2021")
        date
Out[2]: datetime.datetime(2021, 7, 4, 0, 0)
```

Once you have a `datetime` object, you can do things like printing the day of the week:

```
In [3]: date.strftime('%A')
Out[3]: 'Sunday'
```

Here we've used one of the standard string format codes for printing dates (`'%A'`), which you can read about in the strftime section (*https://oreil.ly/bjdsf*) of Python's datetime documentation (*https://oreil.ly/AGVR9*). Documentation of other useful date utilities can be found in `dateutil`'s online documentation (*https://oreil.ly/Y5Rwd*). A related package to be aware of is pytz (*https://oreil.ly/DU9Jr*), which contains tools for working with the most migraine-inducing element of time series data: time zones.

The power of datetime and dateutil lies in their flexibility and easy syntax: you can use these objects and their built-in methods to easily perform nearly any operation you might be interested in. Where they break down is when you wish to work with large arrays of dates and times: just as lists of Python numerical variables are suboptimal compared to NumPy-style typed numerical arrays, lists of Python datetime objects are suboptimal compared to typed arrays of encoded dates.

Typed Arrays of Times: NumPy's datetime64

NumPy's datetime64 dtype encodes dates as 64-bit integers, and thus allows arrays of dates to be represented compactly and operated on in an efficient manner. The datetime64 requires a specific input format:

```
In [4]: import numpy as np
        date = np.array('2021-07-04', dtype=np.datetime64)
        date
Out[4]: array('2021-07-04', dtype='datetime64[D]')
```

Once we have dates in this form, we can quickly do vectorized operations on it:

```
In [5]: date + np.arange(12)
Out[5]: array(['2021-07-04', '2021-07-05', '2021-07-06', '2021-07-07',
               '2021-07-08', '2021-07-09', '2021-07-10', '2021-07-11',
               '2021-07-12', '2021-07-13', '2021-07-14', '2021-07-15'],
              dtype='datetime64[D]')
```

Because of the uniform type in NumPy datetime64 arrays, this kind of operation can be accomplished much more quickly than if we were working directly with Python's datetime objects, especially as arrays get large (we introduced this type of vectorization in Chapter 6).

One detail of the datetime64 and related timedelta64 objects is that they are built on a *fundamental time unit*. Because the datetime64 object is limited to 64-bit precision, the range of encodable times is 2^{64} times this fundamental unit. In other words, datetime64 imposes a trade-off between *time resolution* and *maximum time span*.

For example, if you want a time resolution of 1 nanosecond, you only have enough information to encode a range of 2^{64} nanoseconds, or just under 600 years. NumPy will infer the desired unit from the input; for example, here is a day-based datetime:

```
In [6]: np.datetime64('2021-07-04')
Out[6]: numpy.datetime64('2021-07-04')
```

Here is a minute-based datetime:

```
In [7]: np.datetime64('2021-07-04 12:00')
Out[7]: numpy.datetime64('2021-07-04T12:00')
```

You can force any desired fundamental unit using one of many format codes; for example, here we'll force a nanosecond-based time:

```
In [8]: np.datetime64('2021-07-04 12:59:59.50', 'ns')
Out[8]: numpy.datetime64('2021-07-04T12:59:59.500000000')
```

Table 23-1, drawn from the NumPy datetime64 documentation, lists the available format codes along with the relative and absolute time spans that they can encode.

Table 23-1. Description of date and time codes

Code	Meaning	Time span (relative)	Time span (absolute)
Y	Year	± 9.2e18 years	[9.2e18 BC, 9.2e18 AD]
M	Month	± 7.6e17 years	[7.6e17 BC, 7.6e17 AD]
W	Week	± 1.7e17 years	[1.7e17 BC, 1.7e17 AD]
D	Day	± 2.5e16 years	[2.5e16 BC, 2.5e16 AD]
h	Hour	± 1.0e15 years	[1.0e15 BC, 1.0e15 AD]
m	Minute	± 1.7e13 years	[1.7e13 BC, 1.7e13 AD]
s	Second	± 2.9e12 years	[2.9e9 BC, 2.9e9 AD]
ms	Millisecond	± 2.9e9 years	[2.9e6 BC, 2.9e6 AD]
us	Microsecond	± 2.9e6 years	[290301 BC, 294241 AD]
ns	Nanosecond	± 292 years	[1678 AD, 2262 AD]
ps	Picosecond	± 106 days	[1969 AD, 1970 AD]
fs	Femtosecond	± 2.6 hours	[1969 AD, 1970 AD]
as	Attosecond	± 9.2 seconds	[1969 AD, 1970 AD]

For the types of data we see in the real world, a useful default is `datetime64[ns]`, as it can encode a useful range of modern dates with a suitably fine precision.

Finally, note that while the `datetime64` data type addresses some of the deficiencies of the built-in Python `datetime` type, it lacks many of the convenient methods and functions provided by `datetime` and especially `dateutil`. More information can be found in NumPy's `datetime64` documentation (*https://oreil.ly/XDbck*).

Dates and Times in Pandas: The Best of Both Worlds

Pandas builds upon all the tools just discussed to provide a `Timestamp` object, which combines the ease of use of `datetime` and `dateutil` with the efficient storage and vectorized interface of `numpy.datetime64`. From a group of these `Timestamp` objects, Pandas can construct a `DatetimeIndex` that can be used to index data in a `Series` or `DataFrame`.

For example, we can use Pandas tools to repeat the demonstration from earlier. We can parse a flexibly formatted string date and use format codes to output the day of the week, as follows:

```
In [9]: import pandas as pd
        date = pd.to_datetime("4th of July, 2021")
        date
Out[9]: Timestamp('2021-07-04 00:00:00')

In [10]: date.strftime('%A')
Out[10]: 'Sunday'
```

Additionally, we can do NumPy-style vectorized operations directly on this same object:

```
In [11]: date + pd.to_timedelta(np.arange(12), 'D')
Out[11]: DatetimeIndex(['2021-07-04', '2021-07-05', '2021-07-06', '2021-07-07',
                         '2021-07-08', '2021-07-09', '2021-07-10', '2021-07-11',
                         '2021-07-12', '2021-07-13', '2021-07-14', '2021-07-15'],
                        dtype='datetime64[ns]', freq=None)
```

In the next section, we will take a closer look at manipulating time series data with the tools provided by Pandas.

Pandas Time Series: Indexing by Time

The Pandas time series tools really become useful when you begin to index data by timestamps. For example, we can construct a `Series` object that has time-indexed data:

```
In [12]: index = pd.DatetimeIndex(['2020-07-04', '2020-08-04',
                                   '2021-07-04', '2021-08-04'])
         data = pd.Series([0, 1, 2, 3], index=index)
         data
Out[12]: 2020-07-04    0
         2020-08-04    1
         2021-07-04    2
         2021-08-04    3
         dtype: int64
```

And now that we have this data in a `Series`, we can make use of any of the `Series` indexing patterns we discussed in previous chapters, passing values that can be coerced into dates:

```
In [13]: data['2020-07-04':'2021-07-04']
Out[13]: 2020-07-04    0
         2020-08-04    1
         2021-07-04    2
         dtype: int64
```

There are additional special date-only indexing operations, such as passing a year to obtain a slice of all data from that year:

```
In [14]: data['2021']
Out[14]: 2021-07-04    2
         2021-08-04    3
         dtype: int64
```

Later, we will see additional examples of the convenience of dates-as-indices. But first, let's take a closer look at the available time series data structures.

Pandas Time Series Data Structures

This section will introduce the fundamental Pandas data structures for working with time series data:

- For *timestamps*, Pandas provides the Timestamp type. As mentioned before, this is essentially a replacement for Python's native datetime, but it's based on the more efficient numpy.datetime64 data type. The associated Index structure is DatetimeIndex.

- For *time periods*, Pandas provides the Period type. This encodes a fixed-frequency interval based on numpy.datetime64. The associated index structure is PeriodIndex.

- For *time deltas* or *durations*, Pandas provides the Timedelta type. Timedelta is a more efficient replacement for Python's native datetime.timedelta type, and is based on numpy.timedelta64. The associated index structure is TimedeltaIndex.

The most fundamental of these date/time objects are the Timestamp and Datetime Index objects. While these class objects can be invoked directly, it is more common to use the pd.to_datetime function, which can parse a wide variety of formats. Passing a single date to pd.to_datetime yields a Timestamp; passing a series of dates by default yields a DatetimeIndex, as you can see here:

```
In [15]: dates = pd.to_datetime([datetime(2021, 7, 3), '4th of July, 2021',
                                 '2021-Jul-6', '07-07-2021', '20210708'])
         dates
Out[15]: DatetimeIndex(['2021-07-03', '2021-07-04', '2021-07-06', '2021-07-07',
                        '2021-07-08'],
                       dtype='datetime64[ns]', freq=None)
```

Any DatetimeIndex can be converted to a PeriodIndex with the to_period function, with the addition of a frequency code; here we'll use 'D' to indicate daily frequency:

```
In [16]: dates.to_period('D')
Out[16]: PeriodIndex(['2021-07-03', '2021-07-04', '2021-07-06', '2021-07-07',
                      '2021-07-08'],
                     dtype='period[D]')
```

A TimedeltaIndex is created, for example, when a date is subtracted from another:

```
In [17]: dates - dates[0]
Out[17]: TimedeltaIndex(['0 days', '1 days', '3 days', '4 days', '5 days'],
            > dtype='timedelta64[ns]', freq=None)
```

Regular Sequences: pd.date_range

To make creation of regular date sequences more convenient, Pandas offers a few functions for this purpose: pd.date_range for timestamps, pd.period_range for periods, and pd.timedelta_range for time deltas. We've seen that Python's range and NumPy's np.arange take a start point, end point, and optional step size and return a sequence. Similarly, pd.date_range accepts a start date, an end date, and an optional frequency code to create a regular sequence of dates:

```
In [18]: pd.date_range('2015-07-03', '2015-07-10')
Out[18]: DatetimeIndex(['2015-07-03', '2015-07-04', '2015-07-05', '2015-07-06',
                         '2015-07-07', '2015-07-08', '2015-07-09', '2015-07-10'],
                        dtype='datetime64[ns]', freq='D')
```

Alternatively, the date range can be specified not with a start and end point, but with a start point and a number of periods:

```
In [19]: pd.date_range('2015-07-03', periods=8)
Out[19]: DatetimeIndex(['2015-07-03', '2015-07-04', '2015-07-05', '2015-07-06',
                         '2015-07-07', '2015-07-08', '2015-07-09', '2015-07-10'],
                        dtype='datetime64[ns]', freq='D')
```

The spacing can be modified by altering the freq argument, which defaults to D. For example, here we construct a range of hourly timestamps:

```
In [20]: pd.date_range('2015-07-03', periods=8, freq='H')
Out[20]: DatetimeIndex(['2015-07-03 00:00:00', '2015-07-03 01:00:00',
                         '2015-07-03 02:00:00', '2015-07-03 03:00:00',
                         '2015-07-03 04:00:00', '2015-07-03 05:00:00',
                         '2015-07-03 06:00:00', '2015-07-03 07:00:00'],
                        dtype='datetime64[ns]', freq='H')
```

To create regular sequences of Period or Timedelta values, the similar pd.period_range and pd.timedelta_range functions are useful. Here are some monthly periods:

```
In [21]: pd.period_range('2015-07', periods=8, freq='M')
Out[21]: PeriodIndex(['2015-07', '2015-08', '2015-09',
                       '2015-10', '2015-11', '2015-12',
                       '2016-01', '2016-02'],
                      dtype='period[M]')
```

And a sequence of durations increasing by an hour:

```
In [22]: pd.timedelta_range(0, periods=6, freq='H')
Out[22]: TimedeltaIndex(['0 days 00:00:00', '0 days 01:00:00', '0 days 02:00:00',
                          '0 days 03:00:00', '0 days 04:00:00', '0 days 05:00:00'],
                         dtype='timedelta64[ns]', freq='H')
```

All of these require an understanding of Pandas frequency codes, which are summarized in the next section.

Frequencies and Offsets

Fundamental to these Pandas time series tools is the concept of a *frequency* or *date offset*. The following table summarizes the main codes available; as with the D (day) and H (hour) codes demonstrated in the previous sections, we can use these to specify any desired frequency spacing. Table 23-2 summarizes the main codes available.

Table 23-2. Listing of Pandas frequency codes

Code	Description	Code	Description
D	Calendar day	B	Business day
W	Weekly		
M	Month end	BM	Business month end
Q	Quarter end	BQ	Business quarter end
A	Year end	BA	Business year end
H	Hours	BH	Business hours
T	Minutes		
S	Seconds		
L	Milliseconds		
U	Microseconds		
N	Nanoseconds		

The monthly, quarterly, and annual frequencies are all marked at the end of the specified period. Adding an S suffix to any of these causes them to instead be marked at the beginning (see Table 23-3).

Table 23-3. Listing of start-indexed frequency codes

Code	Description	Code	Description
MS	Month start	BMS	Business month start
QS	Quarter start	BQS	Business quarter start
AS	Year start	BAS	Business year start

Additionally, you can change the month used to mark any quarterly or annual code by adding a three-letter month code as a suffix:

- Q-JAN, BQ-FEB, QS-MAR, BQS-APR, etc.

- A-JAN, BA-FEB, AS-MAR, BAS-APR, etc.

In the same way, the split point of the weekly frequency can be modified by adding a three-letter weekday code: W-SUN, W-MON, W-TUE, W-WED, etc.

On top of this, codes can be combined with numbers to specify other frequencies. For example, for a frequency of 2 hours and 30 minutes, we can combine the hour (H) and minute (T) codes as follows:

```
In [23]: pd.timedelta_range(0, periods=6, freq="2H30T")
Out[23]: TimedeltaIndex(['0 days 00:00:00', '0 days 02:30:00', '0 days 05:00:00',
                          '0 days 07:30:00', '0 days 10:00:00', '0 days 12:30:00'],
                         dtype='timedelta64[ns]', freq='150T')
```

All of these short codes refer to specific instances of Pandas time series offsets, which can be found in the pd.tseries.offsets module. For example, we can create a business day offset directly as follows:

```
In [24]: from pandas.tseries.offsets import BDay
         pd.date_range('2015-07-01', periods=6, freq=BDay())
Out[24]: DatetimeIndex(['2015-07-01', '2015-07-02', '2015-07-03', '2015-07-06',
                        '2015-07-07', '2015-07-08'],
                       dtype='datetime64[ns]', freq='B')
```

For more discussion of the use of frequencies and offsets, see the DateOffset section (*https://oreil.ly/J6JHA*) of the Pandas documentation.

Resampling, Shifting, and Windowing

The ability to use dates and times as indices to intuitively organize and access data is an important aspect of the Pandas time series tools. The benefits of indexed data in general (automatic alignment during operations, intuitive data slicing and access, etc.) still apply, and Pandas provides several additional time series–specific operations.

We will take a look at a few of those here, using some stock price data as an example. Because Pandas was developed largely in a finance context, it includes some very specific tools for financial data. For example, the accompanying pandas-datareader package (installable via pip install pandas-datareader) knows how to import data from various online sources. Here we will load part of the S&P 500 price history:

```
In [25]: from pandas_datareader import data

         sp500 = data.DataReader('^GSPC', start='2018', end='2022',
                                 data_source='yahoo')
         sp500.head()
Out[25]:                    High           Low          Open         Close        Volume \
         Date
         2018-01-02  2695.889893  2682.360107  2683.729980  2695.810059  3367250000
         2018-01-03  2714.370117  2697.770020  2697.850098  2713.060059  3538660000
         2018-01-04  2729.290039  2719.070068  2719.310059  2723.989990  3695260000
         2018-01-05  2743.449951  2727.919922  2731.330078  2743.149902  3236620000
         2018-01-08  2748.510010  2737.600098  2742.669922  2747.709961  3242650000
```

```
          Adj Close
Date
2018-01-02  2695.810059
2018-01-03  2713.060059
2018-01-04  2723.989990
2018-01-05  2743.149902
2018-01-08  2747.709961
```

For simplicity, we'll use just the closing price:

```
In [26]: sp500 = sp500['Close']
```

We can visualize this using the `plot` method, after the normal Matplotlib setup boilerplate (see Part IV); the result is shown in Figure 23-1.

```
In [27]: %matplotlib inline
         import matplotlib.pyplot as plt
         plt.style.use('seaborn-whitegrid')
         sp500.plot();
```

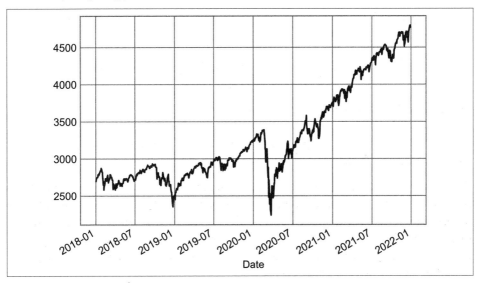

Figure 23-1. S&P500 closing price over time

Resampling and Converting Frequencies

One common need when dealing with time series data is resampling at a higher or lower frequency. This can be done using the `resample` method, or the much simpler `asfreq` method. The primary difference between the two is that `resample` is fundamentally a *data aggregation*, while `asfreq` is fundamentally a *data selection*.

Let's compare what the two return when we downsample the S&P 500 closing price data. Here we will resample the data at the end of business year; Figure 23-2 shows the result.

```
In [28]: sp500.plot(alpha=0.5, style='-')
         sp500.resample('BA').mean().plot(style=':')
         sp500.asfreq('BA').plot(style='--');
         plt.legend(['input', 'resample', 'asfreq'],
                    loc='upper left');
```

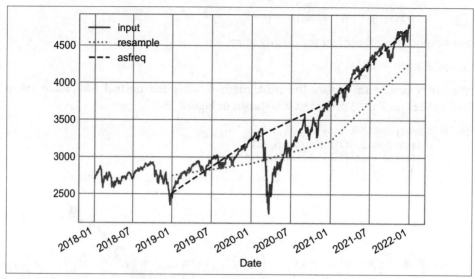

Figure 23-2. Resampling of S&P500 closing price

Notice the difference: at each point, `resample` reports the *average of the previous year*, while `asfreq` reports the *value at the end of the year*.

For upsampling, `resample` and `asfreq` are largely equivalent, though `resample` has many more options available. In this case, the default for both methods is to leave the upsampled points empty; that is, filled with NA values. Like the `pd.fillna` function discussed in Chapter 16, `asfreq` accepts a `method` argument to specify how values are imputed. Here, we will resample the business day data at a daily frequency (i.e., including weekends); Figure 23-3 shows the result.

```
In [29]: fig, ax = plt.subplots(2, sharex=True)
         data = sp500.iloc[:20]

         data.asfreq('D').plot(ax=ax[0], marker='o')

         data.asfreq('D', method='bfill').plot(ax=ax[1], style='-o')
         data.asfreq('D', method='ffill').plot(ax=ax[1], style='--o')
         ax[1].legend(["back-fill", "forward-fill"]);
```

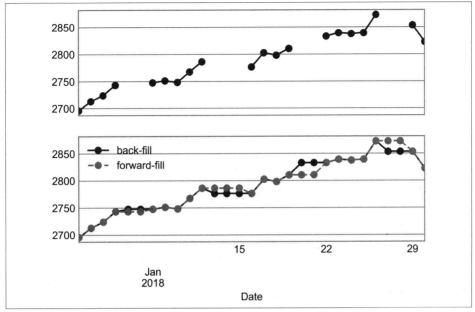

Figure 23-3. Comparison between forward-fill and back-fill interpolation

Because the S&P 500 data only exists for business days, the top panel has gaps representing NA values. The bottom panel shows the differences between two strategies for filling the gaps: forward filling and backward filling.

Time Shifts

Another common time series–specific operation is shifting of data in time. For this, Pandas provides the `shift` method, which can be used to shift data by a given number of entries. With time series data sampled at a regular frequency, this can give us a way to explore trends over time.

For example, here we resample the data to daily values, and shift by 364 to compute the 1-year return on investment for the S&P 500 over time (see Figure 23-4).

```
In [30]: sp500 = sp500.asfreq('D', method='pad')

         ROI = 100 * (sp500.shift(-365) - sp500) / sp500
         ROI.plot()
         plt.ylabel('% Return on Investment after 1 year');
```

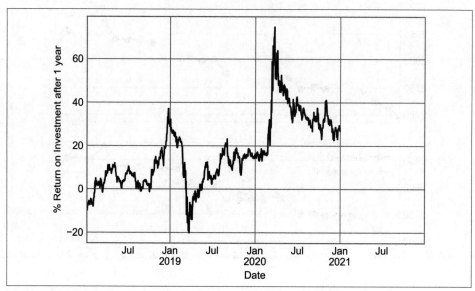

Figure 23-4. Return on investment after one year

The worst one-year return was around March 2019, with the coronavirus-related market crash exactly a year later. As you might expect, the best one-year return was to be found in March 2020, for those with enough foresight or luck to buy low.

Rolling Windows

Calculating rolling statistics is a third type of time series–specific operation implemented by Pandas. This can be accomplished via the `rolling` attribute of `Series` and `DataFrame` objects, which returns a view similar to what we saw with the `groupby` operation (see Chapter 20). This rolling view makes available a number of aggregation operations by default.

For example, we can look at the one-year centered rolling mean and standard deviation of the stock prices (see Figure 23-5).

```
In [31]: rolling = sp500.rolling(365, center=True)

         data = pd.DataFrame({'input': sp500,
                              'one-year rolling_mean': rolling.mean(),
                              'one-year rolling_median': rolling.median()})
         ax = data.plot(style=['-', '--', ':'])
         ax.lines[0].set_alpha(0.3)
```

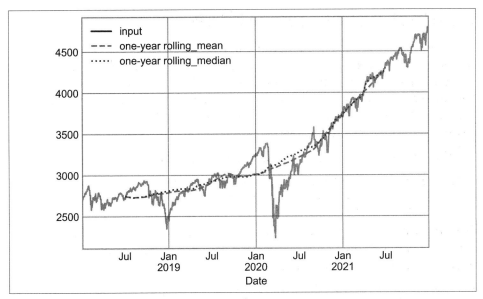

Figure 23-5. Rolling statistics on S&P500 index

As with `groupby` operations, the `aggregate` and `apply` methods can be used for custom rolling computations.

Where to Learn More

This chapter has provided only a brief summary of some of the most essential features of time series tools provided by Pandas; for a more complete discussion, you can refer to the "Time Series/Date Functionality" section (*https://oreil.ly/uC3pB*) of the Pandas online documentation.

Another excellent resource is the book *Python for Data Analysis* (*https://oreil.ly/ik2g7*) by Wes McKinney (O'Reilly). It is an invaluable resource on the use of Pandas. In particular, this book emphasizes time series tools in the context of business and finance, and focuses much more on particular details of business calendars, time zones, and related topics.

As always, you can also use the IPython help functionality to explore and try out further options available to the functions and methods discussed here. I find this often is the best way to learn a new Python tool.

Example: Visualizing Seattle Bicycle Counts

As a more involved example of working with time series data, let's take a look at bicycle counts on Seattle's Fremont Bridge (*https://oreil.ly/6qVBt*). This data comes from an automated bicycle counter installed in late 2012, which has inductive sensors on the east and west sidewalks of the bridge. The hourly bicycle counts can be downloaded from *http://data.seattle.gov*; the Fremont Bridge Bicycle Counter dataset is available under the Transportation category.

The CSV used for this book can be downloaded as follows:

```
In [32]: # url = ('https://raw.githubusercontent.com/jakevdp/'
         #        'bicycle-data/main/FremontBridge.csv')
         # !curl -O {url}
```

Once this dataset is downloaded, we can use Pandas to read the CSV output into a DataFrame. We will specify that we want the Date column as an index, and we want these dates to be automatically parsed:

```
In [33]: data = pd.read_csv('FremontBridge.csv', index_col='Date', parse_dates=True)
         data.head()
Out[33]:                          Fremont Bridge Total  Fremont Bridge East Sidewalk  \
         Date
         2019-11-01 00:00:00                      12.0                           7.0
         2019-11-01 01:00:00                       7.0                           0.0
         2019-11-01 02:00:00                       1.0                           0.0
         2019-11-01 03:00:00                       6.0                           6.0
         2019-11-01 04:00:00                       6.0                           5.0

                              Fremont Bridge West Sidewalk
         Date
         2019-11-01 00:00:00                           5.0
         2019-11-01 01:00:00                           7.0
         2019-11-01 02:00:00                           1.0
         2019-11-01 03:00:00                           0.0
         2019-11-01 04:00:00                           1.0
```

For convenience, we'll shorten the column names:

```
In [34]: data.columns = ['Total', 'East', 'West']
```

Now let's take a look at the summary statistics for this data:

```
In [35]: data.dropna().describe()
Out[35]:                 Total              East           West
         count  147255.000000     147255.000000  147255.000000
         mean      110.341462         50.077763      60.263699
         std       140.422051         64.634038      87.252147
         min         0.000000          0.000000       0.000000
         25%        14.000000          6.000000       7.000000
         50%        60.000000         28.000000      30.000000
```

75%	145.000000	68.000000	74.000000
max	1097.000000	698.000000	850.000000

Visualizing the Data

We can gain some insight into the dataset by visualizing it. Let's start by plotting the raw data (see Figure 23-6).

```
In [36]: data.plot()
         plt.ylabel('Hourly Bicycle Count');
```

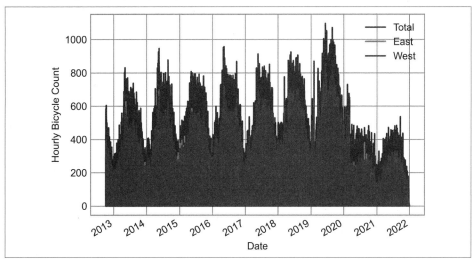

Figure 23-6. Hourly bicycle counts on Seattle's Fremont Bridge

The ~150,000 hourly samples are far too dense for us to make much sense of. We can gain more insight by resampling the data to a coarser grid. Let's resample by week (see Figure 23-7).

```
In [37]: weekly = data.resample('W').sum()
         weekly.plot(style=['-', ':', '--'])
         plt.ylabel('Weekly bicycle count');
```

This reveals some trends: as you might expect, people bicycle more in the summer than in the winter, and even within a particular season the bicycle use varies from week to week (likely dependent on weather; see Chapter 42, where we explore this further). Further, the effect of the COVID-19 pandemic on commuting patterns is quite clear, starting in early 2020.

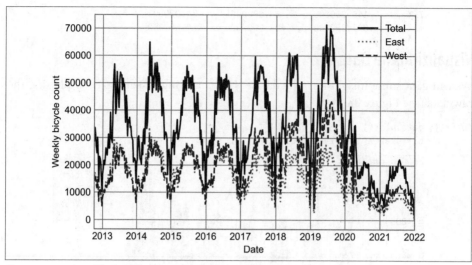

Figure 23-7. Weekly bicycle crossings of Seattle's Fremont Bridge

Another option that comes in handy for aggregating the data is to use a rolling mean, utilizing the `pd.rolling_mean` function. Here we'll examine the 30-day rolling mean of our data, making sure to center the window (see Figure 23-8).

```
In [38]: daily = data.resample('D').sum()
         daily.rolling(30, center=True).sum().plot(style=['-', ':', '--'])
         plt.ylabel('mean hourly count');
```

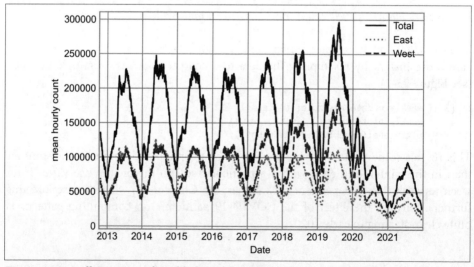

Figure 23-8. Rolling mean of weekly bicycle counts

The jaggedness of the result is due to the hard cutoff of the window. We can get a smoother version of a rolling mean using a window function—for example, a Gaussian window, as shown in Figure 23-9. The following code specifies both the width of the window (here, 50 days) and the width of the Gaussian window (here, 10 days):

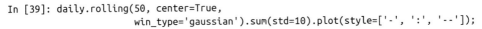

```
In [39]: daily.rolling(50, center=True,
                        win_type='gaussian').sum(std=10).plot(style=['-', ':', '--']);
```

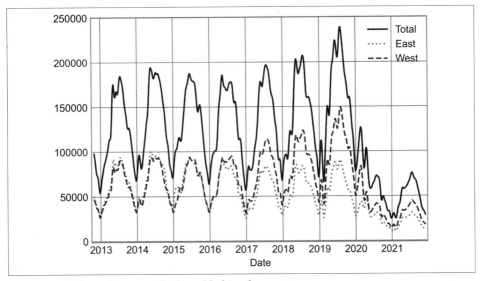

Figure 23-9. Gaussian smoothed weekly bicycle counts

Digging into the Data

While these smoothed data views are useful to get an idea of the general trend in the data, they hide much of the structure. For example, we might want to look at the average traffic as a function of the time of day. We can do this using the groupby functionality discussed in Chapter 20 (see Figure 23-10).

```
In [40]: by_time = data.groupby(data.index.time).mean()
         hourly_ticks = 4 * 60 * 60 * np.arange(6)
         by_time.plot(xticks=hourly_ticks, style=['-', ':', '--']);
```

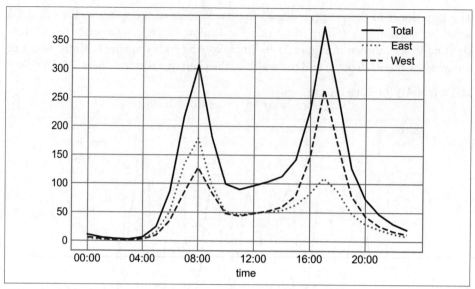

Figure 23-10. Average hourly bicycle counts

The hourly traffic is a strongly bimodal sequence, with peaks around 8:00 a.m. and 5:00 p.m. This is likely evidence of a strong component of commuter traffic crossing the bridge. There is a directional component as well: according to the data, the east sidewalk is used more during the a.m. commute, and the west sidewalk is used more during the p.m. commute.

We also might be curious about how things change based on the day of the week. Again, we can do this with a simple `groupby` (see Figure 23-11).

```
In [41]: by_weekday = data.groupby(data.index.dayofweek).mean()
         by_weekday.index = ['Mon', 'Tues', 'Wed', 'Thurs', 'Fri', 'Sat', 'Sun']
         by_weekday.plot(style=['-', ':', '--']);
```

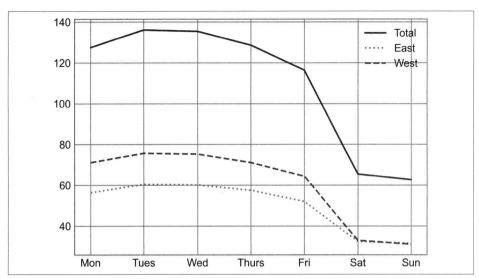

Figure 23-11. Average daily bicycle counts

This shows a strong distinction between weekday and weekend totals, with around twice as many average riders crossing the bridge on Monday through Friday than on Saturday and Sunday.

With this in mind, let's do a compound `groupby` and look at the hourly trends on weekdays versus weekends. We'll start by grouping by flags marking the weekend and the time of day:

```
In [42]: weekend = np.where(data.index.weekday < 5, 'Weekday', 'Weekend')
         by_time = data.groupby([weekend, data.index.time]).mean()
```

Now we'll use some of the Matplotlib tools that will be described in Chapter 31 to plot two panels side by side, as shown in Figure 23-12.

```
In [43]: import matplotlib.pyplot as plt
         fig, ax = plt.subplots(1, 2, figsize=(14, 5))
         by_time.loc['Weekday'].plot(ax=ax[0], title='Weekdays',
                                     xticks=hourly_ticks, style=['-', ':', '--'])
         by_time.loc['Weekend'].plot(ax=ax[1], title='Weekends',
                                     xticks=hourly_ticks, style=['-', ':', '--']);
```

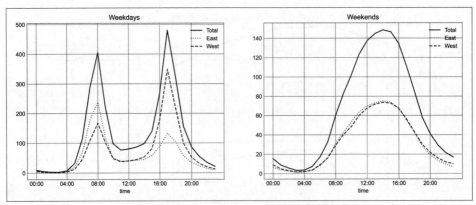

Figure 23-12. Average hourly bicycle counts by weekday and weekend

The result shows a bimodal commuting pattern during the work week, and a unimodal recreational pattern during the weekends. It might be interesting to dig through this data in more detail and examine the effects of weather, temperature, time of year, and other factors on people's commuting patterns; for further discussion, see my blog post "Is Seattle Really Seeing an Uptick in Cycling?" (*https://oreil.ly/j5oEI*), which uses a subset of this data. We will also revisit this dataset in the context of modeling in Chapter 42.

## CHAPTER 24

High-Performance Pandas: eval and query

As we've already seen in previous chapters, the power of the PyData stack is built upon the ability of NumPy and Pandas to push basic operations into lower-level compiled code via an intuitive higher-level syntax: examples are vectorized/broadcasted operations in NumPy, and grouping-type operations in Pandas. While these abstractions are efficient and effective for many common use cases, they often rely on the creation of temporary intermediate objects, which can cause undue overhead in computational time and memory use.

To address this, Pandas includes some methods that allow you to directly access C-speed operations without costly allocation of intermediate arrays: `eval` and `query`, which rely on the NumExpr package (*https://oreil.ly/acvj5*). In this chapter I will walk you through their use and give some rules of thumb about when you might think about using them.

Motivating query and eval: Compound Expressions

We've seen previously that NumPy and Pandas support fast vectorized operations; for example, when adding the elements of two arrays:

```
In [1]: import numpy as np
        rng = np.random.default_rng(42)
        x = rng.random(1000000)
        y = rng.random(1000000)
        %timeit x + y
Out[1]: 2.21 ms ± 142 µs per loop (mean ± std. dev. of 7 runs, 100 loops each)
```

As discussed in Chapter 6, this is much faster than doing the addition via a Python loop or comprehension:

```
In [2]: %timeit np.fromiter((xi + yi for xi, yi in zip(x, y)),
                            dtype=x.dtype, count=len(x))
Out[2]: 263 ms ± 43.4 ms per loop (mean ± std. dev. of 7 runs, 1 loop each)
```

But this abstraction can become less efficient when computing compound expressions. For example, consider the following expression:

```
In [3]: mask = (x > 0.5) & (y < 0.5)
```

Because NumPy evaluates each subexpression, this is roughly equivalent to the following:

```
In [4]: tmp1 = (x > 0.5)
        tmp2 = (y < 0.5)
        mask = tmp1 & tmp2
```

In other words, *every intermediate step is explicitly allocated in memory*. If the x and y arrays are very large, this can lead to significant memory and computational overhead. The NumExpr library gives you the ability to compute this type of compound expression element by element, without the need to allocate full intermediate arrays. The NumExpr documentation (*https://oreil.ly/acvj5*) has more details, but for the time being it is sufficient to say that the library accepts a *string* giving the NumPy-style expression you'd like to compute:

```
In [5]: import numexpr
        mask_numexpr = numexpr.evaluate('(x > 0.5) & (y < 0.5)')
        np.all(mask == mask_numexpr)
Out[5]: True
```

The benefit here is that NumExpr evaluates the expression in a way that avoids temporary arrays where possible, and thus can be much more efficient than NumPy, especially for long sequences of computations on large arrays. The Pandas eval and query tools that we will discuss here are conceptually similar, and are essentially Pandas-specific wrappers of NumExpr functionality.

pandas.eval for Efficient Operations

The eval function in Pandas uses string expressions to efficiently compute operations on DataFrame objects. For example, consider the following data:

```
In [6]: import pandas as pd
        nrows, ncols = 100000, 100
        df1, df2, df3, df4 = (pd.DataFrame(rng.random((nrows, ncols)))
                              for i in range(4))
```

To compute the sum of all four DataFrames using the typical Pandas approach, we can just write the sum:

```
In [7]: %timeit df1 + df2 + df3 + df4
Out[7]: 73.2 ms ± 6.72 ms per loop (mean ± std. dev. of 7 runs, 10 loops each)
```

The same result can be computed via pd.eval by constructing the expression as a string:

```
In [8]: %timeit pd.eval('df1 + df2 + df3 + df4')
Out[8]: 34 ms ± 4.2 ms per loop (mean ± std. dev. of 7 runs, 10 loops each)
```

The eval version of this expression is about 50% faster (and uses much less memory), while giving the same result:

```
In [9]: np.allclose(df1 + df2 + df3 + df4,
                    pd.eval('df1 + df2 + df3 + df4'))
Out[9]: True
```

pd.eval supports a wide range of operations. To demonstrate these, we'll use the following integer data:

```
In [10]: df1, df2, df3, df4, df5 = (pd.DataFrame(rng.integers(0, 1000, (100, 3))))
                                   for i in range(5))
```

Here's a summary of the operations pd.eval supports:

Arithmetic operators

pd.eval supports all arithmetic operators. For example:

```
In [11]: result1 = -df1 * df2 / (df3 + df4) - df5
         result2 = pd.eval('-df1 * df2 / (df3 + df4) - df5')
         np.allclose(result1, result2)
Out[11]: True
```

Comparison operators

pd.eval supports all comparison operators, including chained expressions:

```
In [12]: result1 = (df1 < df2) & (df2 <= df3) & (df3 != df4)
         result2 = pd.eval('df1 < df2 <= df3 != df4')
         np.allclose(result1, result2)
Out[12]: True
```

Bitwise operators

pd.eval supports the & and | bitwise operators:

```
In [13]: result1 = (df1 < 0.5) & (df2 < 0.5) | (df3 < df4)
         result2 = pd.eval('(df1 < 0.5) & (df2 < 0.5) | (df3 < df4)')
         np.allclose(result1, result2)
Out[13]: True
```

In addition, it supports the use of the literal and and or in Boolean expressions:

```
In [14]: result3 = pd.eval('(df1 < 0.5) and (df2 < 0.5) or (df3 < df4)')
         np.allclose(result1, result3)
Out[14]: True
```

Object attributes and indices

pd.eval supports access to object attributes via the obj.attr syntax and indexes via the obj[index] syntax:

```
In [15]: result1 = df2.T[0] + df3.iloc[1]
         result2 = pd.eval('df2.T[0] + df3.iloc[1]')
         np.allclose(result1, result2)
Out[15]: True
```

Other operations

Other operations, such as function calls, conditional statements, loops, and other more involved constructs are currently *not* implemented in pd.eval. If you'd like to execute these more complicated types of expressions, you can use the NumExpr library itself.

DataFrame.eval for Column-Wise Operations

Just as Pandas has a top-level pd.eval function, DataFrame objects have an eval method that works in similar ways. The benefit of the eval method is that columns can be referred to by name. We'll use this labeled array as an example:

```
In [16]: df = pd.DataFrame(rng.random((1000, 3)), columns=['A', 'B', 'C'])
         df.head()
Out[16]:           A         B         C
         0  0.850888  0.966709  0.958690
         1  0.820126  0.385686  0.061402
         2  0.059729  0.831768  0.652259
         3  0.244774  0.140322  0.041711
         4  0.818205  0.753384  0.578851
```

Using pd.eval as in the previous section, we can compute expressions with the three columns like this:

```
In [17]: result1 = (df['A'] + df['B']) / (df['C'] - 1)
         result2 = pd.eval("(df.A + df.B) / (df.C - 1)")
         np.allclose(result1, result2)
Out[17]: True
```

The DataFrame.eval method allows much more succinct evaluation of expressions with the columns:

```
In [18]: result3 = df.eval('(A + B) / (C - 1)')
         np.allclose(result1, result3)
Out[18]: True
```

Notice here that we treat *column names as variables* within the evaluated expression, and the result is what we would wish.

Assignment in DataFrame.eval

In addition to the options just discussed, `DataFrame.eval` also allows assignment to any column. Let's use the `DataFrame` from before, which has columns `'A'`, `'B'`, and `'C'`:

```
In [19]: df.head()
Out[19]:          A         B         C
         0  0.850888  0.966709  0.958690
         1  0.820126  0.385686  0.061402
         2  0.059729  0.831768  0.652259
         3  0.244774  0.140322  0.041711
         4  0.818205  0.753384  0.578851
```

We can use `df.eval` to create a new column `'D'` and assign to it a value computed from the other columns:

```
In [20]: df.eval('D = (A + B) / C', inplace=True)
         df.head()
Out[20]:          A         B         C          D
         0  0.850888  0.966709  0.958690   1.895916
         1  0.820126  0.385686  0.061402  19.638139
         2  0.059729  0.831768  0.652259   1.366782
         3  0.244774  0.140322  0.041711   9.232370
         4  0.818205  0.753384  0.578851   2.715013
```

In the same way, any existing column can be modified:

```
In [21]: df.eval('D = (A - B) / C', inplace=True)
         df.head()
Out[21]:          A         B         C          D
         0  0.850888  0.966709  0.958690  -0.120812
         1  0.820126  0.385686  0.061402   7.075399
         2  0.059729  0.831768  0.652259  -1.183638
         3  0.244774  0.140322  0.041711   2.504142
         4  0.818205  0.753384  0.578851   0.111982
```

Local Variables in DataFrame.eval

The `DataFrame.eval` method supports an additional syntax that lets it work with local Python variables. Consider the following:

```
In [22]: column_mean = df.mean(1)
         result1 = df['A'] + column_mean
         result2 = df.eval('A + @column_mean')
         np.allclose(result1, result2)
Out[22]: True
```

The `@` character here marks a *variable name* rather than a *column name*, and lets you efficiently evaluate expressions involving the two "namespaces": the namespace of columns, and the namespace of Python objects. Notice that this `@` character is only supported by the DataFrame.eval *method*, not by the pandas.eval *function*, because the pandas.eval function only has access to the one (Python) namespace.

The DataFrame.query Method

The DataFrame has another method based on evaluated strings, called query. Consider the following:

```
In [23]: result1 = df[(df.A < 0.5) & (df.B < 0.5)]
         result2 = pd.eval('df[(df.A < 0.5) & (df.B < 0.5)]')
         np.allclose(result1, result2)
Out[23]: True
```

As with the example used in our discussion of DataFrame.eval, this is an expression involving columns of the DataFrame. However, it cannot be expressed using the DataFrame.eval syntax! Instead, for this type of filtering operation, you can use the query method:

```
In [24]: result2 = df.query('A < 0.5 and B < 0.5')
         np.allclose(result1, result2)
Out[24]: True
```

In addition to being a more efficient computation, compared to the masking expression this is much easier to read and understand. Note that the query method also accepts the @ flag to mark local variables:

```
In [25]: Cmean = df['C'].mean()
         result1 = df[(df.A < Cmean) & (df.B < Cmean)]
         result2 = df.query('A < @Cmean and B < @Cmean')
         np.allclose(result1, result2)
Out[25]: True
```

Performance: When to Use These Functions

When considering whether to use eval and query, there are two considerations: *computation time* and *memory use*. Memory use is the most predictable aspect. As already mentioned, every compound expression involving NumPy arrays or Pandas DataFrames will result in implicit creation of temporary arrays. For example, this:

```
In [26]: x = df[(df.A < 0.5) & (df.B < 0.5)]
```

is roughly equivalent to this:

```
In [27]: tmp1 = df.A < 0.5
         tmp2 = df.B < 0.5
         tmp3 = tmp1 & tmp2
         x = df[tmp3]
```

If the size of the temporary `DataFrames` is significant compared to your available system memory (typically several gigabytes), then it's a good idea to use an `eval` or `query` expression. You can check the approximate size of your array in bytes using this:

```
In [28]: df.values.nbytes
Out[28]: 32000
```

On the performance side, `eval` can be faster even when you are not maxing out your system memory. The issue is how your temporary objects compare to the size of the L1 or L2 CPU cache on your system (typically a few megabytes); if they are much bigger, then `eval` can avoid some potentially slow movement of values between the different memory caches. In practice, I find that the difference in computation time between the traditional methods and the `eval`/`query` method is usually not significant—if anything, the traditional method is faster for smaller arrays! The benefit of `eval`/`query` is mainly in the saved memory, and the sometimes cleaner syntax they offer.

We've covered most of the details of `eval` and `query` here; for more information on these, you can refer to the Pandas documentation. In particular, different parsers and engines can be specified for running these queries; for details on this, see the discussion within the "Enhancing Performance" section (*https://oreil.ly/DHNy8*) of the documentation.

Further Resources

In this part of the book, we've covered many of the basics of using Pandas effectively for data analysis. Still, much has been omitted from our discussion. To learn more about Pandas, I recommend the following resources:

Pandas online documentation (http://pandas.pydata.org)
  This is the go-to source for complete documentation of the package. While the examples in the documentation tend to be based on small generated datasets, the description of the options is complete and generally very useful for understanding the use of various functions.

Python for Data Analysis (https://oreil.ly/0hdsf)
  Written by Wes McKinney (the original creator of Pandas), this book contains much more detail on the Pandas package than we had room for in this chapter. In particular, McKinney takes a deep dive into tools for time series, which were his bread and butter as a financial consultant. The book also has many entertaining examples of applying Pandas to gain insight from real-world datasets.

Effective Pandas (https://oreil.ly/cn1ls)

This short ebook by Pandas developer Tom Augspurger provides a succinct outline of using the full power of the Pandas library in an effective and idiomatic way.

Pandas on PyVideo (https://oreil.ly/mh4wI)

From PyCon to SciPy to PyData, many conferences have featured tutorials by Pandas developers and power users. The PyCon tutorials in particular tend to be given by very well-vetted presenters.

Using these resources, combined with the walkthrough given in these chapters, my hope is that you'll be poised to use Pandas to tackle any data analysis problem you come across!

## PART IV

Visualization with Matplotlib

We'll now take an in-depth look at the Matplotlib package for visualization in Python. Matplotlib is a multiplatform data visualization library built on NumPy arrays and designed to work with the broader SciPy stack. It was conceived by John Hunter in 2002, originally as a patch to IPython for enabling interactive MATLAB-style plotting via gnuplot from the IPython command line. IPython's creator, Fernando Perez, was at the time scrambling to finish his PhD, and let John know he wouldn't have time to review the patch for several months. John took this as a cue to set out on his own, and the Matplotlib package was born, with version 0.1 released in 2003. It received an early boost when it was adopted as the plotting package of choice of the Space Telescope Science Institute (the folks behind the Hubble Telescope), which financially supported Matplotlib's development and greatly expanded its capabilities.

One of Matplotlib's most important features is its ability to play well with many operating systems and graphics backends. Matplotlib supports dozens of backends and output types, which means you can count on it to work regardless of which operating system you are using or which output format you desire. This cross-platform, everything-to-everyone approach has been one of the great strengths of Matplotlib. It has led to a large user base, which in turn has led to an active developer base and Matplotlib's powerful tools and ubiquity within the scientific Python world.

In recent years, however, the interface and style of Matplotlib have begun to show their age. Newer tools like ggplot and ggvis in the R language, along with web visualization toolkits based on D3js and HTML5 canvas, often make Matplotlib feel clunky and old-fashioned. Still, I'm of the opinion that we cannot ignore Matplotlib's strength as a well-tested, cross-platform graphics engine. Recent Matplotlib versions make it relatively easy to set new global plotting styles (see Chapter 34), and people have been developing new packages that build on its powerful internals to drive Matplotlib via cleaner, more modern APIs—for example, Seaborn (discussed in Chapter 36), ggpy (*http://yhat.github.io/ggpy*), HoloViews (*http://holoviews.org*), and even Pandas itself can be used as wrappers around Matplotlib's API. Even with wrappers like these, it is still often useful to dive into Matplotlib's syntax to adjust the final plot output. For this reason, I believe that Matplotlib itself will remain a vital piece of the data visualization stack, even if new tools mean the community gradually moves away from using the Matplotlib API directly.

## CHAPTER 25

General Matplotlib Tips

Before we dive into the details of creating visualizations with Matplotlib, there are a few useful things you should know about using the package.

Importing Matplotlib

Just as we use the `np` shorthand for NumPy and the `pd` shorthand for Pandas, we will use some standard shorthands for Matplotlib imports:

```
In [1]: import matplotlib as mpl
        import matplotlib.pyplot as plt
```

The `plt` interface is what we will use most often, as you shall see throughout this part of the book.

Setting Styles

We will use the `plt.style` directive to choose appropriate aesthetic styles for our figures. Here we will set the `classic` style, which ensures that the plots we create use the classic Matplotlib style:

```
In [2]: plt.style.use('classic')
```

Throughout this chapter, we will adjust this style as needed. For more information on stylesheets, see Chapter 34.

show or No show? How to Display Your Plots

A visualization you can't see won't be of much use, but just how you view your Matplotlib plots depends on the context. The best use of Matplotlib differs depending on how you are using it; roughly, the three applicable contexts are using Matplotlib in a script, in an IPython terminal, or in a Jupyter notebook.

Plotting from a Script

If you are using Matplotlib from within a script, the function `plt.show` is your friend. `plt.show` starts an event loop, looks for all currently active `Figure` objects, and opens one or more interactive windows that display your figure or figures.

So, for example, you may have a file called *myplot.py* containing the following:

```
# file: myplot.py
import matplotlib.pyplot as plt
import numpy as np

x = np.linspace(0, 10, 100)

plt.plot(x, np.sin(x))
plt.plot(x, np.cos(x))

plt.show()
```

You can then run this script from the command-line prompt, which will result in a window opening with your figure displayed:

```
$ python myplot.py
```

The `plt.show` command does a lot under the hood, as it must interact with your system's interactive graphical backend. The details of this operation can vary greatly from system to system and even installation to installation, but Matplotlib does its best to hide all these details from you.

One thing to be aware of: the `plt.show` command should be used *only once* per Python session, and is most often seen at the very end of the script. Multiple `show` commands can lead to unpredictable backend-dependent behavior, and should mostly be avoided.

Plotting from an IPython Shell

Matplotlib also works seamlessly within an IPython shell (see Part I). IPython is built to work well with Matplotlib if you specify Matplotlib mode. To enable this mode, you can use the `%matplotlib` magic command after starting `ipython`:

```
In [1]: %matplotlib
Using matplotlib backend: TkAgg

In [2]: import matplotlib.pyplot as plt
```

At this point, any `plt` plot command will cause a figure window to open, and further commands can be run to update the plot. Some changes (such as modifying properties of lines that are already drawn) will not draw automatically: to force an update, use `plt.draw`. Using `plt.show` in IPython's Matplotlib mode is not required.

Plotting from a Jupyter Notebook

The Jupyter notebook is a browser-based interactive data analysis tool that can combine narrative, code, graphics, HTML elements, and much more into a single executable document (see Part I).

Plotting interactively within a Jupyter notebook can be done with the `%matplotlib` command, and works in a similar way to the IPython shell. You also have the option of embedding graphics directly in the notebook, with two possible options:

- `%matplotlib inline` will lead to *static* images of your plot embedded in the notebook.

- `%matplotlib notebook` will lead to *interactive* plots embedded within the notebook.

For this book, we will generally stick with the default, with figures rendered as static images (see Figure 25-1 for the result of this basic plotting example):

```
In [3]: %matplotlib inline

In [4]: import numpy as np
        x = np.linspace(0, 10, 100)

        fig = plt.figure()
        plt.plot(x, np.sin(x), '-')
        plt.plot(x, np.cos(x), '--');
```

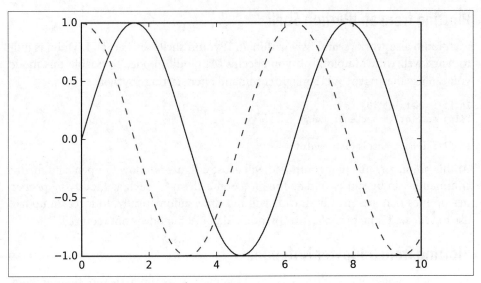

Figure 25-1. Basic plotting example

Saving Figures to File

One nice feature of Matplotlib is the ability to save figures in a wide variety of formats. Saving a figure can be done using the `savefig` command. For example, to save the previous figure as a PNG file, we can run this:

```
In [5]: fig.savefig('my_figure.png')
```

We now have a file called *my_figure.png* in the current working directory:

```
In [6]: !ls -lh my_figure.png
Out[6]: -rw-r--r--  1 jakevdp  staff    26K Feb  1 06:15 my_figure.png
```

To confirm that it contains what we think it contains, let's use the IPython `Image` object to display the contents of this file (see Figure 25-2).

```
In [7]: from IPython.display import Image
        Image('my_figure.png')
```

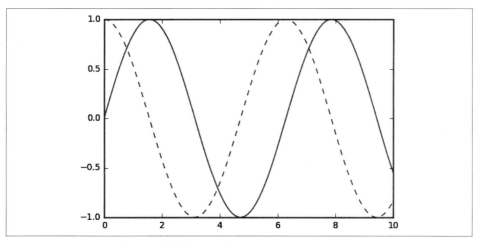

Figure 25-2. PNG rendering of the basic plot

In `savefig`, the file format is inferred from the extension of the given filename. Depending on what backends you have installed, many different file formats are available. The list of supported file types can be found for your system by using the following method of the figure canvas object:

```
In [8]: fig.canvas.get_supported_filetypes()
Out[8]: {'eps': 'Encapsulated Postscript',
         'jpg': 'Joint Photographic Experts Group',
         'jpeg': 'Joint Photographic Experts Group',
         'pdf': 'Portable Document Format',
         'pgf': 'PGF code for LaTeX',
         'png': 'Portable Network Graphics',
         'ps': 'Postscript',
         'raw': 'Raw RGBA bitmap',
         'rgba': 'Raw RGBA bitmap',
         'svg': 'Scalable Vector Graphics',
         'svgz': 'Scalable Vector Graphics',
         'tif': 'Tagged Image File Format',
         'tiff': 'Tagged Image File Format'}
```

Note that when saving your figure, it is not necessary to use `plt.show` or related commands discussed earlier.

Two Interfaces for the Price of One

A potentially confusing feature of Matplotlib is its dual interfaces: a convenient MATLAB-style state-based interface, and a more powerful object-oriented interface. I'll quickly highlight the differences between the two here.

MATLAB-style Interface

Matplotlib was originally conceived as a Python alternative for MATLAB users, and much of its syntax reflects that fact. The MATLAB-style tools are contained in the pyplot (plt) interface. For example, the following code will probably look quite familiar to MATLAB users (Figure 25-3 shows the result).

```
In [9]: plt.figure()  # create a plot figure

        # create the first of two panels and set current axis
        plt.subplot(2, 1, 1) # (rows, columns, panel number)
        plt.plot(x, np.sin(x))

        # create the second panel and set current axis
        plt.subplot(2, 1, 2)
        plt.plot(x, np.cos(x));
```

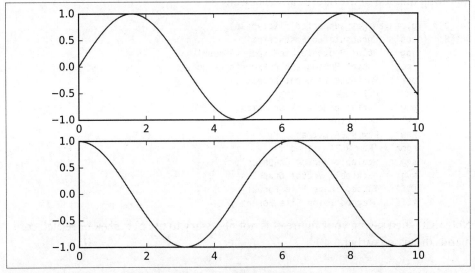

Figure 25-3. Subplots using the MATLAB-style interface

It is important to recognize that this interface is *stateful*: it keeps track of the "current" figure and axes, which are where all plt commands are applied. You can get a reference to these using the plt.gcf (get current figure) and plt.gca (get current axes) routines.

While this stateful interface is fast and convenient for simple plots, it is easy to run into problems. For example, once the second panel is created, how can we go back and add something to the first? This is possible within the MATLAB-style interface, but a bit clunky. Fortunately, there is a better way.

Object-oriented interface

The object-oriented interface is available for these more complicated situations, and for when you want more control over your figure. Rather than depending on some notion of an "active" figure or axes, in the object-oriented interface the plotting functions are *methods* of explicit `Figure` and `Axes` objects. To re-create the previous plot using this style of plotting, as shown in Figure 25-4, you might do the following:

```
In [10]: # First create a grid of plots
         # ax will be an array of two Axes objects
         fig, ax = plt.subplots(2)

         # Call plot() method on the appropriate object
         ax[0].plot(x, np.sin(x))
         ax[1].plot(x, np.cos(x));
```

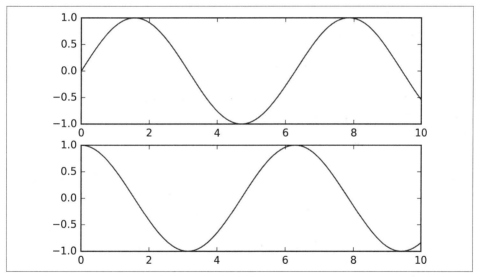

Figure 25-4. Subplots using the object-oriented interface

For simpler plots, the choice of which style to use is largely a matter of preference, but the object-oriented approach can become a necessity as plots become more complicated. Throughout the following chapters, we will switch between the MATLAB-style and object-oriented interfaces, depending on what is most convenient. In most cases, the difference is as small as switching `plt.plot` to `ax.plot`, but there are a few gotchas that I will highlight as they come up in the following chapters.

## CHAPTER 26

Simple Line Plots

Perhaps the simplest of all plots is the visualization of a single function $y = f(x)$. Here we will take a first look at creating a simple plot of this type. As in all the following chapters, we'll start by setting up the notebook for plotting and importing the packages we will use:

```
In [1]: %matplotlib inline
        import matplotlib.pyplot as plt
        plt.style.use('seaborn-whitegrid')
        import numpy as np
```

For all Matplotlib plots, we start by creating a figure and axes. In their simplest form, this can be done as follows (see Figure 26-1).

```
In [2]: fig = plt.figure()
        ax = plt.axes()
```

In Matplotlib, the *figure* (an instance of the class plt.Figure) can be thought of as a single container that contains all the objects representing axes, graphics, text, and labels. The *axes* (an instance of the class plt.Axes) is what we see above: a bounding box with ticks, grids, and labels, which will eventually contain the plot elements that make up our visualization. Throughout this part of the book, I'll commonly use the variable name fig to refer to a figure instance and ax to refer to an axes instance or group of axes instances.

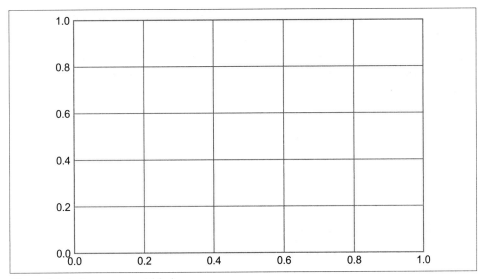

Figure 26-1. An empty gridded axes

Once we have created an axes, we can use the `ax.plot` method to plot some data. Let's start with a simple sinusoid, as shown in Figure 26-2.

```
In [3]: fig = plt.figure()
        ax = plt.axes()

        x = np.linspace(0, 10, 1000)
        ax.plot(x, np.sin(x));
```

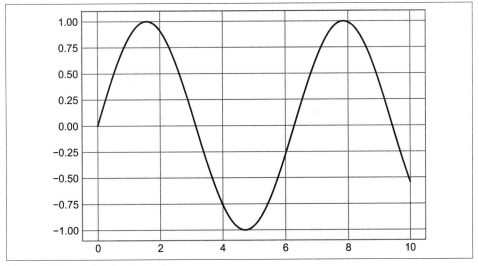

Figure 26-2. A simple sinusoid

Note that the semicolon at the end of the last line is intentional: it suppresses the textual representation of the plot from the output.

Alternatively, we can use the PyLab interface and let the figure and axes be created for us in the background (see Part IV for a discussion of these two interfaces); as Figure 26-3 shows, the result is the same.

```
In [4]: plt.plot(x, np.sin(x));
```

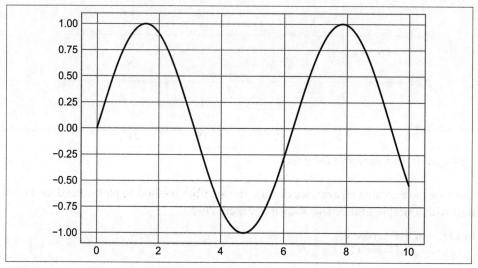

Figure 26-3. A simple sinusoid via the object-oriented interface

If we want to create a single figure with multiple lines (see Figure 26-4), we can simply call the plot function multiple times:

```
In [5]: plt.plot(x, np.sin(x))
        plt.plot(x, np.cos(x));
```

That's all there is to plotting simple functions in Matplotlib! We'll now dive into some more details about how to control the appearance of the axes and lines.

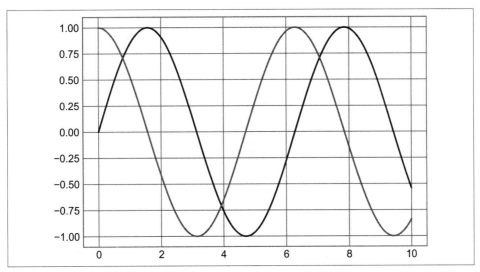

Figure 26-4. Overplotting multiple lines

Adjusting the Plot: Line Colors and Styles

The first adjustment you might wish to make to a plot is to control the line colors and styles. The `plt.plot` function takes additional arguments that can be used to specify these. To adjust the color, you can use the `color` keyword, which accepts a string argument representing virtually any imaginable color. The color can be specified in a variety of ways; see Figure 26-5 for the output of the following examples:

```
In [6]: plt.plot(x, np.sin(x - 0), color='blue')       # specify color by name
        plt.plot(x, np.sin(x - 1), color='g')           # short color code (rgbcmyk)
        plt.plot(x, np.sin(x - 2), color='0.75')        # grayscale between 0 and 1
        plt.plot(x, np.sin(x - 3), color='#FFDD44')     # hex code (RRGGBB, 00 to FF)
        plt.plot(x, np.sin(x - 4), color=(1.0,0.2,0.3)) # RGB tuple, values 0 to 1
        plt.plot(x, np.sin(x - 5), color='chartreuse'); # HTML color names supported
```

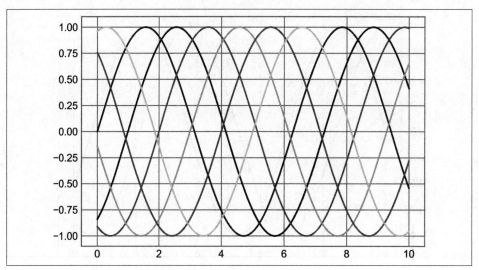

Figure 26-5. Controlling the color of plot elements

If no color is specified, Matplotlib will automatically cycle through a set of default colors for multiple lines.

Similarly, the line style can be adjusted using the linestyle keyword (see Figure 26-6).

```
In [7]: plt.plot(x, x + 0, linestyle='solid')
        plt.plot(x, x + 1, linestyle='dashed')
        plt.plot(x, x + 2, linestyle='dashdot')
        plt.plot(x, x + 3, linestyle='dotted');

        # For short, you can use the following codes:
        plt.plot(x, x + 4, linestyle='-')  # solid
        plt.plot(x, x + 5, linestyle='--') # dashed
        plt.plot(x, x + 6, linestyle='-.') # dashdot
        plt.plot(x, x + 7, linestyle=':'); # dotted
```

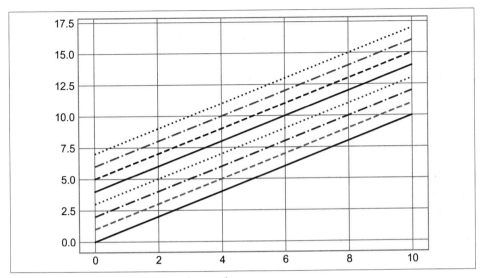

Figure 26-6. Examples of various line styles

Though it may be less clear to someone reading your code, you can save some keystrokes by combining these linestyle and color codes into a single non-keyword argument to the plt.plot function; Figure 26-7 shows the result.

```
In [8]: plt.plot(x, x + 0, '-g')    # solid green
        plt.plot(x, x + 1, '--c')   # dashed cyan
        plt.plot(x, x + 2, '-.k')   # dashdot black
        plt.plot(x, x + 3, ':r');   # dotted red
```

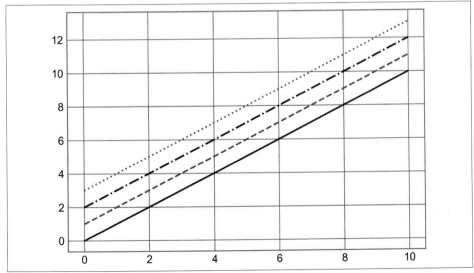

Figure 26-7. Controlling colors and styles with the shorthand syntax

These single-character color codes reflect the standard abbreviations in the RGB (Red/Green/Blue) and CMYK (Cyan/Magenta/Yellow/blacK) color systems, commonly used for digital color graphics.

There are many other keyword arguments that can be used to fine-tune the appearance of the plot; for details, read through the docstring of the plt.plot function using IPython's help tools (see Chapter 1).

Adjusting the Plot: Axes Limits

Matplotlib does a decent job of choosing default axes limits for your plot, but sometimes it's nice to have finer control. The most basic way to adjust the limits is to use the plt.xlim and plt.ylim functions (see Figure 26-8).

```
In [9]: plt.plot(x, np.sin(x))

        plt.xlim(-1, 11)
        plt.ylim(-1.5, 1.5);
```

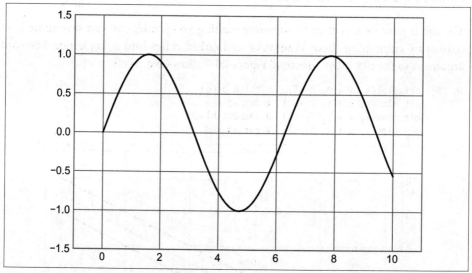

Figure 26-8. Example of setting axis limits

If for some reason you'd like either axis to be displayed in reverse, you can simply reverse the order of the arguments (see Figure 26-9).

```
In [10]: plt.plot(x, np.sin(x))

         plt.xlim(10, 0)
         plt.ylim(1.2, -1.2);
```

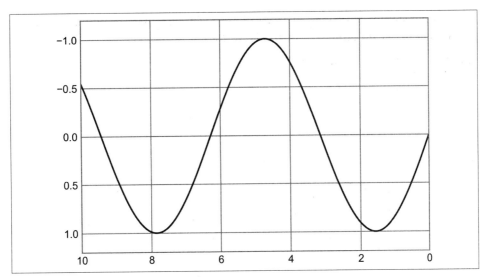

Figure 26-9. Example of reversing the y-axis

A useful related method is `plt.axis` (note here the potential confusion between *axes* with an *e*, and *axis* with an *i*), which allows more qualitative specifications of axis limits. For example, you can automatically tighten the bounds around the current content, as shown in Figure 26-10.

```
In [11]: plt.plot(x, np.sin(x))
         plt.axis('tight');
```

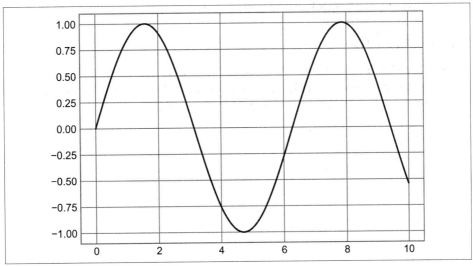

Figure 26-10. Example of a "tight" layout

Or you can specify that you want an equal axis ratio, such that one unit in x is visually equivalent to one unit in y, as seen in Figure 26-11.

```
In [12]: plt.plot(x, np.sin(x))
         plt.axis('equal');
```

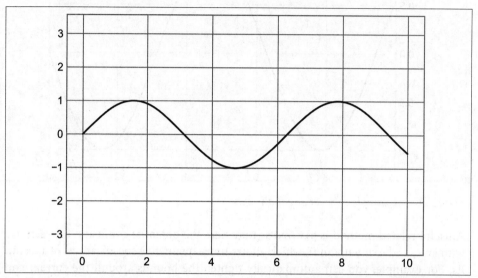

Figure 26-11. Example of an "equal" layout, with units matched to the output resolution

Other axis options include `'on'`, `'off'`, `'square'`, `'image'`, and more. For more information on these, refer to the `plt.axis` docstring.

Labeling Plots

As the last piece of this chapter, we'll briefly look at the labeling of plots: titles, axis labels, and simple legends. Titles and axis labels are the simplest such labels—there are methods that can be used to quickly set them (see Figure 26-12).

```
In [13]: plt.plot(x, np.sin(x))
         plt.title("A Sine Curve")
         plt.xlabel("x")
         plt.ylabel("sin(x)");
```

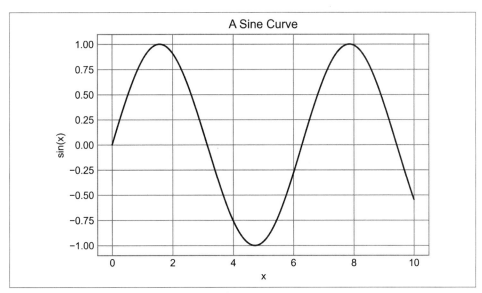

Figure 26-12. Examples of axis labels and title

The position, size, and style of these labels can be adjusted using optional arguments to the functions, described in the docstrings.

When multiple lines are being shown within a single axes, it can be useful to create a plot legend that labels each line type. Again, Matplotlib has a built-in way of quickly creating such a legend; it is done via the (you guessed it) plt.legend method. Though there are several valid ways of using this, I find it easiest to specify the label of each line using the label keyword of the plot function (see Figure 26-13).

```
In [14]: plt.plot(x, np.sin(x), '-g', label='sin(x)')
         plt.plot(x, np.cos(x), ':b', label='cos(x)')
         plt.axis('equal')

         plt.legend();
```

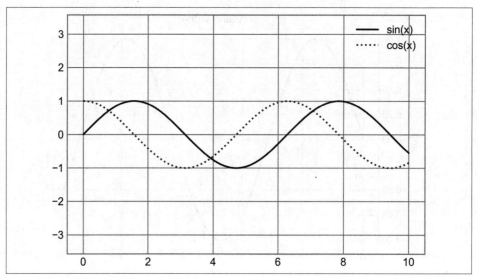

Figure 26-13. Plot legend example

As you can see, the `plt.legend` function keeps track of the line style and color, and matches these with the correct label. More information on specifying and formatting plot legends can be found in the `plt.legend` docstring; additionally, we will cover some more advanced legend options in Chapter 29.

Matplotlib Gotchas

While most `plt` functions translate directly to `ax` methods (`plt.plot` → `ax.plot`, `plt.legend` → `ax.legend`, etc.), this is not the case for all commands. In particular, functions to set limits, labels, and titles are slightly modified. For transitioning between MATLAB-style functions and object-oriented methods, make the following changes:

- `plt.xlabel` → `ax.set_xlabel`
- `plt.ylabel` → `ax.set_ylabel`
- `plt.xlim` → `ax.set_xlim`
- `plt.ylim` → `ax.set_ylim`
- `plt.title` → `ax.set_title`

In the object-oriented interface to plotting, rather than calling these functions individually, it is often more convenient to use the `ax.set` method to set all these properties at once (see Figure 26-14).

```
In [15]: ax = plt.axes()
         ax.plot(x, np.sin(x))
         ax.set(xlim=(0, 10), ylim=(-2, 2),
                xlabel='x', ylabel='sin(x)',
                title='A Simple Plot');
```

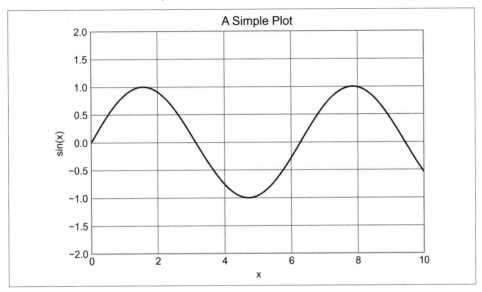

Figure 26-14. Example of using ax.set to set multiple properties at once

## CHAPTER 27

Simple Scatter Plots

Another commonly used plot type is the simple scatter plot, a close cousin of the line plot. Instead of points being joined by line segments, here the points are represented individually with a dot, circle, or other shape. We'll start by setting up the notebook for plotting and importing the packages we will use:

```
In [1]: %matplotlib inline
        import matplotlib.pyplot as plt
        plt.style.use('seaborn-whitegrid')
        import numpy as np
```

Scatter Plots with plt.plot

In the previous chapter we looked at using `plt.plot`/`ax.plot` to produce line plots. It turns out that this same function can produce scatter plots as well (see Figure 27-1).

```
In [2]: x = np.linspace(0, 10, 30)
        y = np.sin(x)

        plt.plot(x, y, 'o', color='black');
```

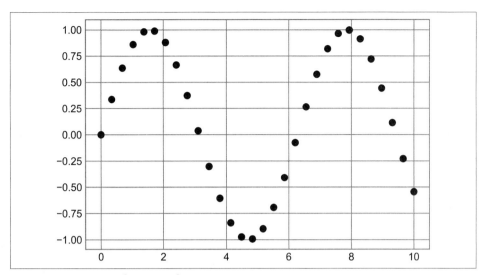

Figure 27-1. Scatter plot example

The third argument in the function call is a character that represents the type of symbol used for the plotting. Just as you can specify options such as `'-'` or `'--'` to control the line style, the marker style has its own set of short string codes. The full list of available symbols can be seen in the documentation of `plt.plot`, or in Matplotlib's online documentation (*https://oreil.ly/tmYIL*). Most of the possibilities are fairly intuitive, and a number of the more common ones are demonstrated here (see Figure 27-2).

```
In [3]: rng = np.random.default_rng(0)
        for marker in ['o', '.', ',', 'x', '+', 'v', '^', '<', '>', 's', 'd']:
            plt.plot(rng.random(2), rng.random(2), marker, color='black',
                    label="marker='{0}'".format(marker))
        plt.legend(numpoints=1, fontsize=13)
        plt.xlim(0, 1.8);
```

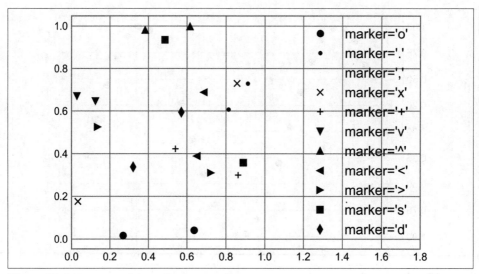

Figure 27-2. Demonstration of point numbers

For even more possibilities, these character codes can be used together with line and color codes to plot points along with a line connecting them (see Figure 27-3).

```
In [4]: plt.plot(x, y, '-ok');
```

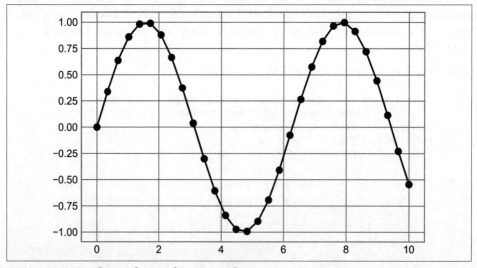

Figure 27-3. Combining line and point markers

Additional keyword arguments to `plt.plot` specify a wide range of properties of the lines and markers, as you can see in Figure 27-4.

```
In [5]: plt.plot(x, y, '-p', color='gray',
                 markersize=15, linewidth=4,
                 markerfacecolor='white',
                 markeredgecolor='gray',
                 markeredgewidth=2)
         plt.ylim(-1.2, 1.2);
```

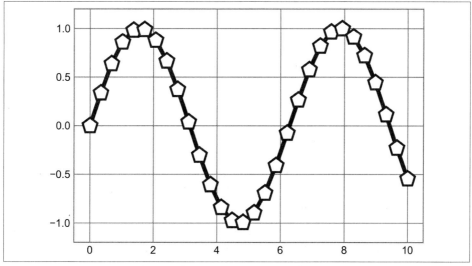

Figure 27-4. Customizing line and point markers

These kinds of options make `plt.plot` the primary workhorse for two-dimensional plots in Matplotlib. For a full description of the options available, refer to the `plt.plot` documentation (*https://oreil.ly/ON1xj*).

Scatter Plots with plt.scatter

A second, more powerful method of creating scatter plots is the `plt.scatter` function, which can be used very similarly to the `plt.plot` function (see Figure 27-5).

```
In [6]: plt.scatter(x, y, marker='o');
```

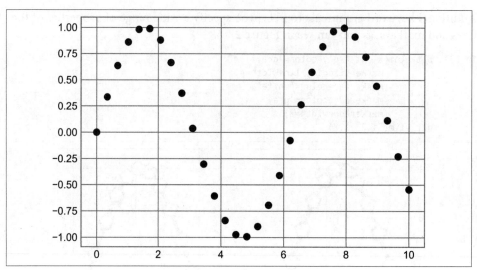

Figure 27-5. A simple scatter plot

The primary difference of `plt.scatter` from `plt.plot` is that it can be used to create scatter plots where the properties of each individual point (size, face color, edge color, etc.) can be individually controlled or mapped to data.

Let's show this by creating a random scatter plot with points of many colors and sizes. In order to better see the overlapping results, we'll also use the `alpha` keyword to adjust the transparency level (see Figure 27-6).

```
In [7]: rng = np.random.default_rng(0)
        x = rng.normal(size=100)
        y = rng.normal(size=100)
        colors = rng.random(100)
        sizes = 1000 * rng.random(100)

        plt.scatter(x, y, c=colors, s=sizes, alpha=0.3)
        plt.colorbar();  # show color scale
```

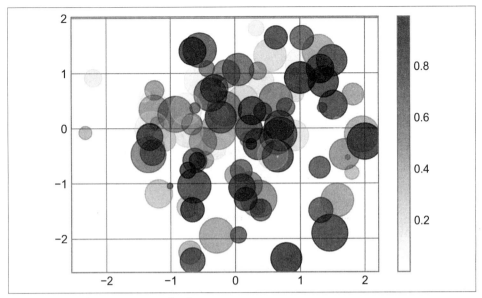

Figure 27-6. Changing size and color in scatter points

Notice that the color argument is automatically mapped to a color scale (shown here by the `colorbar` command), and that the size argument is given in pixels. In this way, the color and size of points can be used to convey information in the visualization, in order to visualize multidimensional data.

For example, we might use the Iris dataset from Scikit-Learn, where each sample is one of three types of flowers that has had the size of its petals and sepals carefully measured (see Figure 27-7).

```
In [8]: from sklearn.datasets import load_iris
        iris = load_iris()
        features = iris.data.T

        plt.scatter(features[0], features[1], alpha=0.4,
                    s=100*features[3], c=iris.target, cmap='viridis')
        plt.xlabel(iris.feature_names[0])
        plt.ylabel(iris.feature_names[1]);
```

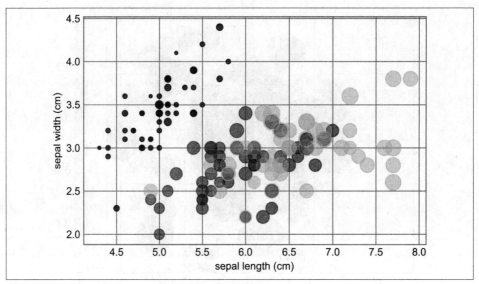

Figure 27-7. Using point properties to encode features of the Iris data[1]

We can see that this scatter plot has given us the ability to simultaneously explore four different dimensions of the data: the (*x, y*) location of each point corresponds to the sepal length and width, the size of the point is related to the petal width, and the color is related to the particular species of flower. Multicolor and multifeature scatter plots like this can be useful for both exploration and presentation of data.

plot Versus scatter: A Note on Efficiency

Aside from the different features available in `plt.plot` and `plt.scatter`, why might you choose to use one over the other? While it doesn't matter as much for small amounts of data, as datasets get larger than a few thousand points, `plt.plot` can be noticeably more efficient than `plt.scatter`. The reason is that `plt.scatter` has the capability to render a different size and/or color for each point, so the renderer must do the extra work of constructing each point individually. With `plt.plot`, on the other hand, the markers for each point are guaranteed to be identical, so the work of determining the appearance of the points is done only once for the entire set of data. For large datasets, this difference can lead to vastly different performance, and for this reason, `plt.plot` should be preferred over `plt.scatter` for large datasets.

---

1 A full-color version of this figure can be found on GitHub (*https://oreil.ly/PDSH_GitHub*).

Visualizing Uncertainties

For any scientific measurement, accurate accounting of uncertainties is nearly as important, if not more so, as accurate reporting of the number itself. For example, imagine that I am using some astrophysical observations to estimate the Hubble Constant, the local measurement of the expansion rate of the Universe. I know that the current literature suggests a value of around 70 (km/s)/Mpc, and I measure a value of 74 (km/s)/Mpc with my method. Are the values consistent? The only correct answer, given this information, is this: there is no way to know.

Suppose I augment this information with reported uncertainties: the current literature suggests a value of 70 ± 2.5 (km/s)/Mpc, and my method has measured a value of 74 ± 5 (km/s)/Mpc. Now are the values consistent? That is a question that can be quantitatively answered.

In visualization of data and results, showing these errors effectively can make a plot convey much more complete information.

Basic Errorbars

One standard way to visualize uncertainties is using an errorbar. A basic errorbar can be created with a single Matplotlib function call, as shown in Figure 27-8.

```
In [1]: %matplotlib inline
        import matplotlib.pyplot as plt
        plt.style.use('seaborn-whitegrid')
        import numpy as np

In [2]: x = np.linspace(0, 10, 50)
        dy = 0.8
        y = np.sin(x) + dy * np.random.randn(50)

        plt.errorbar(x, y, yerr=dy, fmt='.k');
```

Here the `fmt` is a format code controlling the appearance of lines and points, and it has the same syntax as the shorthand used in `plt.plot`, outlined in the previous chapter and earlier in this chapter.

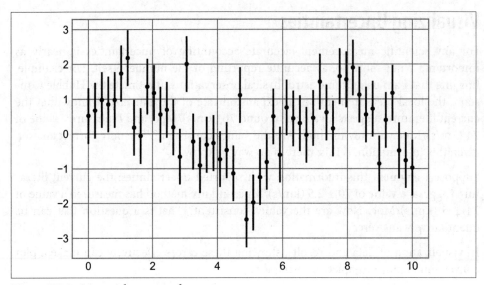

Figure 27-8. An errorbar example

In addition to these basic options, the `errorbar` function has many options to fine-tune the outputs. Using these additional options you can easily customize the aesthetics of your errorbar plot. I often find it helpful, especially in crowded plots, to make the errorbars lighter than the points themselves (see Figure 27-9).

```
In [3]: plt.errorbar(x, y, yerr=dy, fmt='o', color='black',
                      ecolor='lightgray', elinewidth=3, capsize=0);
```

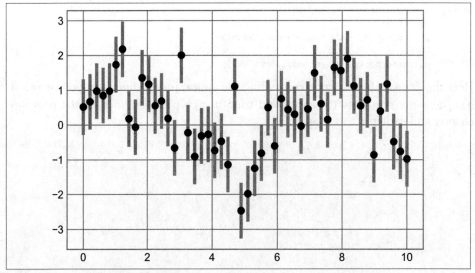

Figure 27-9. Customizing errorbars

In addition to these options, you can also specify horizontal errorbars, one-sided errorbars, and many other variants. For more information on the options available, refer to the docstring of `plt.errorbar`.

Continuous Errors

In some situations it is desirable to show errorbars on continuous quantities. Though Matplotlib does not have a built-in convenience routine for this type of application, it's relatively easy to combine primitives like `plt.plot` and `plt.fill_between` for a useful result.

Here we'll perform a simple *Gaussian process regression*, using the Scikit-Learn API (see Chapter 38 for details). This is a method of fitting a very flexible nonparametric function to data with a continuous measure of the uncertainty. We won't delve into the details of Gaussian process regression at this point, but will focus instead on how you might visualize such a continuous error measurement:

```
In [4]: from sklearn.gaussian_process import GaussianProcessRegressor

        # define the model and draw some data
        model = lambda x: x * np.sin(x)
        xdata = np.array([1, 3, 5, 6, 8])
        ydata = model(xdata)

        # Compute the Gaussian process fit
        gp = GaussianProcessRegressor()
        gp.fit(xdata[:, np.newaxis], ydata)

        xfit = np.linspace(0, 10, 1000)
        yfit, dyfit = gp.predict(xfit[:, np.newaxis], return_std=True)
```

We now have `xfit`, `yfit`, and `dyfit`, which sample the continuous fit to our data. We could pass these to the `plt.errorbar` function as in the previous section, but we don't really want to plot 1,000 points with 1,000 errorbars. Instead, we can use the `plt.fill_between` function with a light color to visualize this continuous error (see Figure 27-10).

```
In [5]: # Visualize the result
        plt.plot(xdata, ydata, 'or')
        plt.plot(xfit, yfit, '-', color='gray')
        plt.fill_between(xfit, yfit - dyfit, yfit + dyfit,
                         color='gray', alpha=0.2)
        plt.xlim(0, 10);
```

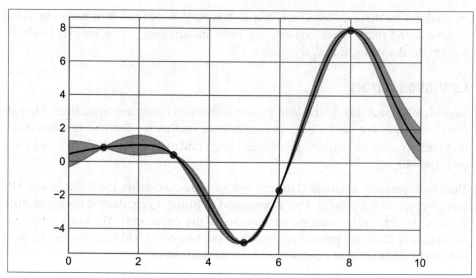

Figure 27-10. Representing continuous uncertainty with filled regions

Take a look at the `fill_between` call signature: we pass an x value, then the lower *y*-bound, then the upper *y*-bound, and the result is that the area between these regions is filled.

The resulting figure gives an intuitive view into what the Gaussian process regression algorithm is doing: in regions near a measured data point, the model is strongly constrained, and this is reflected in the small model uncertainties. In regions far from a measured data point, the model is not strongly constrained, and the model uncertainties increase.

For more information on the options available in `plt.fill_between` (and the closely related `plt.fill` function), see the function docstring or Matplotlib documentation.

Finally, if this seems a bit too low-level for your taste, refer to Chapter 36, where we discuss the Seaborn package, which has a more streamlined API for visualizing this type of continuous errorbar.

## CHAPTER 28

Density and Contour Plots

Sometimes it is useful to display three-dimensional data in two dimensions using contours or color-coded regions. There are three Matplotlib functions that can be helpful for this task: `plt.contour` for contour plots, `plt.contourf` for filled contour plots, and `plt.imshow` for showing images. This chapter looks at several examples of using these. We'll start by setting up the notebook for plotting and importing the functions we will use:

```
In [1]: %matplotlib inline
        import matplotlib.pyplot as plt
        plt.style.use('seaborn-white')
        import numpy as np
```

Visualizing a Three-Dimensional Function

Our first example demonstrates a contour plot using a function $z = f(x, y)$, using the following particular choice for f (we've seen this before in Chapter 8, when we used it as a motivating example for array broadcasting):

```
In [2]: def f(x, y):
            return np.sin(x) ** 10 + np.cos(10 + y * x) * np.cos(x)
```

A contour plot can be created with the `plt.contour` function. It takes three arguments: a grid of x values, a grid of y values, and a grid of z values. The x and y values represent positions on the plot, and the z values will be represented by the contour levels. Perhaps the most straightforward way to prepare such data is to use the `np.meshgrid` function, which builds two-dimensional grids from one-dimensional arrays:

```
In [3]: x = np.linspace(0, 5, 50)
        y = np.linspace(0, 5, 40)
```

```
X, Y = np.meshgrid(x, y)
Z = f(X, Y)
```

Now let's look at this with a standard line-only contour plot (see Figure 28-1).

```
In [4]: plt.contour(X, Y, Z, colors='black');
```

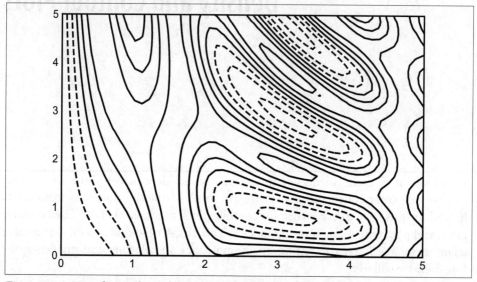

Figure 28-1. Visualizing three-dimensional data with contours

Notice that when a single color is used, negative values are represented by dashed lines and positive values by solid lines. Alternatively, the lines can be color-coded by specifying a colormap with the `cmap` argument. Here we'll also specify that we want more lines to be drawn, at 20 equally spaced intervals within the data range, as shown in Figure 28-2.

```
In [5]: plt.contour(X, Y, Z, 20, cmap='RdGy');
```

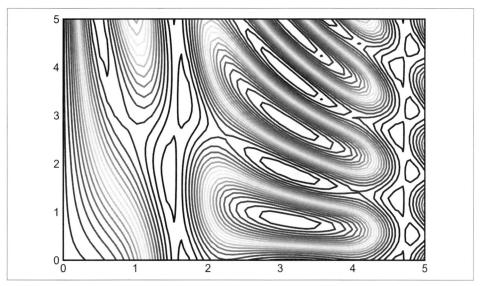

Figure 28-2. Visualizing three-dimensional data with colored contours

Here we chose the `RdGy` (short for *Red–Gray*) colormap, which is a good choice for divergent data: (i.e., data with positive and negative variation around zero). Matplotlib has a wide range of colormaps available, which you can easily browse in IPython by doing a tab completion on the `plt.cm` module:

```
plt.cm.<TAB>
```

Our plot is looking nicer, but the spaces between the lines may be a bit distracting. We can change this by switching to a filled contour plot using the `plt.contourf` function, which uses largely the same syntax as `plt.contour`.

Additionally, we'll add a `plt.colorbar` command, which creates an additional axis with labeled color information for the plot (see Figure 28-3).

```
In [6]: plt.contourf(X, Y, Z, 20, cmap='RdGy')
        plt.colorbar();
```

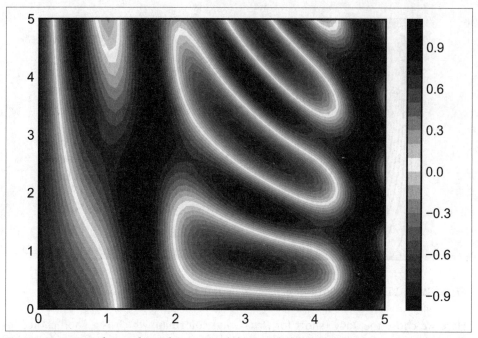

Figure 28-3. Visualizing three-dimensional data with filled contours

The colorbar makes it clear that the black regions are "peaks," while the red regions are "valleys."

One potential issue with this plot is that it is a bit splotchy: the color steps are discrete rather than continuous, which is not always what is desired. This could be remedied by setting the number of contours to a very high number, but this results in a rather inefficient plot: Matplotlib must render a new polygon for each step in the level. A better way to generate a smooth representation is to use the `plt.imshow` function, which offers the `interpolation` argument to generate a smooth two-dimensional representation of the data (see Figure 28-4).

```
In [7]: plt.imshow(Z, extent=[0, 5, 0, 5], origin='lower', cmap='RdGy',
                    interpolation='gaussian', aspect='equal')
        plt.colorbar();
```

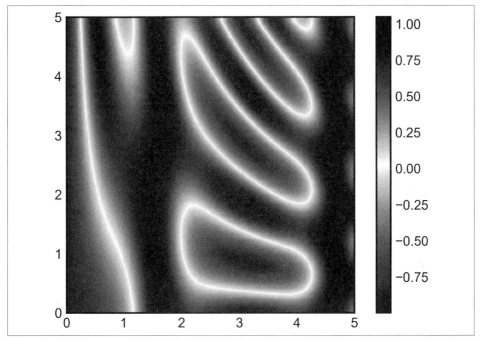

Figure 28-4. Representing three-dimensional data as an image

There are a few potential gotchas with plt.imshow, however:

- It doesn't accept an *x* and *y* grid, so you must manually specify the *extent* [*xmin*, *xmax*, *ymin*, *ymax*] of the image on the plot.
- By default it follows the standard image array definition where the origin is in the upper left, not in the lower left as in most contour plots. This must be changed when showing gridded data.
- It will automatically adjust the axis aspect ratio to match the input data; this can be changed with the aspect argument.

Finally, it can sometimes be useful to combine contour plots and image plots. For example, here we'll use a partially transparent background image (with transparency set via the alpha parameter) and overplot contours with labels on the contours themselves, using the plt.clabel function (see Figure 28-5).

```
In [8]: contours = plt.contour(X, Y, Z, 3, colors='black')
        plt.clabel(contours, inline=True, fontsize=8)

        plt.imshow(Z, extent=[0, 5, 0, 5], origin='lower',
                   cmap='RdGy', alpha=0.5)
        plt.colorbar();
```

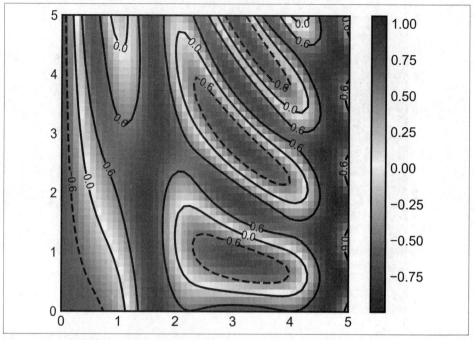

Figure 28-5. Labeled contours on top of an image

The combination of these three functions—`plt.contour`, `plt.contourf`, and `plt.imshow`—gives nearly limitless possibilities for displaying this sort of three-dimensional data within a two-dimensional plot. For more information on the options available in these functions, refer to their docstrings. If you are interested in three-dimensional visualizations of this type of data, see Chapter 35.

Histograms, Binnings, and Density

A simple histogram can be a great first step in understanding a dataset. Earlier, we saw a preview of Matplotlib's histogram function (discussed in Chapter 9), which creates a basic histogram in one line, once the normal boilerplate imports are done (see Figure 28-6).

```
In [1]: %matplotlib inline
        import numpy as np
        import matplotlib.pyplot as plt
        plt.style.use('seaborn-white')

        rng = np.random.default_rng(1701)
        data = rng.normal(size=1000)

In [2]: plt.hist(data);
```

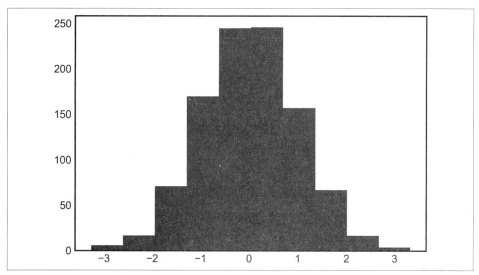

Figure 28-6. A simple histogram

The `hist` function has many options to tune both the calculation and the display; here's an example of a more customized histogram, shown in Figure 28-7.

```
In [3]: plt.hist(data, bins=30, density=True, alpha=0.5,
                histtype='stepfilled', color='steelblue',
                edgecolor='none');
```

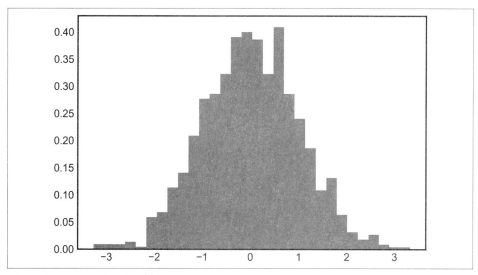

Figure 28-7. A customized histogram

The `plt.hist` docstring has more information on other available customization options. I find this combination of `histtype='stepfilled'` along with some transparency `alpha` to be helpful when comparing histograms of several distributions (see Figure 28-8).

```
In [4]: x1 = rng.normal(0, 0.8, 1000)
        x2 = rng.normal(-2, 1, 1000)
        x3 = rng.normal(3, 2, 1000)

        kwargs = dict(histtype='stepfilled', alpha=0.3, density=True, bins=40)

        plt.hist(x1, **kwargs)
        plt.hist(x2, **kwargs)
        plt.hist(x3, **kwargs);
```

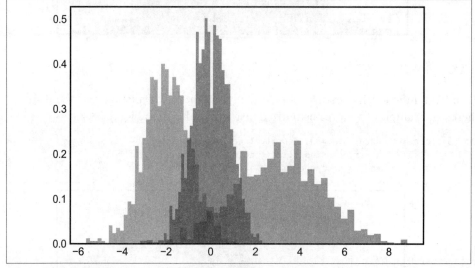

Figure 28-8. Overplotting multiple histograms[1]

If you are interested in computing, but not displaying, the histogram (that is, counting the number of points in a given bin), you can use the `np.histogram` function:

```
In [5]: counts, bin_edges = np.histogram(data, bins=5)
        print(counts)
Out[5]: [ 23 241 491 224  21]
```

---

1 A full-color version of this figure can be found on GitHub (*https://oreil.ly/PDSH_GitHub*).

Two-Dimensional Histograms and Binnings

Just as we create histograms in one dimension by dividing the number line into bins, we can also create histograms in two dimensions by dividing points among two-dimensional bins. We'll take a brief look at several ways to do this. Let's start by defining some data—an x and y array drawn from a multivariate Gaussian distribution:

```
In [6]: mean = [0, 0]
        cov = [[1, 1], [1, 2]]
        x, y = rng.multivariate_normal(mean, cov, 10000).T
```

plt.hist2d: Two-Dimensional Histogram

One straightforward way to plot a two-dimensional histogram is to use Matplotlib's `plt.hist2d` function (see Figure 28-9).

```
In [7]: plt.hist2d(x, y, bins=30)
        cb = plt.colorbar()
        cb.set_label('counts in bin')
```

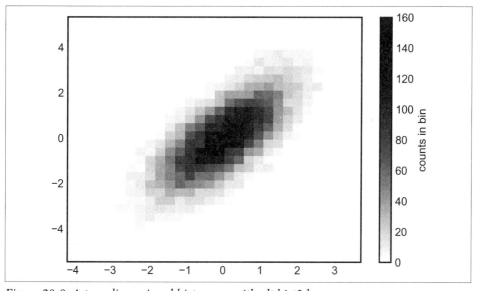

Figure 28-9. A two-dimensional histogram with plt.hist2d

Just like `plt.hist`, `plt.hist2d` has a number of extra options to fine-tune the plot and the binning, which are nicely outlined in the function docstring. Further, just as `plt.hist` has a counterpart in `np.histogram`, `plt.hist2d` has a counterpart in `np.histogram2d`:

```
In [8]: counts, xedges, yedges = np.histogram2d(x, y, bins=30)
        print(counts.shape)
Out[8]: (30, 30)
```

For the generalization of this histogram binning when there are more than two dimensions, see the `np.histogramdd` function.

plt.hexbin: Hexagonal Binnings

The two-dimensional histogram creates a tesselation of squares across the axes. Another natural shape for such a tesselation is the regular hexagon. For this purpose, Matplotlib provides the `plt.hexbin` routine, which represents a two-dimensional dataset binned within a grid of hexagons (see Figure 28-10).

```
In [9]: plt.hexbin(x, y, gridsize=30)
         cb = plt.colorbar(label='count in bin')
```

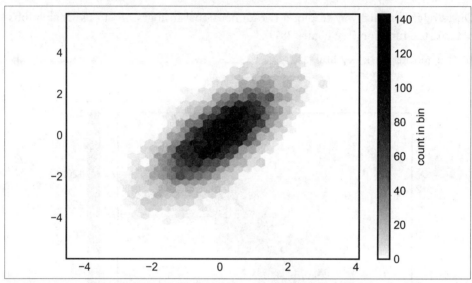

Figure 28-10. A two-dimensional histogram with plt.hexbin

`plt.hexbin` has a number of additional options, including the ability to specify weights for each point and to change the output in each bin to any NumPy aggregate (mean of weights, standard deviation of weights, etc.).

Kernel Density Estimation

Another common method for estimating and representing densities in multiple dimensions is *kernel density estimation* (KDE). This will be discussed more fully in Chapter 49, but for now I'll simply mention that KDE can be thought of as a way to "smear out" the points in space and add up the result to obtain a smooth function. One extremely quick and simple KDE implementation exists in the `scipy.stats` package. Here is a quick example of using KDE (see Figure 28-11).

```
In [10]: from scipy.stats import gaussian_kde

         # fit an array of size [Ndim, Nsamples]
         data = np.vstack([x, y])
         kde = gaussian_kde(data)

         # evaluate on a regular grid
         xgrid = np.linspace(-3.5, 3.5, 40)
         ygrid = np.linspace(-6, 6, 40)
         Xgrid, Ygrid = np.meshgrid(xgrid, ygrid)
         Z = kde.evaluate(np.vstack([Xgrid.ravel(), Ygrid.ravel()]))

         # Plot the result as an image
         plt.imshow(Z.reshape(Xgrid.shape),
                    origin='lower', aspect='auto',
                    extent=[-3.5, 3.5, -6, 6])
         cb = plt.colorbar()
         cb.set_label("density")
```

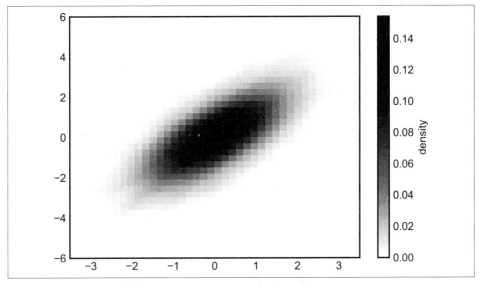

Figure 28-11. A kernel density representation of a distribution

KDE has a smoothing length that effectively slides the knob between detail and smoothness (one example of the ubiquitous bias–variance trade-off). The literature on choosing an appropriate smoothing length is vast; gaussian_kde uses a rule of thumb to attempt to find a nearly optimal smoothing length for the input data.

Other KDE implementations are available within the SciPy ecosystem, each with its own strengths and weaknesses; see, for example, sklearn.neighbors.KernelDensity and statsmodels.nonparametric.KDEMultivariate.

For visualizations based on KDE, using Matplotlib tends to be overly verbose. The Seaborn library, discussed in Chapter 36, provides a much more compact API for creating KDE-based visualizations.

## CHAPTER 29

Customizing Plot Legends

Plot legends give meaning to a visualization, assigning meaning to the various plot elements. We previously saw how to create a simple legend; here we'll take a look at customizing the placement and aesthetics of the legend in Matplotlib.

The simplest legend can be created with the `plt.legend` command, which automatically creates a legend for any labeled plot elements (see Figure 29-1).

```
In [1]: import matplotlib.pyplot as plt
        plt.style.use('seaborn-whitegrid')

In [2]: %matplotlib inline
        import numpy as np

In [3]: x = np.linspace(0, 10, 1000)
        fig, ax = plt.subplots()
        ax.plot(x, np.sin(x), '-b', label='Sine')
        ax.plot(x, np.cos(x), '--r', label='Cosine')
        ax.axis('equal')
        leg = ax.legend()
```

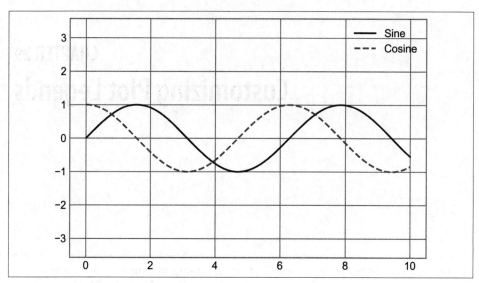

Figure 29-1. A default plot legend

But there are many ways we might want to customize such a legend. For example, we can specify the location and turn on the frame (see Figure 29-2).

```
In [4]: ax.legend(loc='upper left', frameon=True)
        fig
```

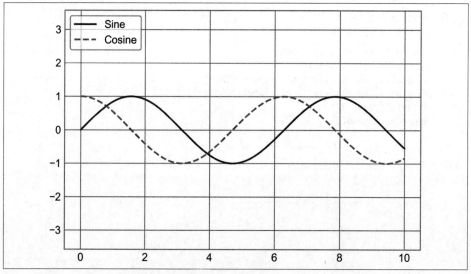

Figure 29-2. A customized plot legend

We can use the ncol command to specify the number of columns in the legend, as shown in Figure 29-3.

```
In [5]: ax.legend(loc='lower center', ncol=2)
        fig
```

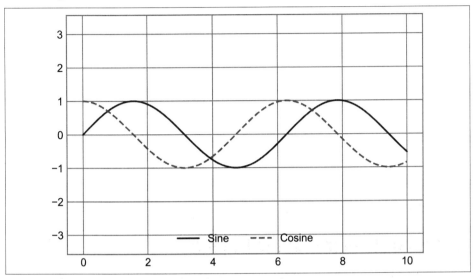

Figure 29-3. A two-column plot legend

And we can use a rounded box (fancybox) or add a shadow, change the transparency (alpha value) of the frame, or change the padding around the text (see Figure 29-4).

```
In [6]: ax.legend(frameon=True, fancybox=True, framealpha=1,
                   shadow=True, borderpad=1)
        fig
```

For more information on available legend options, see the plt.legend docstring.

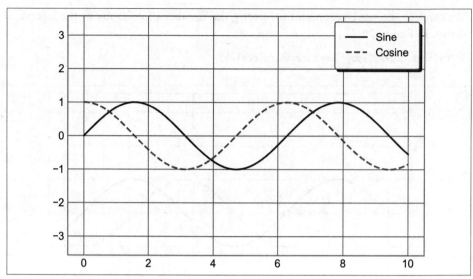

Figure 29-4. A fancybox plot legend

Choosing Elements for the Legend

As we have already seen, by default the legend includes all labeled elements from the plot. If this is not what is desired, we can fine-tune which elements and labels appear in the legend by using the objects returned by plot commands. plt.plot is able to create multiple lines at once, and returns a list of created line instances. Passing any of these to plt.legend will tell it which to identify, along with the labels we'd like to specify (see Figure 29-5).

```
In [7]: y = np.sin(x[:, np.newaxis] + np.pi * np.arange(0, 2, 0.5))
        lines = plt.plot(x, y)

        # lines is a list of plt.Line2D instances
        plt.legend(lines[:2], ['first', 'second'], frameon=True);
```

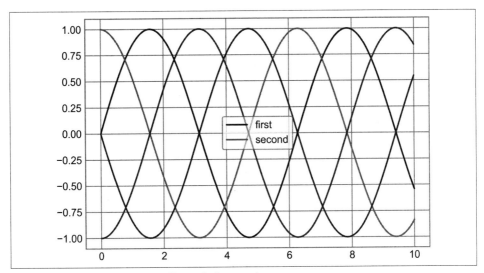

Figure 29-5. Customization of legend elements[1]

I generally find in practice that it is clearer to use the first method, applying labels to the plot elements you'd like to show on the legend (see Figure 29-6).

```
In [8]: plt.plot(x, y[:, 0], label='first')
        plt.plot(x, y[:, 1], label='second')
        plt.plot(x, y[:, 2:])
        plt.legend(frameon=True);
```

Notice that the legend ignores all elements without a label attribute set.

---

1 A full-color version of this figure can be found on GitHub (*https://oreil.ly/PDSH_GitHub*).

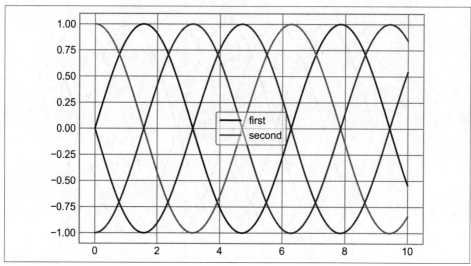

Figure 29-6. Alternative method of customizing legend elements[2]

Legend for Size of Points

Sometimes the legend defaults are not sufficient for the given visualization. For example, perhaps you're using the size of points to mark certain features of the data, and want to create a legend reflecting this. Here is an example where we'll use the size of points to indicate populations of California cities. We'd like a legend that specifies the scale of the sizes of the points, and we'll accomplish this by plotting some labeled data with no entries (see Figure 29-7).

```
In [9]: # Uncomment to download the data
        # url = ('https://raw.githubusercontent.com/jakevdp/
        #         PythonDataScienceHandbook/''master/notebooks/data/
        #         california_cities.csv')
        # !cd data && curl -O {url}

In [10]: import pandas as pd
         cities = pd.read_csv('data/california_cities.csv')

         # Extract the data we're interested in
         lat, lon = cities['latd'], cities['longd']
         population, area = cities['population_total'], cities['area_total_km2']

         # Scatter the points, using size and color but no label
         plt.scatter(lon, lat, label=None,
                     c=np.log10(population), cmap='viridis',
                     s=area, linewidth=0, alpha=0.5)
```

---

2 A full-color version of this figure can be found on GitHub (*https://oreil.ly/PDSH_GitHub*).

```
plt.axis('equal')
plt.xlabel('longitude')
plt.ylabel('latitude')
plt.colorbar(label='log$_{10}$(population)')
plt.clim(3, 7)

# Here we create a legend:
# we'll plot empty lists with the desired size and label
for area in [100, 300, 500]:
    plt.scatter([], [], c='k', alpha=0.3, s=area,
                label=str(area) + ' km$^2$')
plt.legend(scatterpoints=1, frameon=False, labelspacing=1,
           title='City Area')

plt.title('California Cities: Area and Population');
```

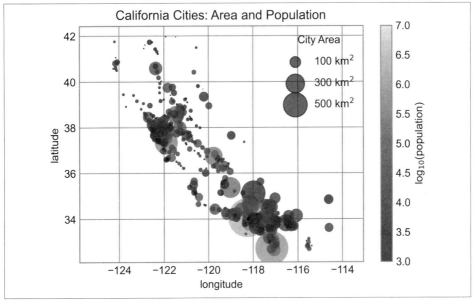

Figure 29-7. Location, geographic size, and population of California cities

The legend will always reference some object that is on the plot, so if we'd like to display a particular shape we need to plot it. In this case, the objects we want (gray circles) are not on the plot, so we fake them by plotting empty lists. Recall that the legend only lists plot elements that have a label specified.

By plotting empty lists, we create labeled plot objects that are picked up by the legend, and now our legend tells us some useful information. This strategy can be useful for creating more sophisticated visualizations.

Multiple Legends

Sometimes when designing a plot you'd like to add multiple legends to the same axes. Unfortunately, Matplotlib does not make this easy: via the standard legend interface, it is only possible to create a single legend for the entire plot. If you try to create a second legend using plt.legend or ax.legend, it will simply override the first one. We can work around this by creating a new legend artist from scratch (Artist is the base class Matplotlib uses for visual attributes), and then using the lower-level ax.add_artist method to manually add the second artist to the plot (see Figure 29-8).

```
In [11]: fig, ax = plt.subplots()

         lines = []
         styles = ['-', '--', '-.', ':']
         x = np.linspace(0, 10, 1000)

         for i in range(4):
             lines += ax.plot(x, np.sin(x - i * np.pi / 2),
                              styles[i], color='black')
         ax.axis('equal')

         # Specify the lines and labels of the first legend
         ax.legend(lines[:2], ['line A', 'line B'], loc='upper right')

         # Create the second legend and add the artist manually
         from matplotlib.legend import Legend
         leg = Legend(ax, lines[2:], ['line C', 'line D'], loc='lower right')
         ax.add_artist(leg);
```

This is a peek into the low-level artist objects that comprise any Matplotlib plot. If you examine the source code of ax.legend (recall that you can do this with within the Jupyter notebook using ax.legend??) you'll see that the function simply consists of some logic to create a suitable Legend artist, which is then saved in the legend_ attribute and added to the figure when the plot is drawn.

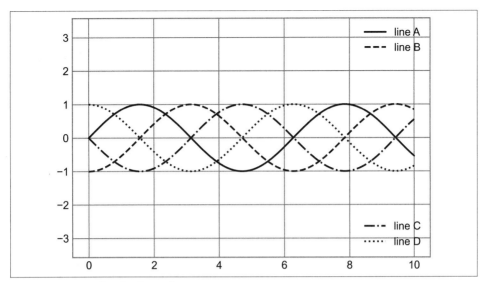

Figure 29-8. A split plot legend

## CHAPTER 30

Customizing Colorbars

Plot legends identify discrete labels of discrete points. For continuous labels based on the color of points, lines, or regions, a labeled colorbar can be a great tool. In Matplotlib, a colorbar is drawn as a separate axes that can provide a key for the meaning of colors in a plot. Because the book is printed in black and white, this chapter has an accompanying online supplement (*https://oreil.ly/PDSH_GitHub*) where you can view the figures in full color. We'll start by setting up the notebook for plotting and importing the functions we will use:

```
In [1]: import matplotlib.pyplot as plt
        plt.style.use('seaborn-white')
```

```
In [2]: %matplotlib inline
        import numpy as np
```

As we have seen several times already, the simplest colorbar can be created with the `plt.colorbar` function (see Figure 30-1).

```
In [3]: x = np.linspace(0, 10, 1000)
        I = np.sin(x) * np.cos(x[:, np.newaxis])

        plt.imshow(I)
        plt.colorbar();
```

 Full-color figures are available in the supplemental materials on GitHub (*https://oreil.ly/PDSH_GitHub*).

We'll now discuss a few ideas for customizing these colorbars and using them effectively in various situations.

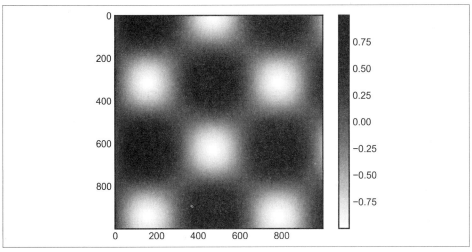

Figure 30-1. A simple colorbar legend

Customizing Colorbars

The colormap can be specified using the `cmap` argument to the plotting function that is creating the visualization (see Figure 30-2).

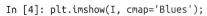

```
In [4]: plt.imshow(I, cmap='Blues');
```

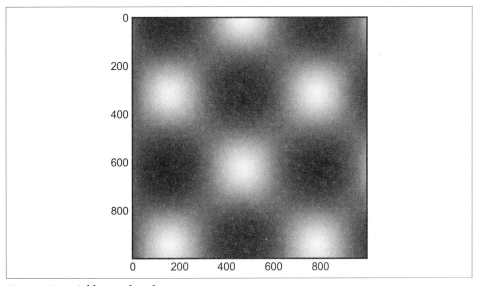

Figure 30-2. A blue-scale colormap

The names of available colormaps are in the `plt.cm` namespace; using IPython's tab completion feature will give you a full list of built-in possibilities:

```
plt.cm.<TAB>
```

But being *able* to choose a colormap is just the first step: more important is how to *decide* among the possibilities! The choice turns out to be much more subtle than you might initially expect.

Choosing the Colormap

A full treatment of color choice within visualizations is beyond the scope of this book, but for entertaining reading on this subject and others, see the article "Ten Simple Rules for Better Figures" (*https://oreil.ly/g4GLV*) by Nicholas Rougier, Michael Droettboom, and Philip Bourne. Matplotlib's online documentation also has an interesting discussion (*https://oreil.ly/Ll1ir*) of colormap choice.

Broadly, you should be aware of three different categories of colormaps:

Sequential colormaps
These are made up of one continuous sequence of colors (e.g., `binary` or `viridis`).

Divergent colormaps
These usually contain two distinct colors, which show positive and negative deviations from a mean (e.g., `RdBu` or `PuOr`).

Qualitative colormaps
These mix colors with no particular sequence (e.g., `rainbow` or `jet`).

The `jet` colormap, which was the default in Matplotlib prior to version 2.0, is an example of a qualitative colormap. Its status as the default was quite unfortunate, because qualitative maps are often a poor choice for representing quantitative data. Among the problems is the fact that qualitative maps usually do not display any uniform progression in brightness as the scale increases.

We can see this by converting the `jet` colorbar into black and white (see Figure 30-3).

```
In [5]: from matplotlib.colors import LinearSegmentedColormap

        def grayscale_cmap(cmap):
            """Return a grayscale version of the given colormap"""
            cmap = plt.cm.get_cmap(cmap)
            colors = cmap(np.arange(cmap.N))

            # Convert RGBA to perceived grayscale luminance
            # cf. http://alienryderflex.com/hsp.html
            RGB_weight = [0.299, 0.587, 0.114]
            luminance = np.sqrt(np.dot(colors[:, :3] ** 2, RGB_weight))
```

```
    colors[:, :3] = luminance[:, np.newaxis]

    return LinearSegmentedColormap.from_list(
        cmap.name + "_gray", colors, cmap.N)


def view_colormap(cmap):
    """Plot a colormap with its grayscale equivalent"""
    cmap = plt.cm.get_cmap(cmap)
    colors = cmap(np.arange(cmap.N))

    cmap = grayscale_cmap(cmap)
    grayscale = cmap(np.arange(cmap.N))

    fig, ax = plt.subplots(2, figsize=(6, 2),
                           subplot_kw=dict(xticks=[], yticks=[]))
    ax[0].imshow([colors], extent=[0, 10, 0, 1])
    ax[1].imshow([grayscale], extent=[0, 10, 0, 1])
```

```
In [6]: view_colormap('jet')
```

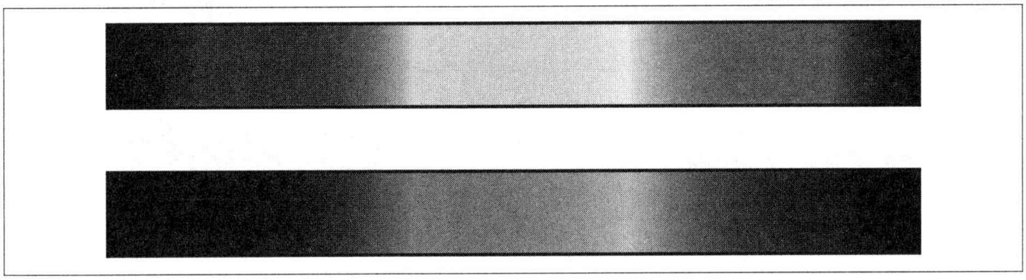

Figure 30-3. The jet colormap and its uneven luminance scale

Notice the bright stripes in the grayscale image. Even in full color, this uneven brightness means that the eye will be drawn to certain portions of the color range, which will potentially emphasize unimportant parts of the dataset. It's better to use a colormap such as viridis (the default as of Matplotlib 2.0), which is specifically constructed to have an even brightness variation across the range; thus, it not only plays well with our color perception, but also will translate well to grayscale printing (see Figure 30-4).

```
In [7]: view_colormap('viridis')
```

Figure 30-4. The viridis colormap and its even luminance scale

For other situations, such as showing positive and negative deviations from some mean, dual-color colorbars such as RdBu (*Red–Blue*) are helpful. However, as you can see in Figure 30-5, it's important to note that the positive/negative information will be lost upon translation to grayscale!

```
In [8]: view_colormap('RdBu')
```

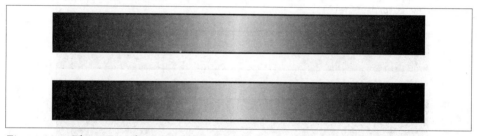

Figure 30-5. The RdBu colormap and its luminance

We'll see examples of using some of these colormaps as we continue.

There are a large number of colormaps available in Matplotlib; to see a list of them, you can use IPython to explore the plt.cm submodule. For a more principled approach to colors in Python, you can refer to the tools and documentation within the Seaborn library (see Chapter 36).

Color Limits and Extensions

Matplotlib allows for a large range of colorbar customization. The colorbar itself is simply an instance of plt.Axes, so all of the axes and tick formatting tricks we've seen so far are applicable. The colorbar has some interesting flexibility: for example, we can narrow the color limits and indicate the out-of-bounds values with a triangular arrow at the top and bottom by setting the extend property. This might come in handy, for example, if displaying an image that is subject to noise (see Figure 30-6).

```
In [9]: # make noise in 1% of the image pixels
        speckles = (np.random.random(I.shape) < 0.01)
        I[speckles] = np.random.normal(0, 3, np.count_nonzero(speckles))
```

```
plt.figure(figsize=(10, 3.5))

plt.subplot(1, 2, 1)
plt.imshow(I, cmap='RdBu')
plt.colorbar()

plt.subplot(1, 2, 2)
plt.imshow(I, cmap='RdBu')
plt.colorbar(extend='both')
plt.clim(-1, 1)
```

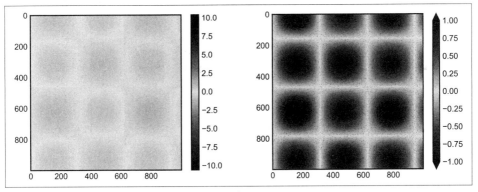

Figure 30-6. Specifying colormap extensions[1]

Notice that in the left panel, the default color limits respond to the noisy pixels, and the range of the noise completely washes out the pattern we are interested in. In the right panel, we manually set the color limits and add extensions to indicate values that are above or below those limits. The result is a much more useful visualization of our data.

Discrete Colorbars

Colormaps are by default continuous, but sometimes you'd like to represent discrete values. The easiest way to do this is to use the `plt.cm.get_cmap` function and pass the name of a suitable colormap along with the number of desired bins (see Figure 30-7).

```
In [10]: plt.imshow(I, cmap=plt.cm.get_cmap('Blues', 6))
         plt.colorbar(extend='both')
         plt.clim(-1, 1);
```

---

1 A full-size version of this figure can be found on GitHub (*https://oreil.ly/PDSH_GitHub*).

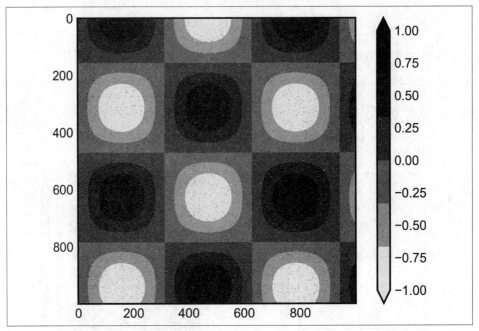

Figure 30-7. A discretized colormap

The discrete version of a colormap can be used just like any other colormap.

Example: Handwritten Digits

As an example of where this can be applied, let's look at an interesting visualization of some handwritten digits from the digits dataset, included in Scikit-Learn; it consists of nearly 2,000 8 × 8 thumbnails showing various handwritten digits.

For now, let's start by downloading the digits dataset and visualizing several of the example images with plt.imshow (see Figure 30-8).

```
In [11]: # load images of the digits 0 through 5 and visualize several of them
         from sklearn.datasets import load_digits
         digits = load_digits(n_class=6)

         fig, ax = plt.subplots(8, 8, figsize=(6, 6))
         for i, axi in enumerate(ax.flat):
             axi.imshow(digits.images[i], cmap='binary')
             axi.set(xticks=[], yticks=[])
```

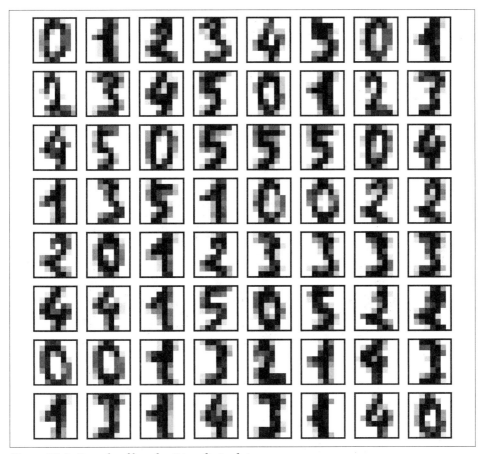

Figure 30-8. Sample of handwritten digits data

Because each digit is defined by the hue of its 64 pixels, we can consider each digit to be a point lying in 64-dimensional space: each dimension represents the brightness of one pixel. Visualizing such high-dimensional data can be difficult, but one way to approach this task is to use a *dimensionality reduction* technique such as manifold learning to reduce the dimensionality of the data while maintaining the relationships of interest. Dimensionality reduction is an example of unsupervised machine learning, and we will discuss it in more detail in Chapter 37.

Deferring the discussion of these details, let's take a look at a two-dimensional manifold learning projection of the digits data (see Chapter 46 for details):

```
In [12]: # project the digits into 2 dimensions using Isomap
         from sklearn.manifold import Isomap
         iso = Isomap(n_components=2, n_neighbors=15)
         projection = iso.fit_transform(digits.data)
```

We'll use our discrete colormap to view the results, setting the `ticks` and `clim` to improve the aesthetics of the resulting colorbar (see Figure 30-9).

```
In [13]: # plot the results
         plt.scatter(projection[:, 0], projection[:, 1], lw=0.1,
                     c=digits.target, cmap=plt.cm.get_cmap('plasma', 6))
         plt.colorbar(ticks=range(6), label='digit value')
         plt.clim(-0.5, 5.5)
```

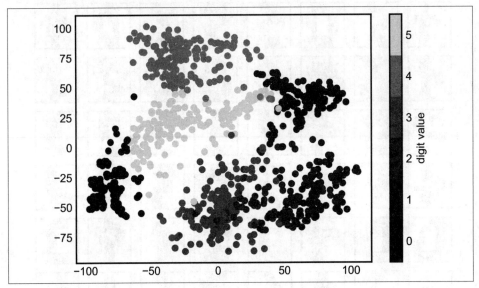

Figure 30-9. Manifold embedding of handwritten digit pixels

The projection also gives us some insights on the relationships within the dataset: for example, the ranges of 2 and 3 nearly overlap in this projection, indicating that some handwritten 2s and 3s are difficult to distinguish, and may be more likely to be confused by an automated classification algorithm. Other values, like 0 and 1, are more distantly separated, and may be less likely to be confused.

We'll return to manifold learning and digit classification in Part V.

## CHAPTER 31

Multiple Subplots

Sometimes it is helpful to compare different views of data side by side. To this end, Matplotlib has the concept of *subplots*: groups of smaller axes that can exist together within a single figure. These subplots might be insets, grids of plots, or other more complicated layouts. In this chapter we'll explore four routines for creating subplots in Matplotlib. We'll start by importing the packages we will use:

```
In [1]: %matplotlib inline
        import matplotlib.pyplot as plt
        plt.style.use('seaborn-white')
        import numpy as np
```

plt.axes: Subplots by Hand

The most basic method of creating an axes is to use the `plt.axes` function. As we've seen previously, by default this creates a standard axes object that fills the entire figure. `plt.axes` also takes an optional argument that is a list of four numbers in the figure coordinate system ([*left, bottom, width, height*]), which ranges from 0 at the bottom left of the figure to 1 at the top right of the figure.

For example, we might create an inset axes at the top-right corner of another axes by setting the *x* and *y* position to 0.65 (that is, starting at 65% of the width and 65% of the height of the figure) and the *x* and *y* extents to 0.2 (that is, the size of the axes is 20% of the width and 20% of the height of the figure). Figure 31-1 shows the result:

```
In [2]: ax1 = plt.axes()  # standard axes
        ax2 = plt.axes([0.65, 0.65, 0.2, 0.2])
```

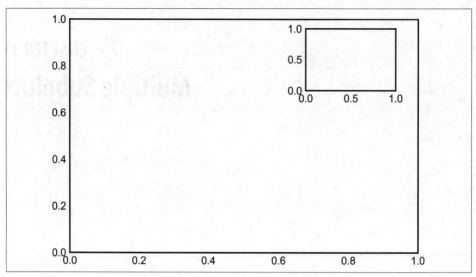

Figure 31-1. Example of an inset axes

The equivalent of this command within the object-oriented interface is `fig.add_axes`. Let's use this to create two vertically stacked axes, as seen in Figure 31-2.

```
In [3]: fig = plt.figure()
        ax1 = fig.add_axes([0.1, 0.5, 0.8, 0.4],
                           xticklabels=[], ylim=(-1.2, 1.2))
        ax2 = fig.add_axes([0.1, 0.1, 0.8, 0.4],
                           ylim=(-1.2, 1.2))

        x = np.linspace(0, 10)
        ax1.plot(np.sin(x))
        ax2.plot(np.cos(x));
```

We now have two axes (the top with no tick labels) that are just touching: the bottom of the upper panel (at position 0.5) matches the top of the lower panel (at position 0.1 + 0.4).

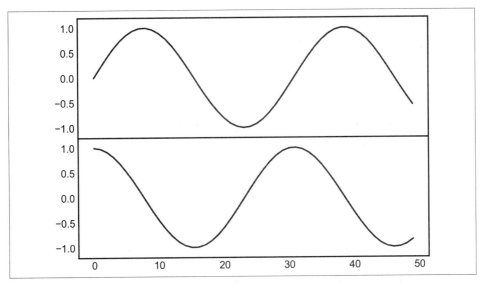

Figure 31-2. Vertically stacked axes example

plt.subplot: Simple Grids of Subplots

Aligned columns or rows of subplots are a common enough need that Matplotlib has several convenience routines that make them easy to create. The lowest level of these is `plt.subplot`, which creates a single subplot within a grid. As you can see, this command takes three integer arguments—the number of rows, the number of columns, and the index of the plot to be created in this scheme, which runs from the upper left to the bottom right (see Figure 31-3).

```
In [4]: for i in range(1, 7):
            plt.subplot(2, 3, i)
            plt.text(0.5, 0.5, str((2, 3, i)),
                     fontsize=18, ha='center')
```

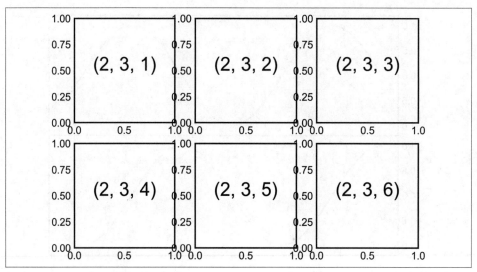

Figure 31-3. A plt.subplot example

The command `plt.subplots_adjust` can be used to adjust the spacing between these plots. The following code uses the equivalent object-oriented command, `fig.add_subplot`; Figure 31-4 shows the result:

```
In [5]: fig = plt.figure()
        fig.subplots_adjust(hspace=0.4, wspace=0.4)
        for i in range(1, 7):
            ax = fig.add_subplot(2, 3, i)
            ax.text(0.5, 0.5, str((2, 3, i)),
                    fontsize=18, ha='center')
```

Here we've used the `hspace` and `wspace` arguments of `plt.subplots_adjust`, which specify the spacing along the height and width of the figure, in units of the subplot size (in this case, the space is 40% of the subplot width and height).

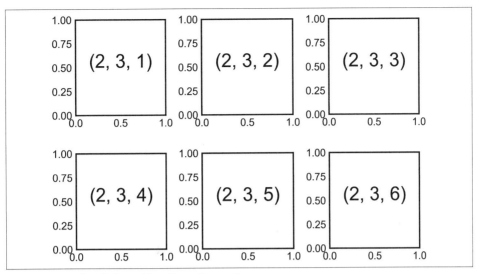

Figure 31-4. plt.subplot with adjusted margins

plt.subplots: The Whole Grid in One Go

The approach just described quickly becomes tedious when creating a large grid of subplots, especially if you'd like to hide the x- and y-axis labels on the inner plots. For this purpose, `plt.subplots` is the easier tool to use (note the s at the end of `subplots`). Rather than creating a single subplot, this function creates a full grid of subplots in a single line, returning them in a NumPy array. The arguments are the number of rows and number of columns, along with optional keywords `sharex` and `sharey`, which allow you to specify the relationships between different axes.

Let's create a 2 × 3 grid of subplots, where all axes in the same row share their y-axis scale, and all axes in the same column share their x-axis scale (see Figure 31-5).

```
In [6]: fig, ax = plt.subplots(2, 3, sharex='col', sharey='row')
```

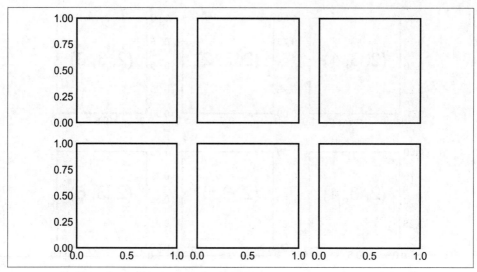

Figure 31-5. Shared x and y axes in plt.subplots

By specifying `sharex` and `sharey`, we've automatically removed inner labels on the grid to make the plot cleaner. The resulting grid of axes instances is returned within a NumPy array, allowing for convenient specification of the desired axes using standard array indexing notation (see Figure 31-6).

```
In [7]: # axes are in a two-dimensional array, indexed by [row, col]
        for i in range(2):
            for j in range(3):
                ax[i, j].text(0.5, 0.5, str((i, j)),
                              fontsize=18, ha='center')
        fig
```

In comparison to `plt.subplot`, `plt.subplots` is more consistent with Python's conventional zero-based indexing, whereas `plt.subplot` uses MATLAB-style one-based indexing.

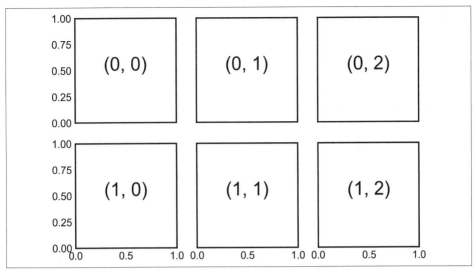

Figure 31-6. Identifying plots in a subplot grid

plt.GridSpec: More Complicated Arrangements

To go beyond a regular grid to subplots that span multiple rows and columns, plt.GridSpec is the best tool. plt.GridSpec does not create a plot by itself; it is rather a convenient interface that is recognized by the plt.subplot command. For example, a GridSpec for a grid of two rows and three columns with some specified width and height space looks like this:

```
In [8]: grid = plt.GridSpec(2, 3, wspace=0.4, hspace=0.3)
```

From this we can specify subplot locations and extents using the familiar Python slicing syntax (see Figure 31-7).

```
In [9]: plt.subplot(grid[0, 0])
        plt.subplot(grid[0, 1:])
        plt.subplot(grid[1, :2])
        plt.subplot(grid[1, 2]);
```

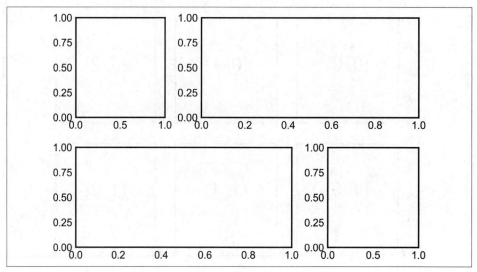

Figure 31-7. Irregular subplots with `plt.GridSpec`

This type of flexible grid alignment has a wide range of uses. I most often use it when creating multiaxes histogram plots like the ones shown in Figure 31-8.

```
In [10]: # Create some normally distributed data
         mean = [0, 0]
         cov = [[1, 1], [1, 2]]
         rng = np.random.default_rng(1701)
         x, y = rng.multivariate_normal(mean, cov, 3000).T

         # Set up the axes with GridSpec
         fig = plt.figure(figsize=(6, 6))
         grid = plt.GridSpec(4, 4, hspace=0.2, wspace=0.2)
         main_ax = fig.add_subplot(grid[:-1, 1:])
         y_hist = fig.add_subplot(grid[:-1, 0], xticklabels=[], sharey=main_ax)
         x_hist = fig.add_subplot(grid[-1, 1:], yticklabels=[], sharex=main_ax)

         # Scatter points on the main axes
         main_ax.plot(x, y, 'ok', markersize=3, alpha=0.2)

         # Histogram on the attached axes
         x_hist.hist(x, 40, histtype='stepfilled',
                     orientation='vertical', color='gray')
         x_hist.invert_yaxis()

         y_hist.hist(y, 40, histtype='stepfilled',
                     orientation='horizontal', color='gray')
         y_hist.invert_xaxis()
```

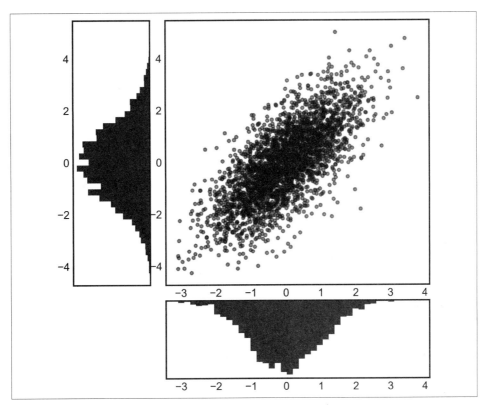

Figure 31-8. Visualizing multidimensional distributions with `plt.GridSpec`

This type of distribution plotted alongside its margins is common enough that it has its own plotting API in the Seaborn package; see Chapter 36 for more details.

## CHAPTER 32

Text and Annotation

Creating a good visualization involves guiding the reader so that the figure tells a story. In some cases, this story can be told in an entirely visual manner, without the need for added text, but in others, small textual cues and labels are necessary. Perhaps the most basic types of annotations you will use are axes labels and titles, but the options go beyond this. Let's take a look at some data and how we might visualize and annotate it to help convey interesting information. We'll start by setting up the notebook for plotting and importing the functions we will use:

```
In [1]: %matplotlib inline
        import matplotlib.pyplot as plt
        import matplotlib as mpl
        plt.style.use('seaborn-whitegrid')
        import numpy as np
        import pandas as pd
```

Example: Effect of Holidays on US Births

Let's return to some data we worked with earlier, in "Example: Birthrate Data" on page 180, where we generated a plot of average births over the course of the calendar year. We'll start with the same cleaning procedure we used there, and plot the results (see Figure 32-1).

```
In [2]: # shell command to download the data:
        # !cd data && curl -O \
        #    https://raw.githubusercontent.com/jakevdp/data-CDCbirths/master/
        #    births.csv
```

```
In [3]: from datetime import datetime

        births = pd.read_csv('data/births.csv')

        quartiles = np.percentile(births['births'], [25, 50, 75])
```

```
mu, sig = quartiles[1], 0.74 * (quartiles[2] - quartiles[0])
births = births.query('(births > @mu - 5 * @sig) &
                       (births < @mu + 5 * @sig)')

births['day'] = births['day'].astype(int)

births.index = pd.to_datetime(10000 * births.year +
                              100 * births.month +
                              births.day, format='%Y%m%d')
births_by_date = births.pivot_table('births',
                                    [births.index.month, births.index.day])
births_by_date.index = [datetime(2012, month, day)
                        for (month, day) in births_by_date.index]
```

```
In [4]: fig, ax = plt.subplots(figsize=(12, 4))
        births_by_date.plot(ax=ax);
```

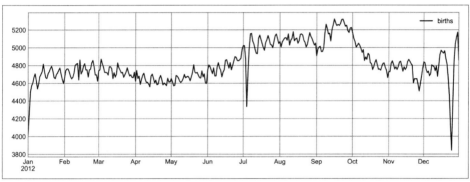

Figure 32-1. Average daily births by date[1]

When we're visualizing data like this, it is often useful to annotate certain features of
the plot to draw the reader's attention. This can be done manually with the `plt.text`/
`ax.text` functions, which will place text at a particular *x/y* value (see Figure 32-2).

```
In [5]: fig, ax = plt.subplots(figsize=(12, 4))
        births_by_date.plot(ax=ax)

        # Add labels to the plot
        style = dict(size=10, color='gray')

        ax.text('2012-1-1', 3950, "New Year's Day", **style)
        ax.text('2012-7-4', 4250, "Independence Day", ha='center', **style)
        ax.text('2012-9-4', 4850, "Labor Day", ha='center', **style)
        ax.text('2012-10-31', 4600, "Halloween", ha='right', **style)
        ax.text('2012-11-25', 4450, "Thanksgiving", ha='center', **style)
        ax.text('2012-12-25', 3850, "Christmas ", ha='right', **style)
```

---

1 A full-size version of this figure can be found on GitHub (*https://oreil.ly/PDSH_GitHub*).

```
# Label the axes
ax.set(title='USA births by day of year (1969-1988)',
       ylabel='average daily births')

# Format the x-axis with centered month labels
ax.xaxis.set_major_locator(mpl.dates.MonthLocator())
ax.xaxis.set_minor_locator(mpl.dates.MonthLocator(bymonthday=15))
ax.xaxis.set_major_formatter(plt.NullFormatter())
ax.xaxis.set_minor_formatter(mpl.dates.DateFormatter('%h'));
```

Figure 32-2. Annotated average daily births by date[2]

The `ax.text` method takes an *x* position, a *y* position, a string, and then optional keywords specifying the color, size, style, alignment, and other properties of the text. Here we used `ha='right'` and `ha='center'`, where `ha` is short for *horizontal alignment*. See the docstrings of `plt.text` and `mpl.text.Text` for more information on the available options.

Transforms and Text Position

In the previous example, we anchored our text annotations to data locations. Sometimes it's preferable to anchor the text to a fixed position on the axes or figure, independent of the data. In Matplotlib, this is done by modifying the *transform*.

Matplotlib makes use of a few different coordinate systems: a data point at $(x, y) = (1, 1)$ corresponds to a certain location on the axes or figure, which in turn corresponds to a particular pixel on the screen. Mathematically, transforming between such coordinate systems is relatively straightforward, and Matplotlib has a well-developed set of tools that it uses internally to perform these transforms (these tools can be explored in the `matplotlib.transforms` submodule).

---

2 A full-size version of this figure can be found on GitHub (*https://oreil.ly/PDSH_GitHub*).

A typical user rarely needs to worry about the details of the transforms, but it is help-ful knowledge to have when considering the placement of text on a figure. There are three predefined transforms that can be useful in this situation:

`ax.transData`
: Transform associated with data coordinates

`ax.transAxes`
: Transform associated with the axes (in units of axes dimensions)

`fig.transFigure`
: Transform associated with the figure (in units of figure dimensions)

Let's look at an example of drawing text at various locations using these transforms (see Figure 32-3).

```
In [6]: fig, ax = plt.subplots(facecolor='lightgray')
        ax.axis([0, 10, 0, 10])

        # transform=ax.transData is the default, but we'll specify it anyway
        ax.text(1, 5, ". Data: (1, 5)", transform=ax.transData)
        ax.text(0.5, 0.1, ". Axes: (0.5, 0.1)", transform=ax.transAxes)
        ax.text(0.2, 0.2, ". Figure: (0.2, 0.2)", transform=fig.transFigure);
```

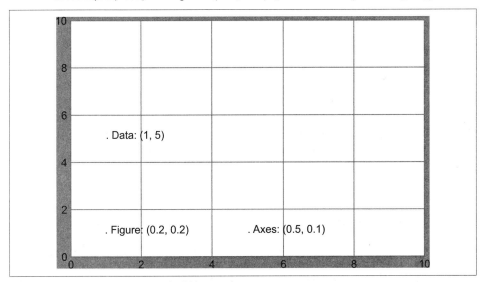

Figure 32-3. Comparing Matplotlib's coordinate systems

Matplotlib's default text alignment is such that the "." at the beginning of each string will approximately mark the specified coordinate location.

The `transData` coordinates give the usual data coordinates associated with the x- and y-axis labels. The `transAxes` coordinates give the location from the bottom-left

corner of the axes (the white box), as a fraction of the total axes size. The `transFig`
`ure` coordinates are similar, but specify the position from the bottom-left corner of
the figure (the gray box) as a fraction of the total figure size.

Notice now that if we change the axes limits, it is only the `transData` coordinates that
will be affected, while the others remain stationary (see Figure 32-4).

```
In [7]: ax.set_xlim(0, 2)
        ax.set_ylim(-6, 6)
        fig
```

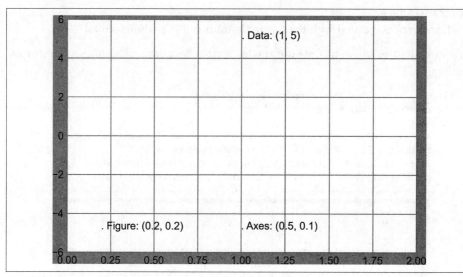

Figure 32-4. Comparing Matplotlib's coordinate systems

This behavior can be seen more clearly by changing the axes limits interactively: if
you are executing this code in a notebook, you can make that happen by changing
`%matplotlib inline` to `%matplotlib notebook` and using each plot's menu to interact with the plot.

Arrows and Annotation

Along with tickmarks and text, another useful annotation mark is the simple arrow.

While there is a `plt.arrow` function available, I wouldn't suggest using it: the arrows
it creates are SVG objects that will be subject to the varying aspect ratio of your plots,
making it tricky to get them right. Instead, I'd suggest using the `plt.annotate` function, which creates some text and an arrow and allows the arrows to be very flexibly
specified.

Here is a demonstration of `annotate` with several of its options (see Figure 32-5).

```
In [8]: fig, ax = plt.subplots()

        x = np.linspace(0, 20, 1000)
        ax.plot(x, np.cos(x))
        ax.axis('equal')

        ax.annotate('local maximum', xy=(6.28, 1), xytext=(10, 4),
                    arrowprops=dict(facecolor='black', shrink=0.05))

        ax.annotate('local minimum', xy=(5 * np.pi, -1), xytext=(2, -6),
                    arrowprops=dict(arrowstyle="->",
                                    connectionstyle="angle3,angleA=0,angleB=-90"));
```

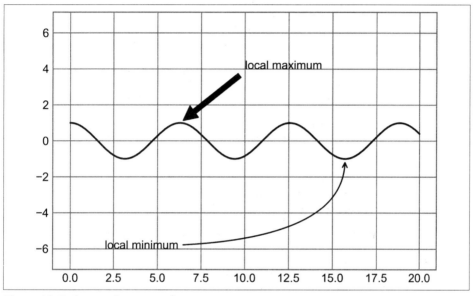

Figure 32-5. Annotation examples

The arrow style is controlled through the `arrowprops` dictionary, which has numerous options available. These options are well documented in Matplotlib's online documentation, so rather than repeating them here it is probably more useful to show some examples. Let's demonstrate several of the possible options using the birthrate plot from before (see Figure 32-6).

```
In [9]: fig, ax = plt.subplots(figsize=(12, 4))
        births_by_date.plot(ax=ax)

        # Add labels to the plot
        ax.annotate("New Year's Day", xy=('2012-1-1', 4100), xycoords='data',
                    xytext=(50, -30), textcoords='offset points',
                    arrowprops=dict(arrowstyle="->",
                                    connectionstyle="arc3,rad=-0.2"))
```

```
ax.annotate("Independence Day", xy=('2012-7-4', 4250),  xycoords='data',
            bbox=dict(boxstyle="round", fc="none", ec="gray"),
            xytext=(10, -40), textcoords='offset points', ha='center',
            arrowprops=dict(arrowstyle="->"))

ax.annotate('Labor Day Weekend', xy=('2012-9-4', 4850), xycoords='data',
            ha='center', xytext=(0, -20), textcoords='offset points')
ax.annotate('', xy=('2012-9-1', 4850), xytext=('2012-9-7', 4850),
            xycoords='data', textcoords='data',
            arrowprops={'arrowstyle': '|-|',widthA=0.2,widthB=0.2', })

ax.annotate('Halloween', xy=('2012-10-31', 4600),  xycoords='data',
            xytext=(-80, -40), textcoords='offset points',
            arrowprops=dict(arrowstyle="fancy",
                            fc="0.6", ec="none",
                            connectionstyle="angle3,angleA=0,angleB=-90"))

ax.annotate('Thanksgiving', xy=('2012-11-25', 4500),  xycoords='data',
            xytext=(-120, -60), textcoords='offset points',
            bbox=dict(boxstyle="round4,pad=.5", fc="0.9"),
            arrowprops=dict(
                arrowstyle="->",
                connectionstyle="angle,angleA=0,angleB=80,rad=20"))


ax.annotate('Christmas', xy=('2012-12-25', 3850),  xycoords='data',
            xytext=(-30, 0), textcoords='offset points',
            size=13, ha='right', va="center",
            bbox=dict(boxstyle="round", alpha=0.1),
            arrowprops=dict(arrowstyle="wedge,tail_width=0.5", alpha=0.1));

# Label the axes
ax.set(title='USA births by day of year (1969-1988)',
       ylabel='average daily births')

# Format the x-axis with centered month labels
ax.xaxis.set_major_locator(mpl.dates.MonthLocator())
ax.xaxis.set_minor_locator(mpl.dates.MonthLocator(bymonthday=15))
ax.xaxis.set_major_formatter(plt.NullFormatter())
ax.xaxis.set_minor_formatter(mpl.dates.DateFormatter('%h'));

ax.set_ylim(3600, 5400);
```

The variety of options make `annotate` powerful and flexible: you can create nearly any arrow style you wish. Unfortunately, it also means that these sorts of features often must be manually tweaked, a process that can be very time-consuming when producing publication-quality graphics! Finally, I'll note that the preceding mix of styles is by no means best practice for presenting data, but rather is included as a demonstration of some of the available options.

More discussion and examples of available arrow and annotation styles can be found in the Matplotlib Annotations tutorial (*https://oreil.ly/abuPw*).

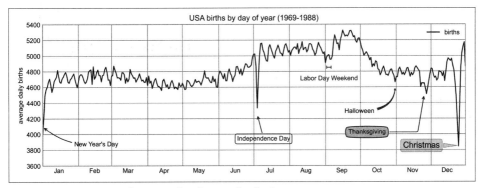

Figure 32-6. Annotated average birth rates by day[3]

---

3 A full-size version of this figure can be found on GitHub (*https://oreil.ly/PDSH_GitHub*).

## CHAPTER 33

Customizing Ticks

Matplotlib's default tick locators and formatters are designed to be generally sufficient in many common situations, but are in no way optimal for every plot. This chapter will give several examples of adjusting the tick locations and formatting for the particular plot type you're interested in.

Before we go into examples, however, let's talk a bit more about the object hierarchy of Matplotlib plots. Matplotlib aims to have a Python object representing everything that appears on the plot: for example, recall that the `Figure` is the bounding box within which plot elements appear. Each Matplotlib object can also act as a container of subobjects: for example, each `Figure` can contain one or more `Axes` objects, each of which in turn contains other objects representing plot contents.

The tickmarks are no exception. Each axes has attributes `xaxis` and `yaxis`, which in turn have attributes that contain all the properties of the lines, ticks, and labels that make up the axes.

Major and Minor Ticks

Within each axes, there is the concept of a *major* tickmark, and a *minor* tickmark. As the names imply, major ticks are usually bigger or more pronounced, while minor ticks are usually smaller. By default, Matplotlib rarely makes use of minor ticks, but one place you can see them is within logarithmic plots (see Figure 33-1).

```
In [1]: import matplotlib.pyplot as plt
        plt.style.use('classic')
        import numpy as np

        %matplotlib inline
```

```
In [2]: ax = plt.axes(xscale='log', yscale='log')
        ax.set(xlim=(1, 1E3), ylim=(1, 1E3))
        ax.grid(True);
```

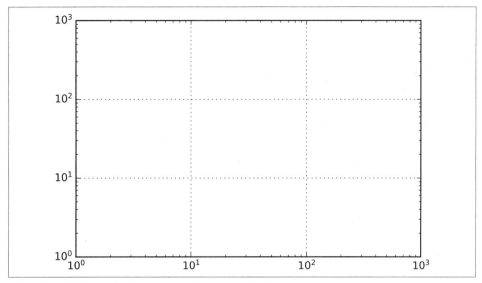

Figure 33-1. Example of logarithmic scales and labels

In this chart each major tick shows a large tickmark, label, and gridline, while each minor tick shows a smaller tickmark with no label or gridline.

These tick properties—locations and labels, that is—can be customized by setting the formatter and locator objects of each axis. Let's examine these for the x-axis of the just-shown plot:

```
In [3]: print(ax.xaxis.get_major_locator())
        print(ax.xaxis.get_minor_locator())
Out[3]: <matplotlib.ticker.LogLocator object at 0x1129b9370>
        <matplotlib.ticker.LogLocator object at 0x1129aaf70>

In [4]: print(ax.xaxis.get_major_formatter())
        print(ax.xaxis.get_minor_formatter())
Out[4]: <matplotlib.ticker.LogFormatterSciNotation object at 0x1129aaa00>
        <matplotlib.ticker.LogFormatterSciNotation object at 0x1129aac10>
```

We see that both major and minor tick labels have their locations specified by a LogLocator (which makes sense for a logarithmic plot). Minor ticks, though, have their labels formatted by a NullFormatter: this says that no labels will be shown.

We'll now look at a few examples of setting these locators and formatters for various plots.

Hiding Ticks or Labels

Perhaps the most common tick/label formatting operation is the act of hiding ticks or labels. This can be done using `plt.NullLocator` and `plt.NullFormatter`, as shown here (see Figure 33-2).

```
In [5]: ax = plt.axes()
        rng = np.random.default_rng(1701)
        ax.plot(rng.random(50))
        ax.grid()

        ax.yaxis.set_major_locator(plt.NullLocator())
        ax.xaxis.set_major_formatter(plt.NullFormatter())
```

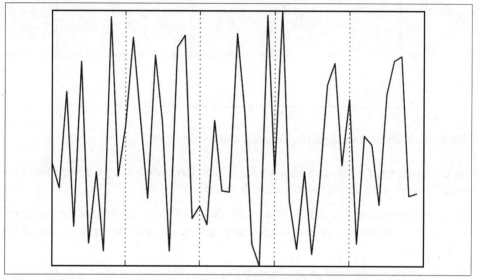

Figure 33-2. Plot with hidden tick labels (x-axis) and hidden ticks (y-axis)

We've removed the labels (but kept the ticks/gridlines) from the x-axis, and removed the ticks (and thus the labels and gridlines as well) from the y-axis. Having no ticks at all can be useful in many situations—for example, when you want to show a grid of images. For instance, consider Figure 33-3, which includes images of different faces, an example often used in supervised machine learning problems (see, for example, Chapter 43):

```
In [6]: fig, ax = plt.subplots(5, 5, figsize=(5, 5))
        fig.subplots_adjust(hspace=0, wspace=0)

        # Get some face data from Scikit-Learn
        from sklearn.datasets import fetch_olivetti_faces
        faces = fetch_olivetti_faces().images
```

```
for i in range(5):
    for j in range(5):
        ax[i, j].xaxis.set_major_locator(plt.NullLocator())
        ax[i, j].yaxis.set_major_locator(plt.NullLocator())
        ax[i, j].imshow(faces[10 * i + j], cmap='binary_r')
```

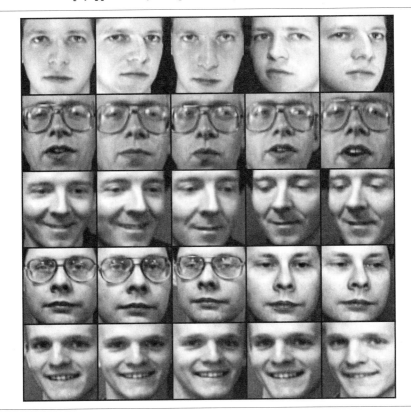

Figure 33-3. Hiding ticks within image plots

Each image is shown in its own axes, and we've set the tick locators to null because the tick values (pixel numbers in this case) do not convey relevant information for this particular visualization.

Reducing or Increasing the Number of Ticks

One common problem with the default settings is that smaller subplots can end up with crowded labels. We can see this in the plot grid shown here (see Figure 33-4).

```
In [7]: fig, ax = plt.subplots(4, 4, sharex=True, sharey=True)
```

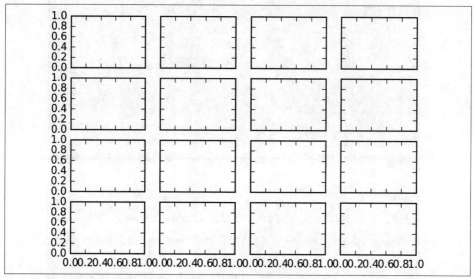

Figure 33-4. A default plot with crowded ticks

Particularly for the x-axis ticks, the numbers nearly overlap, making them quite difficult to decipher. One way to adjust this is with `plt.MaxNLocator`, which allows us to specify the maximum number of ticks that will be displayed. Given this maximum number, Matplotlib will use internal logic to choose the particular tick locations (see Figure 33-5).

```
In [8]: # For every axis, set the x and y major locator
        for axi in ax.flat:
            axi.xaxis.set_major_locator(plt.MaxNLocator(3))
            axi.yaxis.set_major_locator(plt.MaxNLocator(3))
        fig
```

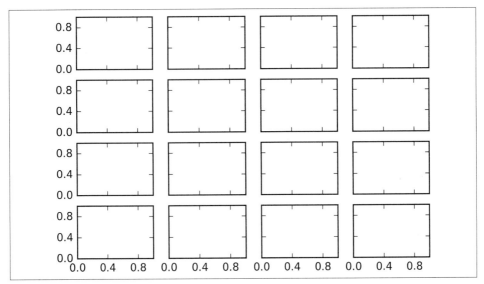

Figure 33-5. Customizing the number of ticks

This makes things much cleaner. If you want even more control over the locations of regularly spaced ticks, you might also use `plt.MultipleLocator`, which we'll discuss in the following section.

Fancy Tick Formats

Matplotlib's default tick formatting can leave a lot to be desired: it works well as a broad default, but sometimes you'd like to do something different. Consider this plot of a sine and a cosine curve (see Figure 33-6).

```
In [9]: # Plot a sine and cosine curve
        fig, ax = plt.subplots()
        x = np.linspace(0, 3 * np.pi, 1000)
        ax.plot(x, np.sin(x), lw=3, label='Sine')
        ax.plot(x, np.cos(x), lw=3, label='Cosine')

        # Set up grid, legend, and limits
        ax.grid(True)
        ax.legend(frameon=False)
        ax.axis('equal')
        ax.set_xlim(0, 3 * np.pi);
```

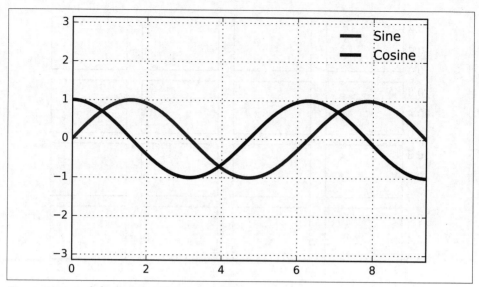

Figure 33-6. A default plot with integer ticks

Full-color figures are available in the supplemental materials on GitHub (*https://oreil.ly/PDSH_GitHub*).

There are a couple of changes we might like to make here. First, it's more natural for this data to space the ticks and gridlines in multiples of π. We can do this by setting a `MultipleLocator`, which locates ticks at a multiple of the number we provide. For good measure, we'll add both major and minor ticks in multiples of $\pi/2$ and $\pi/4$ (see Figure 33-7).

```
In [10]: ax.xaxis.set_major_locator(plt.MultipleLocator(np.pi / 2))
         ax.xaxis.set_minor_locator(plt.MultipleLocator(np.pi / 4))
         fig
```

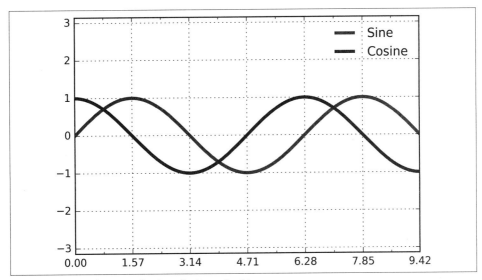

Figure 33-7. Ticks at multiples of π/2 and π/4

But now these tick labels look a little bit silly: we can see that they are multiples of π, but the decimal representation does not immediately convey this. To fix this, we can change the tick formatter. There's no built-in formatter for what we want to do, so we'll instead use plt.FuncFormatter, which accepts a user-defined function giving fine-grained control over the tick outputs (see Figure 33-8).

```
In [11]: def format_func(value, tick_number):
             # find number of multiples of pi/2
             N = int(np.round(2 * value / np.pi))
             if N == 0:
                 return "0"
             elif N == 1:
                 return r"$\pi/2$"
             elif N == 2:
                 return r"$\pi$"
             elif N % 2 > 0:
                 return rf"${N}\pi/2$"
             else:
                 return rf"${N // 2}\pi$"

         ax.xaxis.set_major_formatter(plt.FuncFormatter(format_func))
         fig
```

This is much better! Notice that we've made use of Matplotlib's LaTeX support, specified by enclosing the string within dollar signs. This is very convenient for display of mathematical symbols and formulae: in this case, "π" is rendered as the Greek character π.

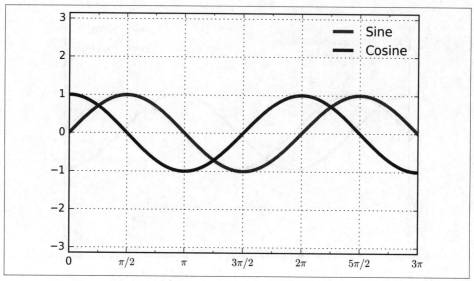

Figure 33-8. Ticks with custom labels

Summary of Formatters and Locators

We've seen a couple of the available formatters and locators; I'll conclude this chapter by listing all of the built-in locator options (Table 33-1) and formatter options (Table 33-2). For more information on of these, refer to the docstrings or to the Matplotlib documentation. Each of the following is available in the `plt` namespace.

Table 33-1. Matplotlib locator options

Locator class	Description
NullLocator	No ticks
FixedLocator	Tick locations are fixed
IndexLocator	Locator for index plots (e.g., where x = range(len(y)))
LinearLocator	Evenly spaced ticks from min to max
LogLocator	Logarithmically spaced ticks from min to max
MultipleLocator	Ticks and range are a multiple of base
MaxNLocator	Finds up to a max number of ticks at nice locations
AutoLocator	(Default) MaxNLocator with simple defaults
AutoMinorLocator	Locator for minor ticks

Table 33-2. Matplotlib formatter options

Formatter class	Description
NullFormatter	No labels on the ticks
IndexFormatter	Set the strings from a list of labels
FixedFormatter	Set the strings manually for the labels
FuncFormatter	User-defined function sets the labels
FormatStrFormatter	Use a format string for each value
ScalarFormatter	Default formatter for scalar values
LogFormatter	Default formatter for log axes

We'll see further examples of these throughout the remainder of the book.

## CHAPTER 34

Customizing Matplotlib: Configurations and Stylesheets

While many of the topics covered in previous chapters involve adjusting the style of plot elements one by one, Matplotlib also offers mechanisms to adjust the overall style of a chart all at once. In this chapter we'll walk through some of Matplotlib's runtime configuration (*rc*) options, and take a look at the *stylesheets* feature, which contains some nice sets of default configurations.

Plot Customization by Hand

Throughout this part of the book, you've seen how it is possible to tweak individual plot settings to end up with something that looks a little nicer than the default. It's also possible to do these customizations for each individual plot. For example, here is a fairly drab default histogram, shown in Figure 34-1.

```
In [1]: import matplotlib.pyplot as plt
        plt.style.use('classic')
        import numpy as np

        %matplotlib inline

In [2]: x = np.random.randn(1000)
        plt.hist(x);
```

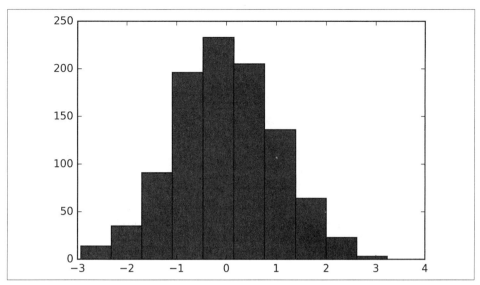

Figure 34-1. A histogram in Matplotlib's default style

We can adjust this by hand to make it a much more visually pleasing plot, as you can see in Figure 34-2.

```
In [3]: # use a gray background
        fig = plt.figure(facecolor='white')
        ax = plt.axes(facecolor='#E6E6E6')
        ax.set_axisbelow(True)

        # draw solid white gridlines
        plt.grid(color='w', linestyle='solid')

        # hide axis spines
        for spine in ax.spines.values():
            spine.set_visible(False)

        # hide top and right ticks
        ax.xaxis.tick_bottom()
        ax.yaxis.tick_left()

        # lighten ticks and labels
        ax.tick_params(colors='gray', direction='out')
        for tick in ax.get_xticklabels():
            tick.set_color('gray')
        for tick in ax.get_yticklabels():
            tick.set_color('gray')

        # control face and edge color of histogram
        ax.hist(x, edgecolor='#E6E6E6', color='#EE6666');
```

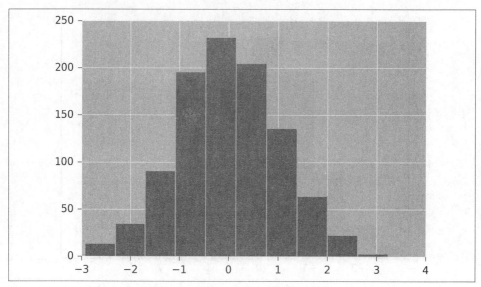

Figure 34-2. A histogram with manual customizations

This looks better, and you may recognize the look as inspired by that of the R language's `ggplot` visualization package. But this took a whole lot of effort! We definitely do not want to have to do all that tweaking each time we create a plot. Fortunately, there is a way to adjust these defaults once in a way that will work for all plots.

Changing the Defaults: rcParams

Each time Matplotlib loads, it defines a runtime configuration containing the default styles for every plot element you create. This configuration can be adjusted at any time using the `plt.rc` convenience routine. Let's see how we can modify the rc parameters so that our default plot will look similar to what we did before.

We can use the `plt.rc` function to change some of these settings:

```
In [4]: from matplotlib import cycler
        colors = cycler('color',
                        ['#EE6666', '#3388BB', '#9988DD',
                         '#EECC55', '#88BB44', '#FFBBBB'])
        plt.rc('figure', facecolor='white')
        plt.rc('axes', facecolor='#E6E6E6', edgecolor='none',
               axisbelow=True, grid=True, prop_cycle=colors)
        plt.rc('grid', color='w', linestyle='solid')
        plt.rc('xtick', direction='out', color='gray')
        plt.rc('ytick', direction='out', color='gray')
        plt.rc('patch', edgecolor='#E6E6E6')
        plt.rc('lines', linewidth=2)
```

With these settings defined, we can now create a plot and see our settings in action (see Figure 34-3).

```
In [5]: plt.hist(x);
```

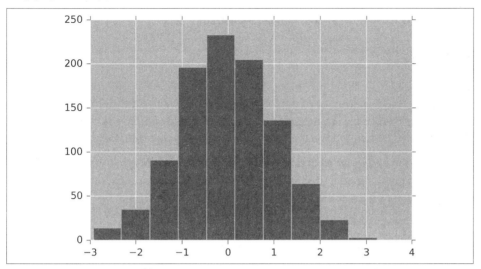

Figure 34-3. A customized histogram using rc settings

Let's see what simple line plots look like with these rc parameters (see Figure 34-4).

```
In [6]: for i in range(4):
            plt.plot(np.random.rand(10))
```

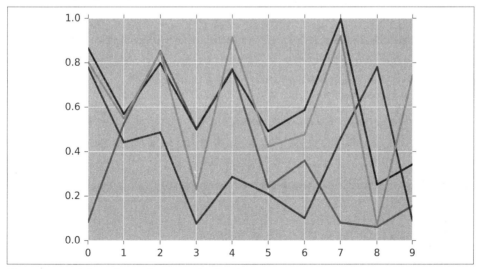

Figure 34-4. A line plot with customized styles

For charts viewed onscreen rather than printed, I find this much more aesthetically pleasing than the default styling. If you disagree with my aesthetic sense, the good news is that you can adjust the rc parameters to suit your own tastes! Optionally, these settings can be saved in a *.matplotlibrc* file, which you can read about in the Matplotlib documentation (*https://oreil.ly/UwM2u*).

Stylesheets

A newer mechanism for adjusting overall chart styles is via Matplotlib's `style` module, which includes a number of default stylesheets, as well as the ability to create and package your own styles. These stylesheets are formatted similarly to the *.matplotlibrc* files mentioned earlier, but must be named with a *.mplstyle* extension.

Even if you don't go as far as creating your own style, you may find what you're looking for in the built-in stylesheets. `plt.style.available` contains a list of the available styles—here I'll list only the first five for brevity:

```
In [7]: plt.style.available[:5]
Out[7]: ['Solarize_Light2', '_classic_test_patch', 'bmh', 'classic',
        >'dark_background']
```

The standard way to switch to a stylesheet is to call `style.use`:

```
    plt.style.use('stylename')
```

But keep in mind that this will change the style for the rest of the Python session! Alternatively, you can use the style context manager, which sets a style temporarily:

```
    with plt.style.context('stylename'):
        make_a_plot()
```

To demonstrate these styles, let's create a function that will make two basic types of plot:

```
In [8]: def hist_and_lines():
            np.random.seed(0)
            fig, ax = plt.subplots(1, 2, figsize=(11, 4))
            ax[0].hist(np.random.randn(1000))
            for i in range(3):
                ax[1].plot(np.random.rand(10))
            ax[1].legend(['a', 'b', 'c'], loc='lower left')
```

We'll use this to explore how these plots look using the various built-in styles.

 Full-color figures are available in the supplemental materials on GitHub (*https://oreil.ly/PDSH_GitHub*).

Default Style

Matplotlib's default style was updated in the version 2.0 release; let's look at this first (see Figure 34-5).

```
In [9]: with plt.style.context('default'):
            hist_and_lines()
```

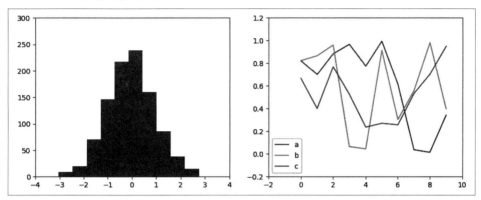

Figure 34-5. Matplotlib's default style

FiveThiryEight Style

The fivethirtyeight style mimics the graphics found on the popular FiveThirtyEight website (*https://fivethirtyeight.com*). As you can see in Figure 34-6, it is typified by bold colors, thick lines, and transparent axes:

```
In [10]: with plt.style.context('fivethirtyeight'):
             hist_and_lines()
```

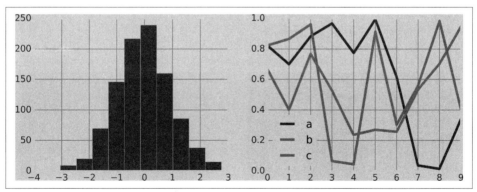

Figure 34-6. The fivethirtyeight style

ggplot Style

The `ggplot` package in the R language is a popular visualization tool among data scientists. Matplotlib's `ggplot` style mimics the default styles from that package (see Figure 34-7).

```
In [11]: with plt.style.context('ggplot'):
             hist_and_lines()
```

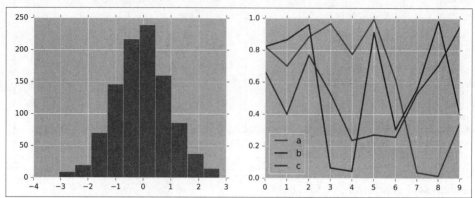

Figure 34-7. The `ggplot` style

Bayesian Methods for Hackers Style

There is a neat short online book called *Probabilistic Programming and Bayesian Methods for Hackers* (*https://oreil.ly/9JIb7*) by Cameron Davidson-Pilon that features figures created with Matplotlib, and uses a nice set of rc parameters to create a consistent and visually appealing style throughout the book. This style is reproduced in the bmh stylesheet (see Figure 34-8).

```
In [12]: with plt.style.context('bmh'):
             hist_and_lines()
```

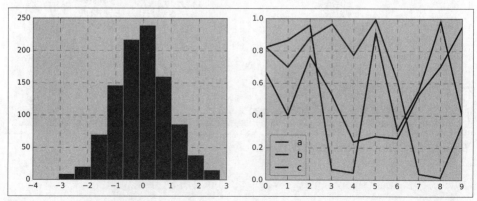

Figure 34-8. The bmh style

Dark Background Style

For figures used within presentations, it is often useful to have a dark rather than light background. The `dark_background` style provides this (see Figure 34-9).

```
In [13]: with plt.style.context('dark_background'):
             hist_and_lines()
```

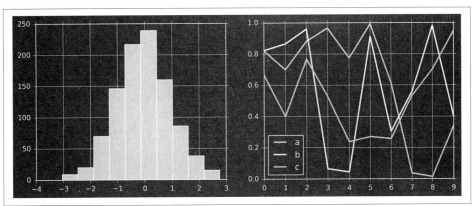

Figure 34-9. The `dark_background` style

Grayscale Style

You might find yourself preparing figures for a print publication that does not accept color figures. For this, the `grayscale` style (see Figure 34-10) can be useful.

```
In [14]: with plt.style.context('grayscale'):
             hist_and_lines()
```

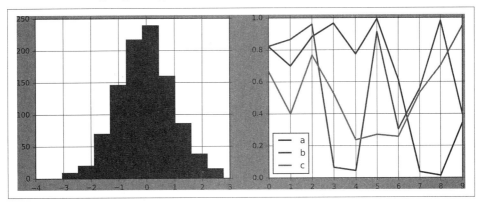

Figure 34-10. The `grayscale` style

Seaborn Style

Matplotlib also has several stylesheets inspired by the Seaborn library (discussed more fully in Chapter 36). I've found these settings to be very nice, and tend to use them as defaults in my own data exploration (see Figure 34-11).

```
In [15]: with plt.style.context('seaborn-whitegrid'):
             hist_and_lines()
```

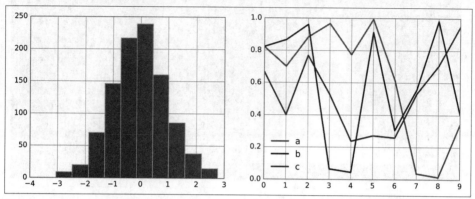

Figure 34-11. The seaborn plotting style

Take some time to explore the built-in options and find one that appeals to you! Throughout this book, I will generally use one or more of these style conventions when creating plots.

## CHAPTER 35

Three-Dimensional Plotting in Matplotlib

Matplotlib was initially designed with only two-dimensional plotting in mind. Around the time of the 1.0 release, some three-dimensional plotting utilities were built on top of Matplotlib's two-dimensional display, and the result is a convenient (if somewhat limited) set of tools for three-dimensional data visualization. Three-dimensional plots are enabled by importing the `mplot3d` toolkit, included with the main Matplotlib installation:

```
In [1]: from mpl_toolkits import mplot3d
```

Once this submodule is imported, a three-dimensional axes can be created by passing the keyword `projection='3d'` to any of the normal axes creation routines, as shown here (see Figure 35-1).

```
In [2]: %matplotlib inline
        import numpy as np
        import matplotlib.pyplot as plt
```

```
In [3]: fig = plt.figure()
        ax = plt.axes(projection='3d')
```

With this three-dimensional axes enabled, we can now plot a variety of three-dimensional plot types. Three-dimensional plotting is one of the functionalities that benefits immensely from viewing figures interactively rather than statically, in the notebook; recall that to use interactive figures, you can use `%matplotlib notebook` rather than `%matplotlib inline` when running this code.

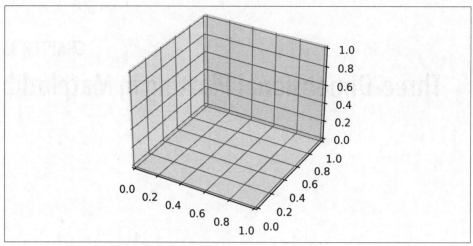

Figure 35-1. An empty three-dimensional axes

Three-Dimensional Points and Lines

The most basic three-dimensional plot is a line or collection of scatter plots created from sets of (x, y, z) triples. In analogy with the more common two-dimensional plots discussed earlier, these can be created using the `ax.plot3D` and `ax.scatter3D` functions. The call signature for these is nearly identical to that of their two-dimensional counterparts, so you can refer to Chapters 26 and 27 for more information on controlling the output. Here we'll plot a trigonometric spiral, along with some points drawn randomly near the line (see Figure 35-2).

```
In [4]: ax = plt.axes(projection='3d')

        # Data for a three-dimensional line
        zline = np.linspace(0, 15, 1000)
        xline = np.sin(zline)
        yline = np.cos(zline)
        ax.plot3D(xline, yline, zline, 'gray')

        # Data for three-dimensional scattered points
        zdata = 15 * np.random.random(100)
        xdata = np.sin(zdata) + 0.1 * np.random.randn(100)
        ydata = np.cos(zdata) + 0.1 * np.random.randn(100)
        ax.scatter3D(xdata, ydata, zdata, c=zdata, cmap='Greens');
```

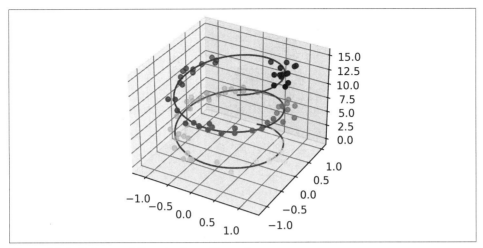

Figure 35-2. Points and lines in three dimensions

Notice that scatter points have their transparency adjusted to give a sense of depth on the page. While the three-dimensional effect is sometimes difficult to see within a static image, an interactive view can lead to some nice intuition about the layout of the points.

Three-Dimensional Contour Plots

Analogous to the contour plots we explored in Chapter 28, `mplot3d` contains tools to create three-dimensional relief plots using the same inputs. Like `ax.contour`, `ax.con tour3D` requires all the input data to be in the form of two-dimensional regular grids, with the *z* data evaluated at each point. Here we'll show a three-dimensional contour diagram of a three-dimensional sinusoidal function (see Figure 35-3).

```
In [5]: def f(x, y):
            return np.sin(np.sqrt(x ** 2 + y ** 2))

        x = np.linspace(-6, 6, 30)
        y = np.linspace(-6, 6, 30)

        X, Y = np.meshgrid(x, y)
        Z = f(X, Y)
In [6]: fig = plt.figure()
        ax = plt.axes(projection='3d')
        ax.contour3D(X, Y, Z, 40, cmap='binary')
        ax.set_xlabel('x')
        ax.set_ylabel('y')
        ax.set_zlabel('z');
```

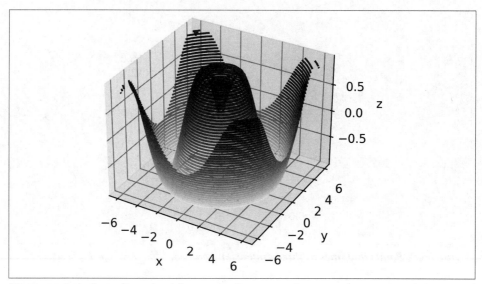

Figure 35-3. A three-dimensional contour plot

Sometimes the default viewing angle is not optimal, in which case we can use the `view_init` method to set the elevation and azimuthal angles. In the following example, visualized in Figure 35-4, we'll use an elevation of 60 degrees (that is, 60 degrees above the x-y plane) and an azimuth of 35 degrees (that is, rotated 35 degrees counter-clockwise about the z-axis):

```
In [7]: ax.view_init(60, 35)
        fig
```

Again, note that this type of rotation can be accomplished interactively by clicking and dragging when using one of Matplotlib's interactive backends.

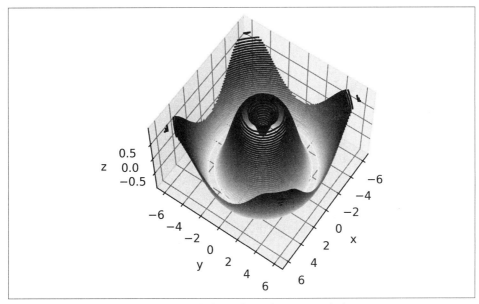

Figure 35-4. Adjusting the view angle for a three-dimensional plot

Wireframes and Surface Plots

Two other types of three-dimensional plots that work on gridded data are wireframes and surface plots. These take a grid of values and project it onto the specified three-dimensional surface, and can make the resulting three-dimensional forms quite easy to visualize. Here's an example of using a wireframe (see Figure 35-5).

```
In [8]: fig = plt.figure()
        ax = plt.axes(projection='3d')
        ax.plot_wireframe(X, Y, Z)
        ax.set_title('wireframe');
```

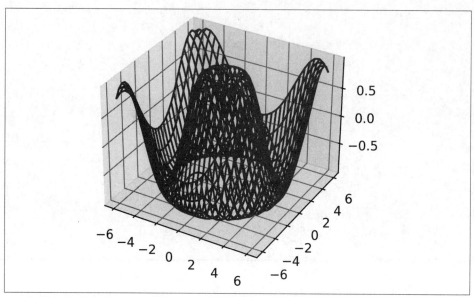

Figure 35-5. A wireframe plot

A surface plot is like a wireframe plot, but each face of the wireframe is a filled polygon. Adding a colormap to the filled polygons can aid perception of the topology of the surface being visualized, as you can see in Figure 35-6.

```
In [9]: ax = plt.axes(projection='3d')
        ax.plot_surface(X, Y, Z, rstride=1, cstride=1,
                        cmap='viridis', edgecolor='none')
        ax.set_title('surface');
```

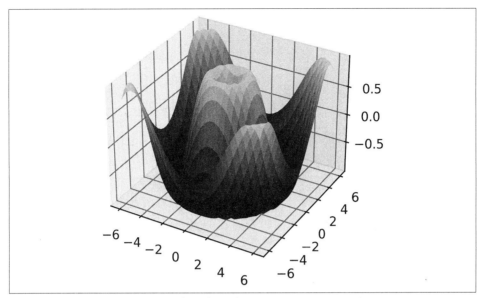

Figure 35-6. A three-dimensional surface plot

Though the grid of values for a surface plot needs to be two-dimensional, it need not be rectilinear. Here is an example of creating a partial polar grid, which when used with the surface3D plot can give us a slice into the function we're visualizing (see Figure 35-7).

```
In [10]: r = np.linspace(0, 6, 20)
         theta = np.linspace(-0.9 * np.pi, 0.8 * np.pi, 40)
         r, theta = np.meshgrid(r, theta)

         X = r * np.sin(theta)
         Y = r * np.cos(theta)
         Z = f(X, Y)

         ax = plt.axes(projection='3d')
         ax.plot_surface(X, Y, Z, rstride=1, cstride=1,
                         cmap='viridis', edgecolor='none');
```

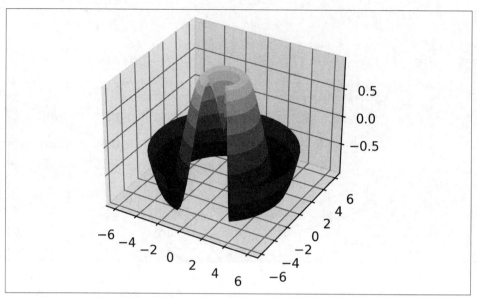

Figure 35-7. A polar surface plot

Surface Triangulations

For some applications, the evenly sampled grids required by the preceding routines are too restrictive. In these situations, triangulation-based plots can come in handy. What if rather than an even draw from a Cartesian or a polar grid, we instead have a set of random draws?

```
In [11]: theta = 2 * np.pi * np.random.random(1000)
         r = 6 * np.random.random(1000)
         x = np.ravel(r * np.sin(theta))
         y = np.ravel(r * np.cos(theta))
         z = f(x, y)
```

We could create a scatter plot of the points to get an idea of the surface we're sampling from, as shown in Figure 35-8.

```
In [12]: ax = plt.axes(projection='3d')
         ax.scatter(x, y, z, c=z, cmap='viridis', linewidth=0.5);
```

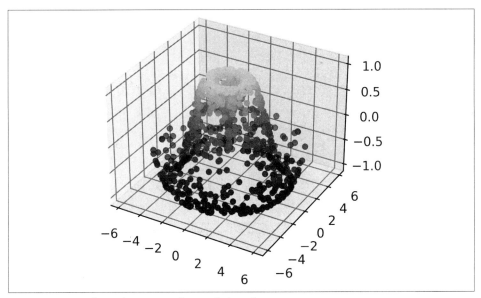

Figure 35-8. A three-dimensional sampled surface

This point cloud leaves a lot to be desired. The function that will help us in this case is `ax.plot_trisurf`, which creates a surface by first finding a set of triangles formed between adjacent points (remember that x, y, and z here are one-dimensional arrays); Figure 35-9 shows the result:

```
In [13]: ax = plt.axes(projection='3d')
         ax.plot_trisurf(x, y, z,
                         cmap='viridis', edgecolor='none');
```

The result is certainly not as clean as when it is plotted with a grid, but the flexibility of such a triangulation allows for some really interesting three-dimensional plots. For example, it is actually possible to plot a three-dimensional Möbius strip using this, as we'll see next.

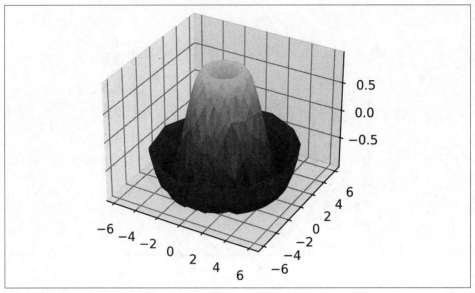

Figure 35-9. A triangulated surface plot

Example: Visualizing a Möbius Strip

A Möbius strip is similar to a strip of paper glued into a loop with a half-twist, result-ing in an object with only a single side! Here we will visualize such an object using Matplotlib's three-dimensional tools. The key to creating the Möbius strip is to think about its parametrization: it's a two-dimensional strip, so we need two intrinsic dimensions. Let's call them θ, which ranges from 0 to 2π around the loop, and w, which ranges from –1 to 1 across the width of the strip:

```
In [14]: theta = np.linspace(0, 2 * np.pi, 30)
         w = np.linspace(-0.25, 0.25, 8)
         w, theta = np.meshgrid(w, theta)
```

Now from this parametrization, we must determine the (x, y, z) positions of the embedded strip.

Thinking about it, we might realize that there are two rotations happening: one is the position of the loop about its center (what we've called θ), while the other is the twist-ing of the strip about its axis (we'll call this φ). For a Möbius strip, we must have the strip make half a twist during a full loop, or $\Delta\varphi = \Delta\theta/2$:

```
In [15]: phi = 0.5 * theta
```

Now we use our recollection of trigonometry to derive the three-dimensional embed-ding. We'll define r, the distance of each point from the center, and use this to find the embedded (x, y, z) coordinates:

```
In [16]: # radius in x-y plane
         r = 1 + w * np.cos(phi)

         x = np.ravel(r * np.cos(theta))
         y = np.ravel(r * np.sin(theta))
         z = np.ravel(w * np.sin(phi))
```

Finally, to plot the object, we must make sure the triangulation is correct. The best way to do this is to define the triangulation *within the underlying parametrization*, and then let Matplotlib project this triangulation into the three-dimensional space of the Möbius strip. This can be accomplished as follows (see Figure 35-10).

```
In [17]: # triangulate in the underlying parametrization
         from matplotlib.tri import Triangulation
         tri = Triangulation(np.ravel(w), np.ravel(theta))

         ax = plt.axes(projection='3d')
         ax.plot_trisurf(x, y, z, triangles=tri.triangles,
                         cmap='Greys', linewidths=0.2);

         ax.set_xlim(-1, 1); ax.set_ylim(-1, 1); ax.set_zlim(-1, 1)
         ax.axis('off');
```

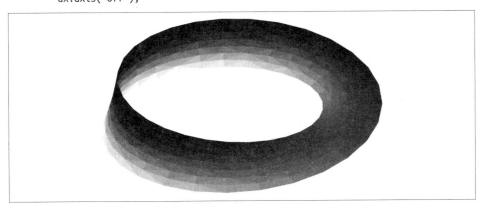

Figure 35-10. Visualizing a Möbius strip

Combining all of these techniques, it is possible to create and display a wide variety of three-dimensional objects and patterns in Matplotlib.

## CHAPTER 36

Visualization with Seaborn

Matplotlib has been at the core of scientific visualization in Python for decades, but even avid users will admit it often leaves much to be desired. There are several complaints about Matplotlib that often come up:

- A common early complaint, which is now outdated: prior to version 2.0, Matplotlib's color and style defaults were at times poor and looked dated.

- Matplotlib's API is relatively low-level. Doing sophisticated statistical visualization is possible, but often requires a *lot* of boilerplate code.

- Matplotlib predated Pandas by more than a decade, and thus is not designed for use with Pandas `DataFrame` objects. In order to visualize data from a `DataFrame`, you must extract each `Series` and often concatenate them together into the right format. It would be nicer to have a plotting library that can intelligently use the `DataFrame` labels in a plot.

An answer to these problems is Seaborn (*http://seaborn.pydata.org*). Seaborn provides an API on top of Matplotlib that offers sane choices for plot style and color defaults, defines simple high-level functions for common statistical plot types, and integrates with the functionality provided by Pandas.

To be fair, the Matplotlib team has adapted to the changing landscape: it added the `plt.style` tools discussed in Chapter 34, and Matplotlib is starting to handle Pandas data more seamlessly. But for all the reasons just discussed, Seaborn remains a useful add-on.

By convention, Seaborn is often imported as sns:

```
In [1]: %matplotlib inline
        import matplotlib.pyplot as plt
        import seaborn as sns
        import numpy as np
        import pandas as pd

        sns.set()  # seaborn's method to set its chart style
```

Full-color figures are available in the supplemental materials on GitHub (*https://oreil.ly/PDSH_GitHub*).

Exploring Seaborn Plots

The main idea of Seaborn is that it provides high-level commands to create a variety of plot types useful for statistical data exploration, and even some statistical model fitting.

Let's take a look at a few of the datasets and plot types available in Seaborn. Note that all of the following *could* be done using raw Matplotlib commands (this is, in fact, what Seaborn does under the hood), but the Seaborn API is much more convenient.

Histograms, KDE, and Densities

Often in statistical data visualization, all you want is to plot histograms and joint distributions of variables. We have seen that this is relatively straightforward in Matplotlib (see Figure 36-1).

```
In [2]: data = np.random.multivariate_normal([0, 0], [[5, 2], [2, 2]], size=2000)
        data = pd.DataFrame(data, columns=['x', 'y'])

        for col in 'xy':
            plt.hist(data[col], density=True, alpha=0.5)
```

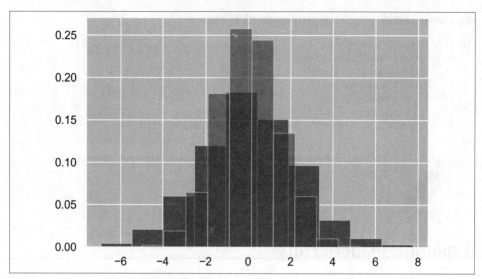

Figure 36-1. Histograms for visualizing distributions

Rather than just providing a histogram as a visual output, we can get a smooth esti-mate of the distribution using kernel density estimation (introduced in Chapter 28), which Seaborn does with sns.kdeplot (see Figure 36-2).

In [3]: sns.kdeplot(data=data, shade=True);

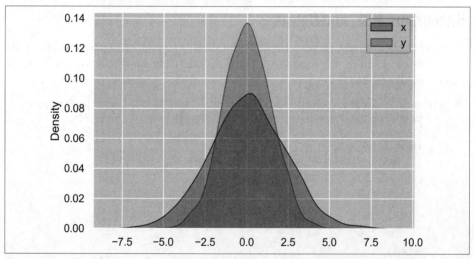

Figure 36-2. Kernel density estimates for visualizing distributions

If we pass x and y columns to kdeplot, we instead get a two-dimensional visualization of the joint density (see Figure 36-3).

```
In [4]: sns.kdeplot(data=data, x='x', y='y');
```

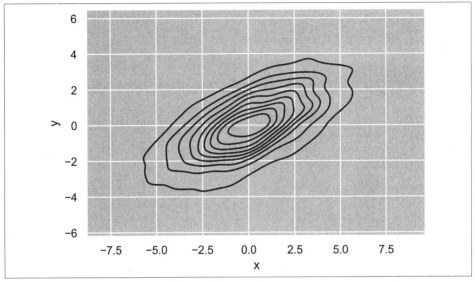

Figure 36-3. A two-dimensional kernel density plot

We can see the joint distribution and the marginal distributions together using sns.jointplot, which we'll explore further later in this chapter.

Pair Plots

When you generalize joint plots to datasets of larger dimensions, you end up with *pair plots*. These are very useful for exploring correlations between multidimensional data, when you'd like to plot all pairs of values against each other.

We'll demo this with the well-known Iris dataset, which lists measurements of petals and sepals of three Iris species:

```
In [5]: iris = sns.load_dataset("iris")
        iris.head()
Out[5]:    sepal_length  sepal_width  petal_length  petal_width species
        0           5.1          3.5           1.4          0.2  setosa
        1           4.9          3.0           1.4          0.2  setosa
        2           4.7          3.2           1.3          0.2  setosa
        3           4.6          3.1           1.5          0.2  setosa
        4           5.0          3.6           1.4          0.2  setosa
```

Visualizing the multidimensional relationships among the samples is as easy as calling sns.pairplot (see Figure 36-4).

In [6]: sns.pairplot(iris, hue='species', height=2.5);

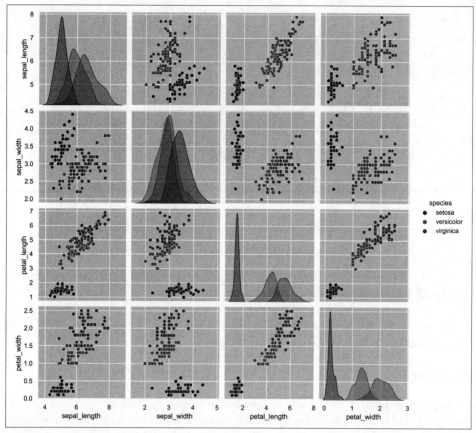

Figure 36-4. A pair plot showing the relationships between four variables

Faceted Histograms

Sometimes the best way to view data is via histograms of subsets, as shown in Figure 36-5. Seaborn's FacetGrid makes this simple. We'll take a look at some data that shows the amount that restaurant staff receive in tips based on various indicator data:[1]

---

1 The restaurant staff data used in this section divides employees into two sexes: female and male. Biological sex isn't binary, but the following discussion and visualizations are limited by this data.

```
In [7]: tips = sns.load_dataset('tips')
        tips.head()
Out[7]:    total_bill   tip     sex smoker  day    time  size
        0      16.99  1.01  Female     No  Sun  Dinner     2
        1      10.34  1.66    Male     No  Sun  Dinner     3
        2      21.01  3.50    Male     No  Sun  Dinner     3
        3      23.68  3.31    Male     No  Sun  Dinner     2
        4      24.59  3.61  Female     No  Sun  Dinner     4

In [8]: tips['tip_pct'] = 100 * tips['tip'] / tips['total_bill']

        grid = sns.FacetGrid(tips, row="sex", col="time", margin_titles=True)
        grid.map(plt.hist, "tip_pct", bins=np.linspace(0, 40, 15));
```

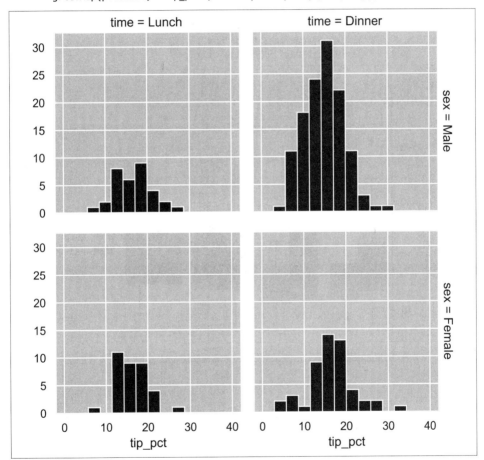

Figure 36-5. An example of a faceted histogram

The faceted chart gives us some quick insights into the dataset: for example, we see that it contains far more data on male servers during the dinner hour than other categories, and typical tip amounts appear to range from approximately 10% to 20%, with some outliers on either end.

Categorical Plots

Categorical plots can be useful for this kind of visualization as well. These allow you to view the distribution of a parameter within bins defined by any other parameter, as shown in Figure 36-6.

```
In [9]: with sns.axes_style(style='ticks'):
            g = sns.catplot(x="day", y="total_bill", hue="sex",
                            data=tips, kind="box")
            g.set_axis_labels("Day", "Total Bill");
```

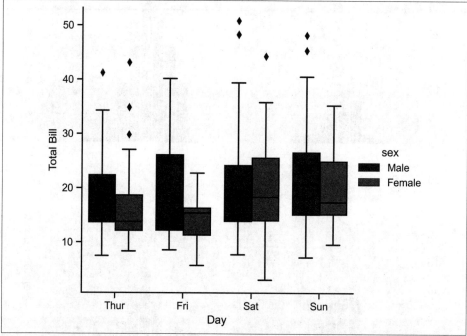

Figure 36-6. An example of a factor plot, comparing distributions given various discrete factors

Joint Distributions

Similar to the pair plot we saw earlier, we can use `sns.jointplot` to show the joint distribution between different datasets, along with the associated marginal distributions (see Figure 36-7).

```
In [10]: with sns.axes_style('white'):
             sns.jointplot(x="total_bill", y="tip", data=tips, kind='hex')
```

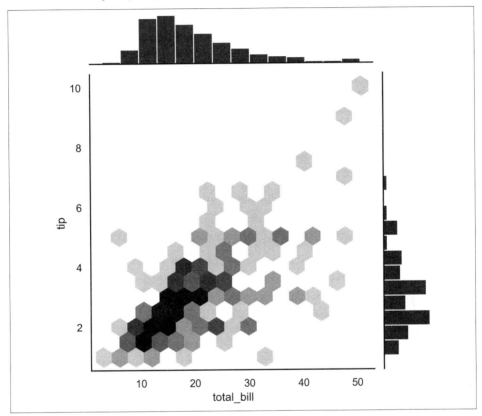

Figure 36-7. A joint distribution plot

The joint plot can even do some automatic kernel density estimation and regression, as shown in Figure 36-8.

```
In [11]: sns.jointplot(x="total_bill", y="tip", data=tips, kind='reg');
```

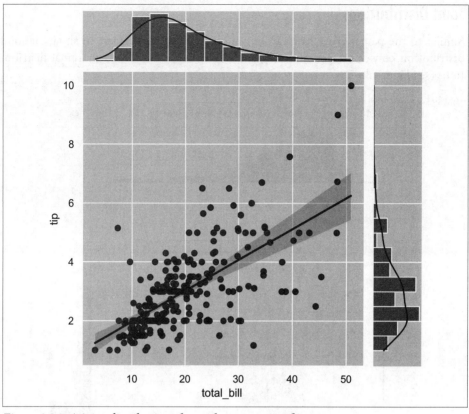

Figure 36-8. A joint distribution plot with a regression fit

Bar Plots

Time series can be plotted using `sns.factorplot`. In the following example, we'll use
the Planets dataset that we first saw in Chapter 20; see Figure 36-9 for the result.

```
In [12]: planets = sns.load_dataset('planets')
         planets.head()
Out[12]:            method  number  orbital_period   mass  distance  year
         0  Radial Velocity       1         269.300   7.10     77.40  2006
         1  Radial Velocity       1         874.774   2.21     56.95  2008
         2  Radial Velocity       1         763.000   2.60     19.84  2011
         3  Radial Velocity       1         326.030  19.40    110.62  2007
         4  Radial Velocity       1         516.220  10.50    119.47  2009

In [13]: with sns.axes_style('white'):
             g = sns.catplot(x="year", data=planets, aspect=2,
                             kind="count", color='steelblue')
             g.set_xticklabels(step=5)
```

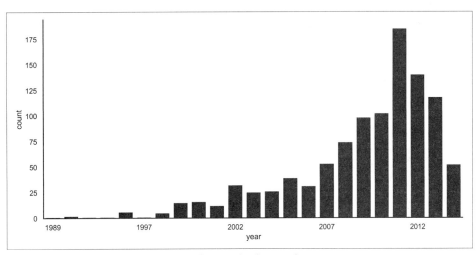

Figure 36-9. A histogram as a special case of a factor plot

We can learn more by looking at the *method* of discovery of each of these planets (see Figure 36-10).

```
In [14]: with sns.axes_style('white'):
            g = sns.catplot(x="year", data=planets, aspect=4.0, kind='count',
                            hue='method', order=range(2001, 2015))
            g.set_ylabels('Number of Planets Discovered')
```

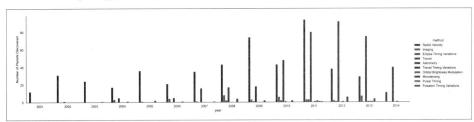

Figure 36-10. Number of planets discovered by year and type

For more information on plotting with Seaborn, see the Seaborn documentation (*https://oreil.ly/fCHxn*), and particularly the example gallery (*https://oreil.ly/08xGE*).

Example: Exploring Marathon Finishing Times

Here we'll look at using Seaborn to help visualize and understand finishing results from a marathon.[2] I've scraped the data from sources on the web, aggregated it and removed any identifying information, and put it on GitHub, where it can be downloaded.[3]

We will start by downloading the data and loading it into Pandas:

```
In [15]: # url = ('https://raw.githubusercontent.com/jakevdp/'
         #        'marathon-data/master/marathon-data.csv')
         # !cd data && curl -O {url}

In [16]: data = pd.read_csv('data/marathon-data.csv')
         data.head()
Out[16]:    age gender    split    final
         0   33      M  01:05:38 02:08:51
         1   32      M  01:06:26 02:09:28
         2   31      M  01:06:49 02:10:42
         3   38      M  01:06:16 02:13:45
         4   31      M  01:06:32 02:13:59
```

Notice that Pandas loaded the time columns as Python strings (type `object`); we can see this by looking at the `dtypes` attribute of the `DataFrame`:

```
In [17]: data.dtypes
Out[17]: age        int64
         gender    object
         split     object
         final     object
         dtype: object
```

Let's fix this by providing a converter for the times:

```
In [18]: import datetime

         def convert_time(s):
             h, m, s = map(int, s.split(':'))
             return datetime.timedelta(hours=h, minutes=m, seconds=s)

         data = pd.read_csv('data/marathon-data.csv',
                            converters={'split':convert_time, 'final':convert_time})
         data.head()
Out[18]:    age gender          split           final
         0   33      M 0 days 01:05:38 0 days 02:08:51
         1   32      M 0 days 01:06:26 0 days 02:09:28
```

---

2 The marathon data used in this section divides runners into two genders: men and women. While gender is a spectrum, the following discussion and visualizations use this binary because they depend on the data.

3 If you are interested in using Python for web scraping, I would recommend *Web Scraping with Python* by Ryan Mitchell, also from O'Reilly.

```
        2   31      M 0 days 01:06:49 0 days 02:10:42
        3   38      M 0 days 01:06:16 0 days 02:13:45
        4   31      M 0 days 01:06:32 0 days 02:13:59
In [19]: data.dtypes
Out[19]: age                   int64
         gender               object
         split      timedelta64[ns]
         final      timedelta64[ns]
         dtype: object
```

That will make it easier to manipulate the temporal data. For the purpose of our Seaborn plotting utilities, let's next add columns that give the times in seconds:

```
In [20]: data['split_sec'] = data['split'].view(int) / 1E9
         data['final_sec'] = data['final'].view(int) / 1E9
         data.head()
Out[20]:    age gender          split          final  split_sec  final_sec
         0   33      M 0 days 01:05:38 0 days 02:08:51     3938.0     7731.0
         1   32      M 0 days 01:06:26 0 days 02:09:28     3986.0     7768.0
         2   31      M 0 days 01:06:49 0 days 02:10:42     4009.0     7842.0
         3   38      M 0 days 01:06:16 0 days 02:13:45     3976.0     8025.0
         4   31      M 0 days 01:06:32 0 days 02:13:59     3992.0     8039.0
```

To get an idea of what the data looks like, we can plot a jointplot over the data; Figure 36-11 shows the result.

```
In [21]: with sns.axes_style('white'):
             g = sns.jointplot(x='split_sec', y='final_sec', data=data, kind='hex')
             g.ax_joint.plot(np.linspace(4000, 16000),
                             np.linspace(8000, 32000), ':k')
```

The dotted line shows where someone's time would lie if they ran the marathon at a perfectly steady pace. The fact that the distribution lies above this indicates (as you might expect) that most people slow down over the course of the marathon. If you have run competitively, you'll know that those who do the opposite—run faster during the second half of the race—are said to have "negative-split" the race.

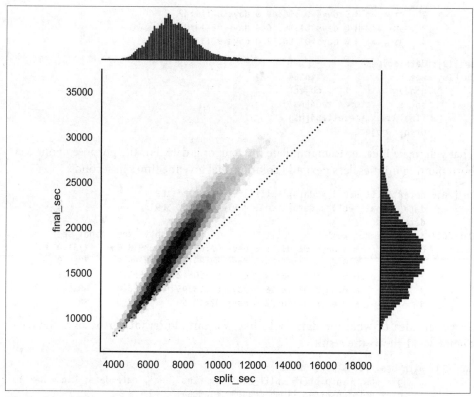

Figure 36-11. The relationship between the split for the first half-marathon and the fin-ishing time for the full marathon

Let's create another column in the data, the split fraction, which measures the degree to which each runner negative-splits or positive-splits the race:

```
In [22]: data['split_frac'] = 1 - 2 * data['split_sec'] / data['final_sec']
         data.head()
Out[22]:    age gender          split             final  split_sec  final_sec  \
         0   33      M 0 days 01:05:38 0 days 02:08:51     3938.0     7731.0
         1   32      M 0 days 01:06:26 0 days 02:09:28     3986.0     7768.0
         2   31      M 0 days 01:06:49 0 days 02:10:42     4009.0     7842.0
         3   38      M 0 days 01:06:16 0 days 02:13:45     3976.0     8025.0
         4   31      M 0 days 01:06:32 0 days 02:13:59     3992.0     8039.0

            split_frac
         0   -0.018756
         1   -0.026262
         2   -0.022443
         3    0.009097
         4    0.006842
```

Where this split difference is less than zero, the person negative-split the race by that fraction. Let's do a distribution plot of this split fraction (see Figure 36-12).

```
In [23]: sns.displot(data['split_frac'], kde=False)
         plt.axvline(0, color="k", linestyle="--");
```

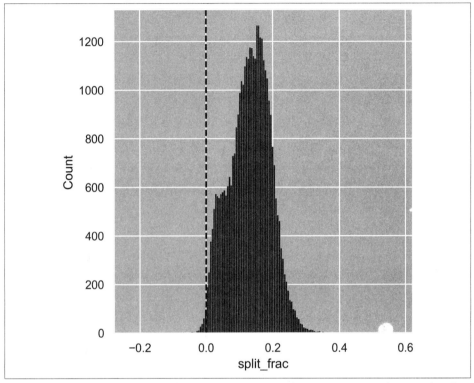

Figure 36-12. The distribution of split fractions; 0.0 indicates a runner who completed the first and second halves in identical times

```
In [24]: sum(data.split_frac < 0)
Out[24]: 251
```

Out of nearly 40,000 participants, there were only 250 people who negative-split their marathon.

Let's see whether there is any correlation between this split fraction and other variables. We'll do this using a `PairGrid`, which draws plots of all these correlations (see Figure 36-13).

```
In [25]: g = sns.PairGrid(data, vars=['age', 'split_sec', 'final_sec', 'split_frac'],
                        hue='gender', palette='RdBu_r')
         g.map(plt.scatter, alpha=0.8)
         g.add_legend();
```

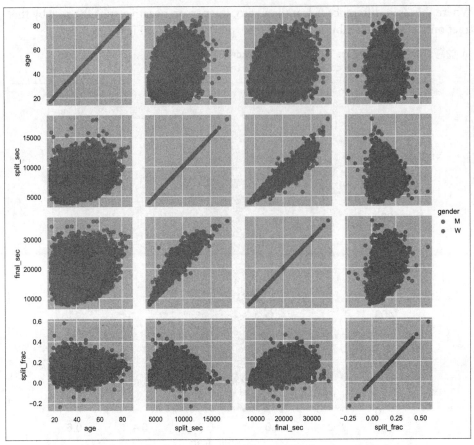

Figure 36-13. The relationship between quantities within the marathon dataset

It looks like the split fraction does not correlate particularly with age, but does correlate with the final time: faster runners tend to have closer to even splits on their marathon time. Let's zoom in on the histogram of split fractions separated by gender, shown in Figure 36-14.

```
In [26]: sns.kdeplot(data.split_frac[data.gender=='M'], label='men', shade=True)
         sns.kdeplot(data.split_frac[data.gender=='W'], label='women', shade=True)
         plt.xlabel('split_frac');
```

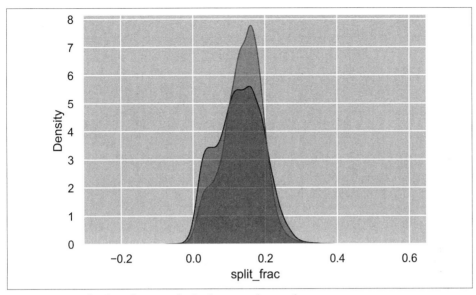

Figure 36-14. The distribution of split fractions by gender

The interesting thing here is that there are many more men than women who are running close to an even split! It almost looks like a bimodal distribution among the men and women. Let's see if we can suss out what's going on by looking at the distributions as a function of age.

A nice way to compare distributions is to use a *violin plot*, shown in Figure 36-15.

```
In [27]: sns.violinplot(x="gender", y="split_frac", data=data,
                         palette=["lightblue", "lightpink"]);
```

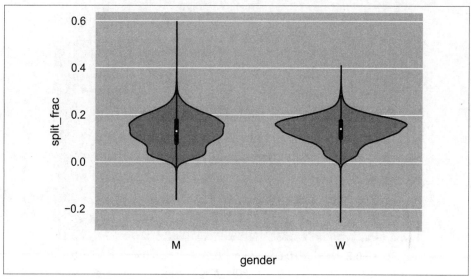

Figure 36-15. A violin plot showing the split fraction by gender

Let's look a little deeper, and compare these violin plots as a function of age (see Figure 36-16). We'll start by creating a new column in the array that specifies the age range that each person is in, by decade:

```
In [28]: data['age_dec'] = data.age.map(lambda age: 10 * (age // 10))
         data.head()
Out[28]:    age gender          split           final  split_sec  final_sec  \
         0   33      M  0 days 01:05:38  0 days 02:08:51     3938.0     7731.0
         1   32      M  0 days 01:06:26  0 days 02:09:28     3986.0     7768.0
         2   31      M  0 days 01:06:49  0 days 02:10:42     4009.0     7842.0
         3   38      M  0 days 01:06:16  0 days 02:13:45     3976.0     8025.0
         4   31      M  0 days 01:06:32  0 days 02:13:59     3992.0     8039.0

            split_frac  age_dec
         0   -0.018756       30
         1   -0.026262       30
         2   -0.022443       30
         3    0.009097       30
         4    0.006842       30

In [29]: men = (data.gender == 'M')
         women = (data.gender == 'W')

         with sns.axes_style(style=None):
             sns.violinplot(x="age_dec", y="split_frac", hue="gender", data=data,
                            split=True, inner="quartile",
                            palette=["lightblue", "lightpink"]);
```

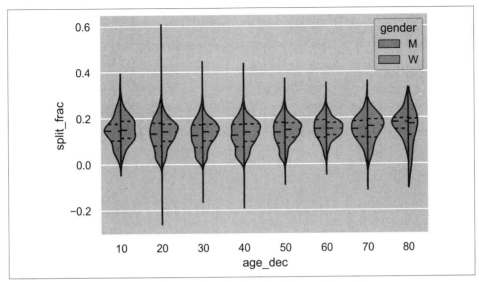

Figure 36-16. A violin plot showing the split fraction by gender and age

We can see where the distributions among men and women differ: the split distributions of men in their 20s to 50s show a pronounced overdensity toward lower splits when compared to women of the same age (or of any age, for that matter).

Also surprisingly, it appears that the 80-year-old women seem to outperform *everyone* in terms of their split time, although this is likely a small number effect, as there are only a handful of runners in that range:

```
In [30]: (data.age > 80).sum()
Out[30]: 7
```

Back to the men with negative splits: who are these runners? Does this split fraction correlate with finishing quickly? We can plot this very easily. We'll use `regplot`, which will automatically fit a linear regression model to the data (see Figure 36-17).

```
In [31]: g = sns.lmplot(x='final_sec', y='split_frac', col='gender', data=data,
                        markers=".", scatter_kws=dict(color='c'))
         g.map(plt.axhline, y=0.0, color="k", ls=":");
```

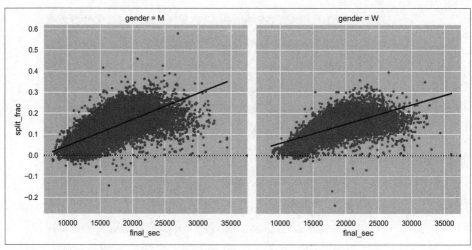

Figure 36-17. Split fraction versus finishing time by gender

Apparently, among both men and women, the people with fast splits tend to be faster runners who are finishing within ~15,000 seconds, or about 4 hours. People slower than that are much less likely to have a fast second split.

Further Resources

A single part of a book can never hope to cover all the available features and plot types available in Matplotlib. As with other packages we've seen, liberal use of IPython's tab completion and help functions (see Chapter 1) can be very helpful when exploring Matplotlib's API. In addition, Matplotlib's online documentation (*http://matplotlib.org*) can be a helpful reference. See in particular the Matplotlib gallery (*https://oreil.ly/WNiHP*), which shows thumbnails of hundreds of different plot types, each one linked to a page with the Python code snippet used to generate it. This allows you to visually inspect and learn about a wide range of different plotting styles and visualization techniques.

For a book-length treatment of Matplotlib, I would recommend *Interactive Applications Using Matplotlib* (Packt), written by Matplotlib core developer Ben Root.

Other Python Visualization Libraries

Although Matplotlib is the most prominent Python visualization library, there are other more modern tools that are worth exploring as well. I'll mention a few of them briefly here:

- Bokeh (*http://bokeh.pydata.org*) is a JavaScript visualization library with a Python frontend that creates highly interactive visualizations capable of handling very large and/or streaming datasets.
- Plotly (*http://plot.ly*) is the eponymous open source product of the Plotly company, and is similar in spirit to Bokeh. It is actively developed and provides a wide range of interactive chart types.
- HoloViews (*https://holoviews.org*) is a more declarative, unified API for generating charts in a variety of backends, including Bokeh and Matplotlib.
- Vega (*https://vega.github.io*) and Vega-Lite (*https://vega.github.io/vega-lite*) are declarative graphics representations, and are the product of years of research into how to think about data visualization and interaction. The reference rendering implementation is JavaScript, and the Altair package (*https://altair-viz.github.io*) provides a Python API to generate these charts.

The visualization landscape in the Python world is constantly evolving, and I expect that this list may be out of date by the time this book is published. Additionally, because Python is used in so many domains, you'll find many other visualization tools built for more specific use cases. It can be hard to keep track of all of them, but a good resource for learning about this wide variety of visualization tools is PyViz (*https://pyviz.org*), an open, community-driven site containing tutorials and examples of many different visualization tools.

Other Python Visualization Libraries

PART V

Machine Learning

This final part is an introduction to the very broad topic of machine learning, mainly via Python's Scikit-Learn package (*http://scikit-learn.org*). You can think of machine learning as a class of algorithms that allow a program to detect particular patterns in a dataset, and thus "learn" from the data to draw inferences from it. This is not meant to be a comprehensive introduction to the field of machine learning; that is a large subject and necessitates a more technical approach than we take here. Nor is it meant to be a comprehensive manual for the use of the Scikit-Learn package (for this, you can refer to the resources listed in "Further Machine Learning Resources" on page 550). Rather, the goals here are:

- To introduce the fundamental vocabulary and concepts of machine learning
- To introduce the Scikit-Learn API and show some examples of its use
- To take a deeper dive into the details of several of the more important classical machine learning approaches, and develop an intuition into how they work and when and where they are applicable

Much of this material is drawn from the Scikit-Learn tutorials and workshops I have given on several occasions at PyCon, SciPy, PyData, and other conferences. Any clarity in the following pages is likely due to the many workshop participants and co-instructors who have given me valuable feedback on this material over the years!

## CHAPTER 37

What Is Machine Learning?

Before we take a look at the details of several machine learning methods, let's start by looking at what machine learning is, and what it isn't. Machine learning is often categorized as a subfield of artificial intelligence, but I find that categorization can be misleading. The study of machine learning certainly arose from research in this context, but in the data science application of machine learning methods, it's more helpful to think of machine learning as a means of *building models of data*.

In this context, "learning" enters the fray when we give these models *tunable parameters* that can be adapted to observed data; in this way the program can be considered to be "learning" from the data. Once these models have been fit to previously seen data, they can be used to predict and understand aspects of newly observed data. I'll leave to the reader the more philosophical digression regarding the extent to which this type of mathematical, model-based "learning" is similar to the "learning" exhibited by the human brain.

Understanding the problem setting in machine learning is essential to using these tools effectively, and so we will start with some broad categorizations of the types of approaches we'll discuss here.

All of the figures in this chapter are generated based on actual machine learning computations; the code behind them can be found in the online appendix (*https://oreil.ly/o1Zya*).

Categories of Machine Learning

Machine learning can be categorized into two main types: supervised learning and unsupervised learning.

Supervised learning involves somehow modeling the relationship between measured features of data and some labels associated with the data; once this model is determined, it can be used to apply labels to new, unknown data. This is sometimes further subdivided into classification tasks and regression tasks: in *classification*, the labels are discrete categories, while in *regression*, the labels are continuous quantities. You will see examples of both types of supervised learning in the following section.

Unsupervised learning involves modeling the features of a dataset without reference to any label. These models include tasks such as *clustering* and *dimensionality reduction*. Clustering algorithms identify distinct groups of data, while dimensionality reduction algorithms search for more succinct representations of the data. You will also see examples of both types of unsupervised learning in the following section.

In addition, there are so-called *semi-supervised learning* methods, which fall somewhere between supervised learning and unsupervised learning. Semi-supervised learning methods are often useful when only incomplete labels are available.

Qualitative Examples of Machine Learning Applications

To make these ideas more concrete, let's take a look at a few very simple examples of a machine learning task. These examples are meant to give an intuitive, non-quantitative overview of the types of machine learning tasks we will be looking at in this part of the book. In later chapters, we will go into more depth regarding the particular models and how they are used. For a preview of these more technical aspects, you can find the Python source that generates the figures in the online appendix (*https://oreil.ly/o1Zya*).

Classification: Predicting Discrete Labels

We will first take a look at a simple classification task, in which we are given a set of labeled points and want to use these to classify some unlabeled points.

Imagine that we have the data shown in Figure 37-1. This data is two-dimensional: that is, we have two *features* for each point, represented by the (x,y) positions of the points on the plane. In addition, we have one of two *class labels* for each point, here represented by the colors of the points. From these features and labels, we would like to create a model that will let us decide whether a new point should be labeled "blue" or "red."

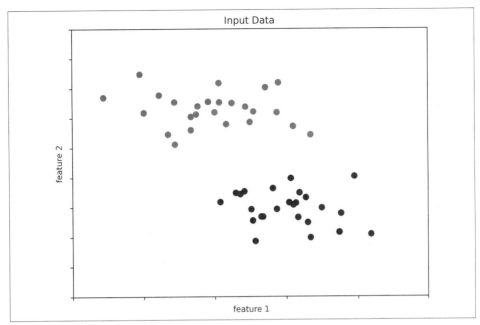

Input Data

Figure 37-1. A simple dataset for classification

There are a number of possible models for such a classification task, but we will start with a very simple one. We will make the assumption that the two groups can be separated by drawing a straight line through the plane between them, such that points on each side of the line all fall in the same group. Here the *model* is a quantitative version of the statement "a straight line separates the classes," while the *model parameters* are the particular numbers describing the location and orientation of that line for our data. The optimal values for these model parameters are learned from the data (this is the "learning" in machine learning), which is often called *training the model*.

Figure 37-2 shows a visual representation of what the trained model looks like for this data.

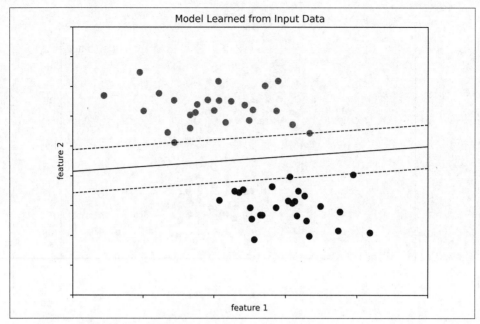

Figure 37-2. A simple classification model

Now that this model has been trained, it can be generalized to new, unlabeled data. In other words, we can take a new set of data, draw this line through it, and assign labels to the new points based on this model (see Figure 37-3). This stage is usually called *prediction*.

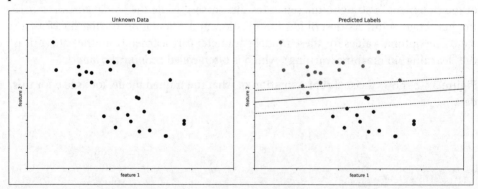

Figure 37-3. Applying a classification model to new data

This is the basic idea of a classification task in machine learning, where "classification" indicates that the data has discrete class labels. At first glance this may seem trivial: it's easy to look at our data and draw such a discriminatory line to accomplish this classification. A benefit of the machine learning approach, however, is that it can generalize to much larger datasets in many more dimensions. For example, this is similar to the task of automated spam detection for email. In this case, we might use the following features and labels:

- *feature 1*, *feature 2*, etc. → normalized counts of important words or phrases ("Viagra", "Extended warranty", etc.)
- *label* → "spam" or "not spam"

For the training set, these labels might be determined by individual inspection of a small representative sample of emails; for the remaining emails, the label would be determined using the model. For a suitably trained classification algorithm with enough well-constructed features (typically thousands or millions of words or phrases), this type of approach can be very effective. We will see an example of such text-based classification in Chapter 41.

Some important classification algorithms that we will discuss in more detail are Gaussian naive Bayes (see Chapter 41), support vector machines (see Chapter 43), and random forest classification (see Chapter 44).

Regression: Predicting Continuous Labels

In contrast with the discrete labels of a classification algorithm, we will next look at a simple regression task in which the labels are continuous quantities.

Consider the data shown in Figure 37-4, which consists of a set of points each with a continuous label.

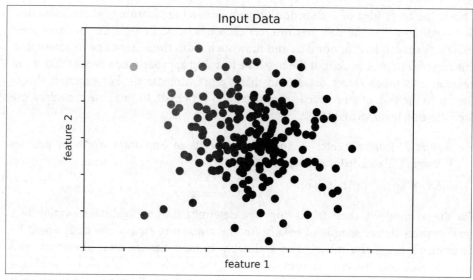

Figure 37-4. A simple dataset for regression

As with the classification example, we have two-dimensional data: that is, there are two features describing each data point. The color of each point represents the continuous label for that point.

There are a number of possible regression models we might use for this type of data, but here we will use a simple linear regression model to predict the points. This simple model assumes that if we treat the label as a third spatial dimension, we can fit a plane to the data. This is a higher-level generalization of the well-known problem of fitting a line to data with two coordinates.

We can visualize this setup as shown in Figure 37-5.

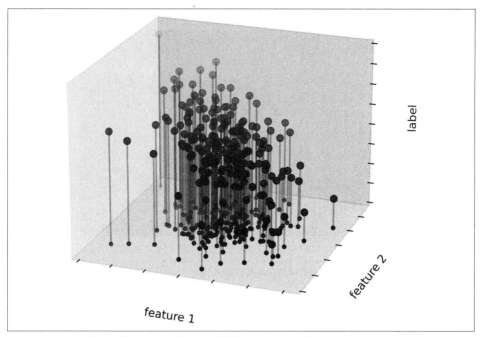

Figure 37-5. A three-dimensional view of the regression data

Notice that the *feature 1–feature 2* plane here is the same as in the two-dimensional plot in Figure 37-4; in this case, however, we have represented the labels by both color and three-dimensional axis position. From this view, it seems reasonable that fitting a plane through this three-dimensional data would allow us to predict the expected label for any set of input parameters. Returning to the two-dimensional projection, when we fit such a plane we get the result shown in Figure 37-6.

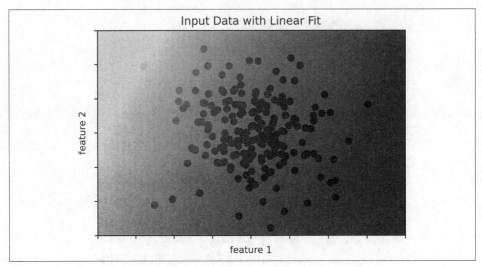

Figure 37-6. A representation of the regression model

This plane of fit gives us what we need to predict labels for new points. Visually, we find the results shown in Figure 37-7.

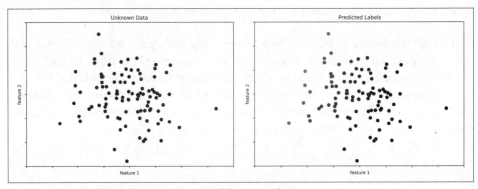

Figure 37-7. Applying the regression model to new data

As with the classification example, this task may seem trivial in a low number of dimensions. But the power of these methods is that they can be straightforwardly applied and evaluated in the case of data with many, many features. For example, this is similar to the task of computing the distance to galaxies observed through a telescope—in this case, we might use the following features and labels:

- *feature 1*, *feature 2*, etc. → brightness of each galaxy at one of several wavelengths or colors
- *label* → distance or redshift of the galaxy

The distances for a small number of these galaxies might be determined through an independent set of (typically more expensive or complex) observations. Distances to remaining galaxies could then be estimated using a suitable regression model, without the need to employ the more expensive observation across the entire set. In astronomy circles, this is known as the "photometric redshift" problem.

Some important regression algorithms that we will discuss are linear regression (see Chapter 42), support vector machines (see Chapter 43), and random forest regression (see Chapter 44).

Clustering: Inferring Labels on Unlabeled Data

The classification and regression illustrations we just saw are examples of supervised learning algorithms, in which we are trying to build a model that will predict labels for new data. Unsupervised learning involves models that describe data without reference to any known labels.

One common case of unsupervised learning is "clustering," in which data is automatically assigned to some number of discrete groups. For example, we might have some two-dimensional data like that shown in Figure 37-8.

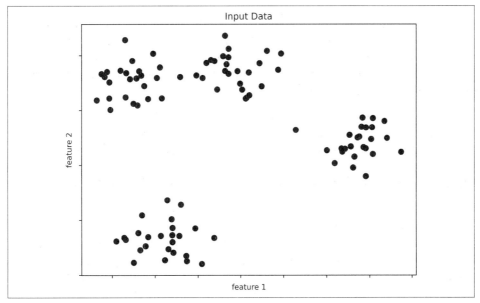

Figure 37-8. Example data for clustering

By eye, it is clear that each of these points is part of a distinct group. Given this input, a clustering model will use the intrinsic structure of the data to determine which points are related. Using the very fast and intuitive *k*-means algorithm (see Chapter 47), we find the clusters shown in Figure 37-9.

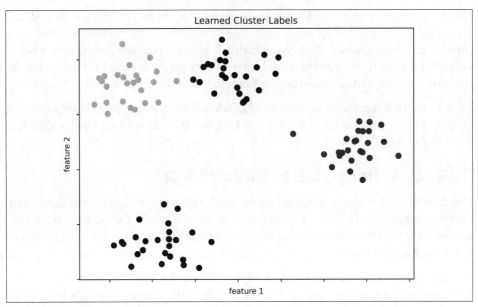

Figure 37-9. Data labeled with a k-means clustering model

k-means fits a model consisting of *k* cluster centers; the optimal centers are assumed to be those that minimize the distance of each point from its assigned center. Again, this might seem like a trivial exercise in two dimensions, but as our data becomes larger and more complex such clustering algorithms can continue to be employed to extract useful information from the dataset.

We will discuss the *k*-means algorithm in more depth in Chapter 47. Other important clustering algorithms include Gaussian mixture models (see Chapter 48) and spectral clustering (see Scikit-Learn's clustering documentation (*https://oreil.ly/9FHKO*)).

Dimensionality Reduction: Inferring Structure of Unlabeled Data

Dimensionality reduction is another example of an unsupervised algorithm, in which labels or other information are inferred from the structure of the dataset itself. Dimensionality reduction is a bit more abstract than the examples we looked at before, but generally it seeks to pull out some low-dimensional representation of data that in some way preserves relevant qualities of the full dataset. Different dimensionality reduction routines measure these relevant qualities in different ways, as we will see in Chapter 46.

As an example of this, consider the data shown in Figure 37-10.

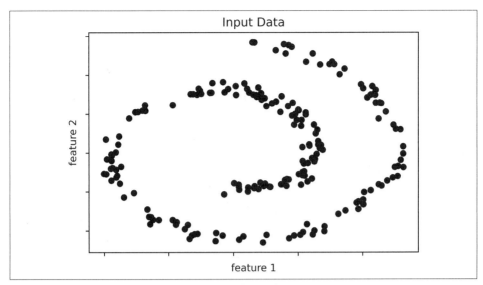

Figure 37-10. Example data for dimensionality reduction

Visually, it is clear that there is some structure in this data: it is drawn from a one-dimensional line that is arranged in a spiral within this two-dimensional space. In a sense, you could say that this data is "intrinsically" only one-dimensional, though this one-dimensional data is embedded in two-dimensional space. A suitable dimensionality reduction model in this case would be sensitive to this nonlinear embedded structure and be able to detect this lower-dimensionality representation.

Figure 37-11 shows a visualization of the results of the Isomap algorithm, a manifold learning algorithm that does exactly this.

Notice that the colors (which represent the extracted one-dimensional latent variable) change uniformly along the spiral, which indicates that the algorithm did in fact detect the structure we saw by eye. As with the previous examples, the power of dimensionality reduction algorithms becomes clearer in higher-dimensional cases. For example, we might wish to visualize important relationships within a dataset that has 100 or 1,000 features. Visualizing 1,000-dimensional data is a challenge, and one way we can make this more manageable is to use a dimensionality reduction technique to reduce the data to 2 or 3 dimensions.

Some important dimensionality reduction algorithms that we will discuss are principal component analysis (see Chapter 45) and various manifold learning algorithms, including Isomap and locally linear embedding (see Chapter 46).

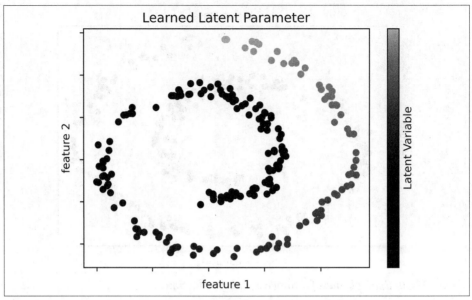

Figure 37-11. Data with labels learned via dimensionality reduction

Summary

Here we have seen a few simple examples of some of the basic types of machine learning approaches. Needless to say, there are a number of important practical details that we have glossed over, but this chapter was designed to give you a basic idea of what types of problems machine learning approaches can solve.

In short, we saw the following:

- *Supervised learning*: Models that can predict labels based on labeled training data
  - *Classification*: Models that predict labels as two or more discrete categories
  - *Regression*: Models that predict continuous labels
- *Unsupervised learning*: Models that identify structure in unlabeled data
  - *Clustering*: Models that detect and identify distinct groups in the data
  - *Dimensionality reduction*: Models that detect and identify lower-dimensional structure in higher-dimensional data

In the following chapters, we will go into much greater depth within these categories, and see some more interesting examples of where these concepts can be useful.

## CHAPTER 38

Introducing Scikit-Learn

Several Python libraries provide solid implementations of a range of machine learning algorithms. One of the best known is Scikit-Learn (*http://scikit-learn.org*), a package that provides efficient versions of a large number of common algorithms. Scikit-Learn is characterized by a clean, uniform, and streamlined API, as well as by very useful and complete documentation. A benefit of this uniformity is that once you understand the basic use and syntax of Scikit-Learn for one type of model, switching to a new model or algorithm is straightforward.

This chapter provides an overview of the Scikit-Learn API. A solid understanding of these API elements will form the foundation for understanding the deeper practical discussion of machine learning algorithms and approaches in the following chapters.

We will start by covering data representation in Scikit-Learn, then delve into the Estimator API, and finally go through a more interesting example of using these tools for exploring a set of images of handwritten digits.

Data Representation in Scikit-Learn

Machine learning is about creating models from data; for that reason, we'll start by discussing how data can be represented. The best way to think about data within Scikit-Learn is in terms of *tables*.

A basic table is a two-dimensional grid of data, in which the rows represent individual elements of the dataset, and the columns represent quantities related to each of these elements. For example, consider the Iris dataset (*https://oreil.ly/TeWYs*), famously analyzed by Ronald Fisher in 1936. We can download this dataset in the form of a Pandas DataFrame using the Seaborn library (*http://seaborn.pydata.org*), and take a look at the first few items:

```
In [1]: import seaborn as sns
        iris = sns.load_dataset('iris')
        iris.head()
Out[1]:    sepal_length  sepal_width  petal_length  petal_width  species
        0           5.1          3.5           1.4          0.2  setosa
        1           4.9          3.0           1.4          0.2  setosa
        2           4.7          3.2           1.3          0.2  setosa
        3           4.6          3.1           1.5          0.2  setosa
        4           5.0          3.6           1.4          0.2  setosa
```

Here each row of the data refers to a single observed flower, and the number of rows is the total number of flowers in the dataset. In general, we will refer to the rows of the matrix as *samples*, and the number of rows as n_samples.

Likewise, each column of the data refers to a particular quantitative piece of information that describes each sample. In general, we will refer to the columns of the matrix as *features*, and the number of columns as n_features.

The Features Matrix

The table layout makes clear that the information can be thought of as a two-dimensional numerical array or matrix, which we will call the *features matrix*. By convention, this matrix is often stored in a variable named X. The features matrix is assumed to be two-dimensional, with shape [n_samples, n_features], and is most often contained in a NumPy array or a Pandas DataFrame, though some Scikit-Learn models also accept SciPy sparse matrices.

The samples (i.e., rows) always refer to the individual objects described by the dataset. For example, a sample might represent a flower, a person, a document, an image, a sound file, a video, an astronomical object, or anything else you can describe with a set of quantitative measurements.

The features (i.e., columns) always refer to the distinct observations that describe each sample in a quantitative manner. Features are often real-valued, but may be Boolean or discrete-valued in some cases.

The Target Array

In addition to the feature matrix X, we also generally work with a *label* or *target* array, which by convention we will usually call y. The target array is usually one-dimensional, with length n_samples, and is generally contained in a NumPy array or Pandas Series. The target array may have continuous numerical values, or discrete classes/labels. While some Scikit-Learn estimators do handle multiple target values in the form of a two-dimensional, [n_samples, n_targets] target array, we will primarily be working with the common case of a one-dimensional target array.

A common point of confusion is how the target array differs from the other feature columns. The distinguishing characteristic of the target array is that it is usually the quantity we want to *predict from the features*: in statistical terms, it is the dependent variable. For example, given the preceding data we may wish to construct a model that can predict the species of flower based on the other measurements; in this case, the `species` column would be considered the target array.

With this target array in mind, we can use Seaborn (discussed in Chapter 36) to conveniently visualize the data (see Figure 38-1).

```
In [2]: %matplotlib inline
        import seaborn as sns
        sns.pairplot(iris, hue='species', height=1.5);
```

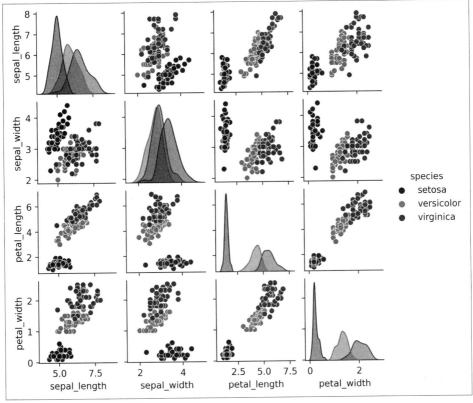

Figure 38-1. A visualization of the Iris dataset[1]

---

1 A full-size, full-color version of this figure can be found on GitHub (*https://oreil.ly/PDSH_GitHub*).

For use in Scikit-Learn, we will extract the features matrix and target array from the DataFrame, which we can do using some of the Pandas DataFrame operations discussed in Part III:

```
In [3]: X_iris = iris.drop('species', axis=1)
        X_iris.shape
Out[3]: (150, 4)

In [4]: y_iris = iris['species']
        y_iris.shape
Out[4]: (150,)
```

To summarize, the expected layout of features and target values is visualized in Figure 38-2.

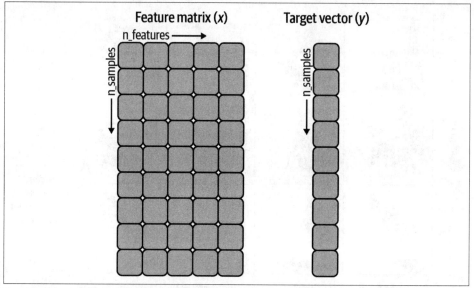

Figure 38-2. Scikit-Learn's data layout[2]

With this data properly formatted, we can move on to consider Scikit-Learn's Estimator API.

The Estimator API

The Scikit-Learn API is designed with the following guiding principles in mind, as outlined in the Scikit-Learn API paper (*http://arxiv.org/abs/1309.0238*):

---

2 Code to produce this figure can be found in the online appendix (*https://oreil.ly/J8V6U*).

Consistency

All objects share a common interface drawn from a limited set of methods, with consistent documentation.

Inspection

All specified parameter values are exposed as public attributes.

Limited object hierarchy

Only algorithms are represented by Python classes; datasets are represented in standard formats (NumPy arrays, Pandas `DataFrame` objects, SciPy sparse matrices) and parameter names use standard Python strings.

Composition

Many machine learning tasks can be expressed as sequences of more fundamental algorithms, and Scikit-Learn makes use of this wherever possible.

Sensible defaults

When models require user-specified parameters, the library defines an appropriate default value.

In practice, these principles make Scikit-Learn very easy to use, once the basic principles are understood. Every machine learning algorithm in Scikit-Learn is implemented via the Estimator API, which provides a consistent interface for a wide range of machine learning applications.

Basics of the API

Most commonly, the steps in using the Scikit-Learn Estimator API are as follows:

1. Choose a class of model by importing the appropriate estimator class from Scikit-Learn.
2. Choose model hyperparameters by instantiating this class with desired values.
3. Arrange data into a features matrix and target vector, as outlined earlier in this chapter.
4. Fit the model to your data by calling the `fit` method of the model instance.
5. Apply the model to new data:
   - For supervised learning, often we predict labels for unknown data using the `predict` method.
   - For unsupervised learning, we often transform or infer properties of the data using the `transform` or `predict` method.

We will now step through several simple examples of applying supervised and unsupervised learning methods.

Supervised Learning Example: Simple Linear Regression

As an example of this process, let's consider a simple linear regression—that is, the common case of fitting a line to (x, y) data. We will use the following simple data for our regression example (see Figure 38-3).

```
In [5]: import matplotlib.pyplot as plt
        import numpy as np

        rng = np.random.RandomState(42)
        x = 10 * rng.rand(50)
        y = 2 * x - 1 + rng.randn(50)
        plt.scatter(x, y);
```

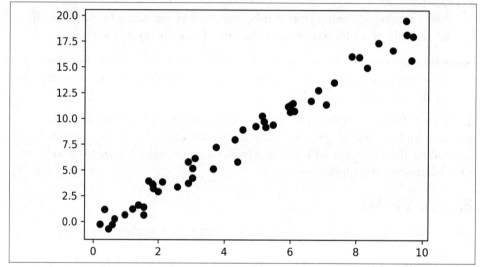

Figure 38-3. Data for linear regression

With this data in place, we can use the recipe outlined earlier. We'll walk through the process in the following sections.

1. Choose a class of model

In Scikit-Learn, every class of model is represented by a Python class. So, for example, if we would like to compute a simple `LinearRegression` model, we can import the linear regression class:

```
In [6]: from sklearn.linear_model import LinearRegression
```

Note that other more general linear regression models exist as well; you can read more about them in the `sklearn.linear_model` module documentation (*https://oreil.ly/YVOFd*).

2. Choose model hyperparameters

An important point is that *a class of model is not the same as an instance of a model.*

Once we have decided on our model class, there are still some options open to us. Depending on the model class we are working with, we might need to answer one or more questions like the following:

- Would we like to fit for the offset (i.e., *y*-intercept)?
- Would we like the model to be normalized?
- Would we like to preprocess our features to add model flexibility?
- What degree of regularization would we like to use in our model?
- How many model components would we like to use?

These are examples of the important choices that must be made *once the model class is selected.* These choices are often represented as *hyperparameters*, or parameters that must be set before the model is fit to data. In Scikit-Learn, hyperparameters are chosen by passing values at model instantiation. We will explore how you can quantitatively choose hyperparameters in Chapter 39.

For our linear regression example, we can instantiate the `LinearRegression` class and specify that we'd like to fit the intercept using the `fit_intercept` hyperparameter:

```
In [7]: model = LinearRegression(fit_intercept=True)
        model
Out[7]: LinearRegression()
```

Keep in mind that when the model is instantiated, the only action is the storing of these hyperparameter values. In particular, we have not yet applied the model to any data: the Scikit-Learn API makes very clear the distinction between *choice of model* and *application of model to data.*

3. Arrange data into a features matrix and target vector

Previously we examined the Scikit-Learn data representation, which requires a two-dimensional features matrix and a one-dimensional target array. Here our target variable y is already in the correct form (a length-n_samples array), but we need to massage the data x to make it a matrix of size [n_samples, n_features].

In this case, this amounts to a simple reshaping of the one-dimensional array:

```
In [8]: X = x[:, np.newaxis]
        X.shape
Out[8]: (50, 1)
```

4. Fit the model to the data

Now it is time to apply our model to the data. This can be done with the `fit` method of the model:

```
In [9]: model.fit(X, y)
Out[9]: LinearRegression()
```

This `fit` command causes a number of model-dependent internal computations to take place, and the results of these computations are stored in model-specific attributes that the user can explore. In Scikit-Learn, by convention all model parameters that were learned during the `fit` process have trailing underscores; for example in this linear model, we have the following:

```
In [10]: model.coef_
Out[10]: array([1.9776566])

In [11]: model.intercept_
Out[11]: -0.9033107255311146
```

These two parameters represent the slope and intercept of the simple linear fit to the data. Comparing the results to the data definition, we see that they are close to the values used to generate the data: a slope of 2 and intercept of –1.

One question that frequently comes up regards the uncertainty in such internal model parameters. In general, Scikit-Learn does not provide tools to draw conclusions from internal model parameters themselves: interpreting model parameters is much more a *statistical modeling* question than a *machine learning* question. Machine learning instead focuses on what the model *predicts*. If you would like to dive into the meaning of fit parameters within the model, other tools are available, including the `statsmodels` Python package (*https://oreil.ly/adDFZ*).

5. Predict labels for unknown data

Once the model is trained, the main task of supervised machine learning is to evaluate it based on what it says about new data that was not part of the training set. In Scikit-Learn, this can be done using the `predict` method. For the sake of this example, our "new data" will be a grid of x values, and we will ask what y values the model predicts:

```
In [12]: xfit = np.linspace(-1, 11)
```

As before, we need to coerce these x values into a `[n_samples, n_features]` features matrix, after which we can feed it to the model:

```
In [13]: Xfit = xfit[:, np.newaxis]
         yfit = model.predict(Xfit)
```

Finally, let's visualize the results by plotting first the raw data, and then this model fit (see Figure 38-4).

```
In [14]: plt.scatter(x, y)
         plt.plot(xfit, yfit);
```

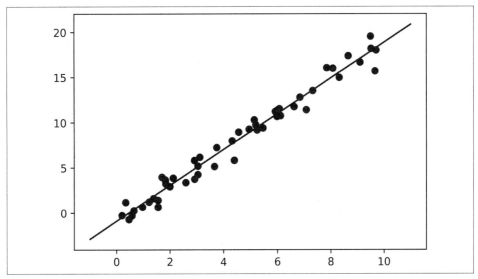

Figure 38-4. A simple linear regression fit to the data

Typically the efficacy of the model is evaluated by comparing its results to some known baseline, as we will see in the next example.

Supervised Learning Example: Iris Classification

Let's take a look at another example of this process, using the Iris dataset we discussed earlier. Our question will be this: given a model trained on a portion of the Iris data, how well can we predict the remaining labels?

For this task, we will use a simple generative model known as *Gaussian naive Bayes*, which proceeds by assuming each class is drawn from an axis-aligned Gaussian distribution (see Chapter 41 for more details). Because it is so fast and has no hyperparameters to choose, Gaussian naive Bayes is often a good model to use as a baseline classification, before exploring whether improvements can be found through more sophisticated models.

We would like to evaluate the model on data it has not seen before, so we will split the data into a *training set* and a *testing set*. This could be done by hand, but it is more convenient to use the train_test_split utility function:

```
In [15]: from sklearn.model_selection import train_test_split
         Xtrain, Xtest, ytrain, ytest = train_test_split(X_iris, y_iris,
                                                         random_state=1)
```

With the data arranged, we can follow our recipe to predict the labels:

```
In [16]: from sklearn.naive_bayes import GaussianNB # 1. choose model class
         model = GaussianNB()                       # 2. instantiate model
         model.fit(Xtrain, ytrain)                  # 3. fit model to data
         y_model = model.predict(Xtest)             # 4. predict on new data
```

Finally, we can use the `accuracy_score` utility to see the fraction of predicted labels that match their true values:

```
In [17]: from sklearn.metrics import accuracy_score
         accuracy_score(ytest, y_model)
Out[17]: 0.9736842105263158
```

With an accuracy topping 97%, we see that even this very naive classification algorithm is effective for this particular dataset!

Unsupervised Learning Example: Iris Dimensionality

As an example of an unsupervised learning problem, let's take a look at reducing the dimensionality of the Iris data so as to more easily visualize it. Recall that the Iris data is four-dimensional: there are four features recorded for each sample.

The task of dimensionality reduction centers around determining whether there is a suitable lower-dimensional representation that retains the essential features of the data. Often dimensionality reduction is used as an aid to visualizing data: after all, it is much easier to plot data in two dimensions than in four dimensions or more!

Here we will use *principal component analysis* (PCA; see Chapter 45), which is a fast linear dimensionality reduction technique. We will ask the model to return two components—that is, a two-dimensional representation of the data.

Following the sequence of steps outlined earlier, we have:

```
In [18]: from sklearn.decomposition import PCA # 1. choose model class
         model = PCA(n_components=2)           # 2. instantiate model
         model.fit(X_iris)                     # 3. fit model to data
         X_2D = model.transform(X_iris)        # 4. transform the data
```

Now let's plot the results. A quick way to do this is to insert the results into the original Iris `DataFrame`, and use Seaborn's `lmplot` to show the results (see Figure 38-5).

```
In [19]: iris['PCA1'] = X_2D[:, 0]
         iris['PCA2'] = X_2D[:, 1]
         sns.lmplot(x="PCA1", y="PCA2", hue='species', data=iris, fit_reg=False);
```

We see that in the two-dimensional representation, the species are fairly well separated, even though the PCA algorithm had no knowledge of the species labels! This suggests to us that a relatively straightforward classification will probably be effective on the dataset, as we saw before.

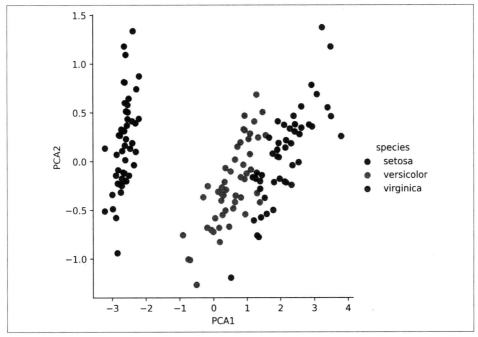

Figure 38-5. The Iris data projected to two dimensions[3]

Unsupervised Learning Example: Iris Clustering

Let's next look at applying clustering to the Iris data. A clustering algorithm attempts to find distinct groups of data without reference to any labels. Here we will use a powerful clustering method called a *Gaussian mixture model* (GMM), discussed in more detail in Chapter 48. A GMM attempts to model the data as a collection of Gaussian blobs.

We can fit the Gaussian mixture model as follows:

```
In [20]: from sklearn.mixture import GaussianMixture      # 1. choose model class
         model = GaussianMixture(n_components=3,
                                 covariance_type='full')   # 2. instantiate model
         model.fit(X_iris)                                 # 3. fit model to data
         y_gmm = model.predict(X_iris)                     # 4. determine labels
```

As before, we will add the cluster label to the Iris `DataFrame` and use Seaborn to plot the results (see Figure 38-6).

---

3 A full-color version of this figure can be found on GitHub (*https://oreil.ly/PDSH_GitHub*).

```
In [21]: iris['cluster'] = y_gmm
         sns.lmplot(x="PCA1", y="PCA2", data=iris, hue='species',
                    col='cluster', fit_reg=False);
```

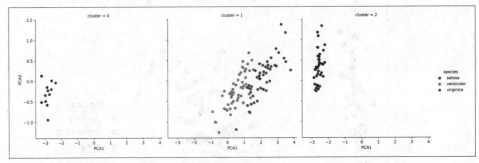

Figure 38-6. k-means clusters within the Iris data[4]

By splitting the data by cluster number, we see exactly how well the GMM algorithm has recovered the underlying labels: the *setosa* species is separated perfectly within cluster 0, while there remains a small amount of mixing between *versicolor* and *virginica*. This means that even without an expert to tell us the species labels of the individual flowers, the measurements of these flowers are distinct enough that we could *automatically* identify the presence of these different groups of species with a simple clustering algorithm! This sort of algorithm might further give experts in the field clues as to the relationships between the samples they are observing.

Application: Exploring Handwritten Digits

To demonstrate these principles on a more interesting problem, let's consider one piece of the optical character recognition problem: the identification of handwritten digits. In the wild, this problem involves both locating and identifying characters in an image. Here we'll take a shortcut and use Scikit-Learn's set of preformatted digits, which is built into the library.

Loading and Visualizing the Digits Data

We can use Scikit-Learn's data access interface to take a look at this data:

```
In [22]: from sklearn.datasets import load_digits
         digits = load_digits()
         digits.images.shape
Out[22]: (1797, 8, 8)
```

---

4 A full-size, full-color version of this figure can be found on GitHub (*https://oreil.ly/PDSH_GitHub*).

The images data is a three-dimensional array: 1,797 samples each consisting of an 8 × 8 grid of pixels. Let's visualize the first hundred of these (see Figure 38-7).

```
In [23]: import matplotlib.pyplot as plt

         fig, axes = plt.subplots(10, 10, figsize=(8, 8),
                               subplot_kw={'xticks':[], 'yticks':[]},
                               gridspec_kw=dict(hspace=0.1, wspace=0.1))

         for i, ax in enumerate(axes.flat):
             ax.imshow(digits.images[i], cmap='binary', interpolation='nearest')
             ax.text(0.05, 0.05, str(digits.target[i]),
                   transform=ax.transAxes, color='green')
```

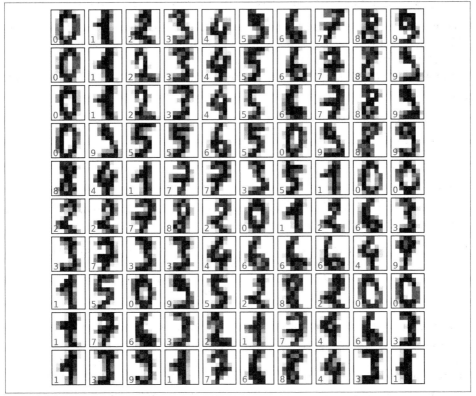

Figure 38-7. The handwritten digits data; each sample is represented by one 8 × 8 grid of pixels

In order to work with this data within Scikit-Learn, we need a two-dimensional, [n_samples, n_features] representation. We can accomplish this by treating each pixel in the image as a feature: that is, by flattening out the pixel arrays so that we have a length-64 array of pixel values representing each digit. Additionally, we need the target array, which gives the previously determined label for each digit. These two quantities are built into the digits dataset under the `data` and `target` attributes, respectively:

```
In [24]: X = digits.data
         X.shape
Out[24]: (1797, 64)

In [25]: y = digits.target
         y.shape
Out[25]: (1797,)
```

We see here that there are 1,797 samples and 64 features.

Unsupervised Learning Example: Dimensionality Reduction

We'd like to visualize our points within the 64-dimensional parameter space, but it's difficult to effectively visualize points in such a high-dimensional space. Instead, we'll reduce the number of dimensions, using an unsupervised method. Here, we'll make use of a manifold learning algorithm called Isomap (see Chapter 46) and transform the data to two dimensions:

```
In [26]: from sklearn.manifold import Isomap
         iso = Isomap(n_components=2)
         iso.fit(digits.data)
         data_projected = iso.transform(digits.data)
         print(data_projected.shape)
Out[26]: (1797, 2)
```

We see that the projected data is now two-dimensional. Let's plot this data to see if we can learn anything from its structure (see Figure 38-8).

```
In [27]: plt.scatter(data_projected[:, 0], data_projected[:, 1], c=digits.target,
                      edgecolor='none', alpha=0.5,
                      cmap=plt.cm.get_cmap('viridis', 10))
         plt.colorbar(label='digit label', ticks=range(10))
         plt.clim(-0.5, 9.5);
```

This plot gives us some good intuition into how well various numbers are separated in the larger 64-dimensional space. For example, zeros and ones have very little overlap in the parameter space. Intuitively, this makes sense: a zero is empty in the middle of the image, while a one will generally have ink in the middle. On the other hand, there seems to be a more or less continuous spectrum between ones and fours: we can understand this by realizing that some people draw ones with "hats" on them, which causes them to look similar to fours.

Overall, however, despite some mixing at the edges, the different groups appear to be fairly well localized in the parameter space: this suggests that even a very straightforward supervised classification algorithm should perform suitably on the full high-dimensional dataset. Let's give it a try.

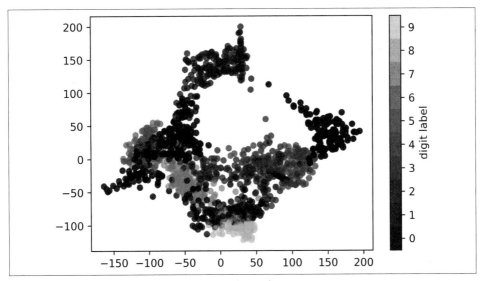

Figure 38-8. An Isomap embedding of the digits data

Classification on Digits

Let's apply a classification algorithm to the digits data. As we did with the Iris data previously, we will split the data into training and testing sets and fit a Gaussian naive Bayes model:

```
In [28]: Xtrain, Xtest, ytrain, ytest = train_test_split(X, y, random_state=0)
```

```
In [29]: from sklearn.naive_bayes import GaussianNB
         model = GaussianNB()
         model.fit(Xtrain, ytrain)
         y_model = model.predict(Xtest)
```

Now that we have the model's predictions, we can gauge its accuracy by comparing the true values of the test set to the predictions:

```
In [30]: from sklearn.metrics import accuracy_score
         accuracy_score(ytest, y_model)
Out[30]: 0.8333333333333334
```

With even this very simple model, we find about 83% accuracy for classification of the digits! However, this single number doesn't tell us where we've gone wrong. One nice way to do this is to use the *confusion matrix*, which we can compute with Scikit-Learn and plot with Seaborn (see Figure 38-9).

```
In [31]: from sklearn.metrics import confusion_matrix

         mat = confusion_matrix(ytest, y_model)

         sns.heatmap(mat, square=True, annot=True, cbar=False, cmap='Blues')
```

```
plt.xlabel('predicted value')
plt.ylabel('true value');
```

This shows us where the mislabeled points tend to be: for example, many of the twos here are misclassified as either ones or eights.

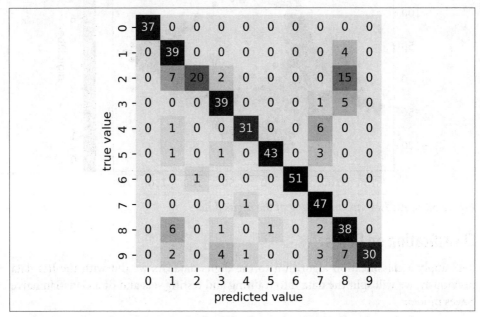

Figure 38-9. A confusion matrix showing the frequency of misclassifications by our classifier

Another way to gain intuition into the characteristics of the model is to plot the inputs again, with their predicted labels. We'll use green for correct labels and red for incorrect labels; see Figure 38-10.

```
In [32]: fig, axes = plt.subplots(10, 10, figsize=(8, 8),
                                   subplot_kw={'xticks':[], 'yticks':[]},
                                   gridspec_kw=dict(hspace=0.1, wspace=0.1))

         test_images = Xtest.reshape(-1, 8, 8)

         for i, ax in enumerate(axes.flat):
             ax.imshow(test_images[i], cmap='binary', interpolation='nearest')
             ax.text(0.05, 0.05, str(y_model[i]),
                     transform=ax.transAxes,
                     color='green' if (ytest[i] == y_model[i]) else 'red')
```

Examining this subset of the data can give us some insight into where the algorithm might be not performing optimally. To go beyond our 83% classification success rate, we might switch to a more sophisticated algorithm such as support vector machines (see Chapter 43), random forests (see Chapter 44), or another classification approach.

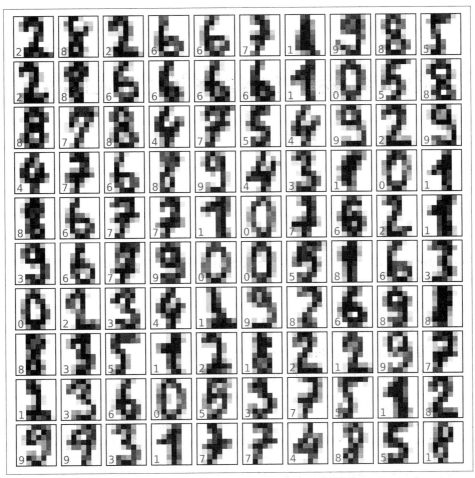

Figure 38-10. Data showing correct (green) and incorrect (red) labels; for a color version of this plot, see the online version of the book (https://oreil.ly/PDSH_GitHub)

Summary

In this chapter we covered the essential features of the Scikit-Learn data representation and the Estimator API. Regardless of the type of estimator used, the same import/instantiate/fit/predict pattern holds. Armed with this information, you can explore the Scikit-Learn documentation and try out various models on your data.

In the next chapter, we will explore perhaps the most important topic in machine learning: how to select and validate your model.

## CHAPTER 39

Hyperparameters and Model Validation

In the previous chapter, we saw the basic recipe for applying a supervised machine learning model:

1. Choose a class of model.
2. Choose model hyperparameters.
3. Fit the model to the training data.
4. Use the model to predict labels for new data.

The first two pieces of this—the choice of model and choice of hyperparameters—are perhaps the most important part of using these tools and techniques effectively. In order to make informed choices, we need a way to *validate* that our model and our hyperparameters are a good fit to the data. While this may sound simple, there are some pitfalls that you must avoid to do this effectively.

Thinking About Model Validation

In principle, model validation is very simple: after choosing a model and its hyperparameters, we can estimate how effective it is by applying it to some of the training data and comparing the predictions to the known values.

This section will first show a naive approach to model validation and why it fails, before exploring the use of holdout sets and cross-validation for more robust model evaluation.

Model Validation the Wrong Way

Let's start with the naive approach to validation using the Iris dataset, which we saw in the previous chapter. We will start by loading the data:

```
In [1]: from sklearn.datasets import load_iris
        iris = load_iris()
        X = iris.data
        y = iris.target
```

Next, we choose a model and hyperparameters. Here we'll use a *k*-nearest neighbors classifier with n_neighbors=1. This is a very simple and intuitive model that says "the label of an unknown point is the same as the label of its closest training point":

```
In [2]: from sklearn.neighbors import KNeighborsClassifier
        model = KNeighborsClassifier(n_neighbors=1)
```

Then we train the model, and use it to predict labels for data whose labels we already know:

```
In [3]: model.fit(X, y)
        y_model = model.predict(X)
```

Finally, we compute the fraction of correctly labeled points:

```
In [4]: from sklearn.metrics import accuracy_score
        accuracy_score(y, y_model)
Out[4]: 1.0
```

We see an accuracy score of 1.0, which indicates that 100% of points were correctly labeled by our model! But is this truly measuring the expected accuracy? Have we really come upon a model that we expect to be correct 100% of the time?

As you may have gathered, the answer is no. In fact, this approach contains a fundamental flaw: *it trains and evaluates the model on the same data*. Furthermore, this nearest neighbor model is an *instance-based* estimator that simply stores the training data, and predicts labels by comparing new data to these stored points: except in contrived cases, it will get 100% accuracy every time!

Model Validation the Right Way: Holdout Sets

So what can be done? A better sense of a model's performance can be found by using what's known as a *holdout set*: that is, we hold back some subset of the data from the training of the model, and then use this holdout set to check the model's performance. This splitting can be done using the train_test_split utility in Scikit-Learn:

```
In [5]: from sklearn.model_selection import train_test_split
        # split the data with 50% in each set
        X1, X2, y1, y2 = train_test_split(X, y, random_state=0,
                                          train_size=0.5)
```

```
        # fit the model on one set of data
        model.fit(X1, y1)

        # evaluate the model on the second set of data
        y2_model = model.predict(X2)
        accuracy_score(y2, y2_model)
Out[5]: 0.9066666666666666
```

We see here a more reasonable result: the one-nearest-neighbor classifier is about 90% accurate on this holdout set. The holdout set is similar to unknown data, because the model has not "seen" it before.

Model Validation via Cross-Validation

One disadvantage of using a holdout set for model validation is that we have lost a portion of our data to the model training. In the preceding case, half the dataset does not contribute to the training of the model! This is not optimal, especially if the initial set of training data is small.

One way to address this is to use *cross-validation*; that is, to do a sequence of fits where each subset of the data is used both as a training set and as a validation set. Visually, it might look something like Figure 39-1.

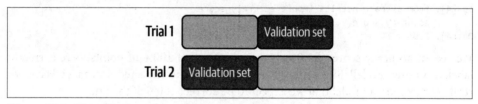

Figure 39-1. Visualization of two-fold cross-validation[1]

Here we do two validation trials, alternately using each half of the data as a holdout set. Using the split data from earlier, we could implement it like this:

```
In [6]: y2_model = model.fit(X1, y1).predict(X2)
        y1_model = model.fit(X2, y2).predict(X1)
        accuracy_score(y1, y1_model), accuracy_score(y2, y2_model)
Out[6]: (0.96, 0.9066666666666666)
```

What comes out are two accuracy scores, which we could combine (by, say, taking the mean) to get a better measure of the global model performance. This particular form of cross-validation is a *two-fold cross-validation*—that is, one in which we have split the data into two sets and used each in turn as a validation set.

---

1 Code to produce this figure can be found in the online appendix (*https://oreil.ly/jv0wb*).

We could expand on this idea to use even more trials, and more folds in the data—for example, Figure 39-2 shows a visual depiction of five-fold cross-validation.

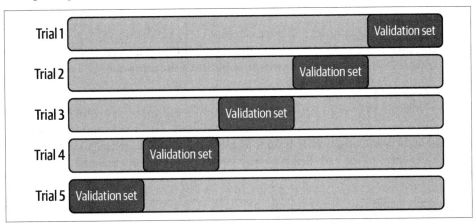

Figure 39-2. *Visualization of five-fold cross-validation*[2]

Here we split the data into five groups and use each in turn to evaluate the model fit on the other four-fifths of the data. This would be rather tedious to do by hand, but we can use Scikit-Learn's `cross_val_score` convenience routine to do it succinctly:

```
In [7]: from sklearn.model_selection import cross_val_score
        cross_val_score(model, X, y, cv=5)
Out[7]: array([0.96666667, 0.96666667, 0.93333333, 0.93333333, 1.       ])
```

Repeating the validation across different subsets of the data gives us an even better idea of the performance of the algorithm.

Scikit-Learn implements a number of cross-validation schemes that are useful in particular situations; these are implemented via iterators in the `model_selection` module. For example, we might wish to go to the extreme case in which our number of folds is equal to the number of data points: that is, we train on all points but one in each trial. This type of cross-validation is known as *leave-one-out* cross validation, and can be used as follows:

```
In [8]: from sklearn.model_selection import LeaveOneOut
        scores = cross_val_score(model, X, y, cv=LeaveOneOut())
        scores
Out[8]: array([1., 1., 1., 1., 1., 1., 1., 1., 1., 1., 1., 1., 1., 1., 1., 1., 1.,
               1., 1., 1., 1., 1., 1., 1., 1., 1., 1., 1., 1., 1., 1., 1., 1.,
               1., 1., 1., 1., 1., 1., 1., 1., 1., 1., 1., 1., 1., 1., 1., 1.,
               1., 1., 1., 1., 1., 1., 1., 1., 1., 1., 1., 1., 1., 1., 1., 1.,
               1., 1., 0., 1., 0., 1., 1., 1., 1., 1., 1., 1., 1., 1., 0., 1.,
```

---

2 Code to produce this figure can be found in the online appendix (*https://oreil.ly/2BP2o*).

```
         1., 1., 1., 1., 1., 1., 1., 1., 1., 1., 1., 1., 1., 1., 1., 1., 1.,
         1., 1., 1., 1., 0., 1., 1., 1., 1., 1., 1., 1., 1., 1., 1., 1., 1.,
         0., 1., 1., 1., 1., 1., 1., 1., 1., 1., 1., 1., 1., 1., 0., 1., 1.,
         1., 1., 1., 1., 1., 1., 1., 1., 1., 1., 1., 1., 1., 1.])
```

Because we have 150 samples, the leave-one-out cross-validation yields scores for 150 trials, and each score indicates either a successful (1.0) or an unsuccessful (0.0) prediction. Taking the mean of these gives an estimate of the error rate:

```
In [9]: scores.mean()
Out[9]: 0.96
```

Other cross-validation schemes can be used similarly. For a description of what is available in Scikit-Learn, use IPython to explore the `sklearn.model_selection` submodule, or take a look at Scikit-Learn's cross-validation documentation (*https:// oreil.ly/rITkn*).

Selecting the Best Model

Now that we've explored the basics of validation and cross-validation, we will go into a little more depth regarding model selection and selection of hyperparameters. These issues are some of the most important aspects of the practice of machine learning, but I find that this information is often glossed over in introductory machine learning tutorials.

Of core importance is the following question: *if our estimator is underperforming, how should we move forward?* There are several possible answers:

- Use a more complicated/more flexible model.
- Use a less complicated/less flexible model.
- Gather more training samples.
- Gather more data to add features to each sample.

The answer to this question is often counterintuitive. In particular, sometimes using a more complicated model will give worse results, and adding more training samples may not improve your results! The ability to determine what steps will improve your model is what separates the successful machine learning practitioners from the unsuccessful.

The Bias-Variance Trade-off

Fundamentally, finding "the best model" is about finding a sweet spot in the trade-off between *bias* and *variance*. Consider Figure 39-3, which presents two regression fits to the same dataset.

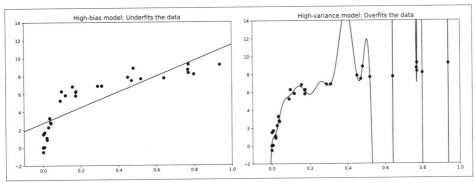

Figure 39-3. High-bias and high-variance regression models[3]

It is clear that neither of these models is a particularly good fit to the data, but they fail in different ways.

The model on the left attempts to find a straight-line fit through the data. Because in this case a straight line cannot accurately split the data, the straight-line model will never be able to describe this dataset well. Such a model is said to *underfit* the data: that is, it does not have enough flexibility to suitably account for all the features in the data. Another way of saying this is that the model has high bias.

The model on the right attempts to fit a high-order polynomial through the data. Here the model fit has enough flexibility to nearly perfectly account for the fine features in the data, but even though it very accurately describes the training data, its precise form seems to be more reflective of the particular noise properties of the data than of the intrinsic properties of whatever process generated that data. Such a model is said to *overfit* the data: that is, it has so much flexibility that the model ends up accounting for random errors as well as the underlying data distribution. Another way of saying this is that the model has high variance.

To look at this in another light, consider what happens if we use these two models to predict the *y*-values for some new data. In the plots in Figure 39-4, the red/lighter points indicate data that is omitted from the training set.

---

3 Code to produce this figure can be found in the online appendix (*https://oreil.ly/j9G96*).

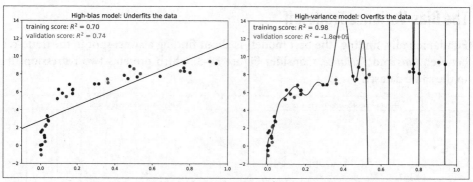

Figure 39-4. Training and validation scores in high-bias and high-variance models[4]

The score here is the R^2 score, or coefficient of determination (*https://oreil.ly/2AtV8*), which measures how well a model performs relative to a simple mean of the target values. $R^2 = 1$ indicates a perfect match, $R^2 = 0$ indicates the model does no better than simply taking the mean of the data, and negative values mean even worse models. From the scores associated with these two models, we can make an observation that holds more generally:

- For high-bias models, the performance of the model on the validation set is similar to the performance on the training set.

- For high-variance models, the performance of the model on the validation set is far worse than the performance on the training set.

If we imagine that we have some ability to tune the model complexity, we would expect the training score and validation score to behave as illustrated in Figure 39-5, often called a *validation curve*, and we see the following features:

- The training score is everywhere higher than the validation score. This is generally the case: the model will be a better fit to data it has seen than to data it has not seen.

- For very low model complexity (a high-bias model), the training data is underfit, which means that the model is a poor predictor both for the training data and for any previously unseen data.

- For very high model complexity (a high-variance model), the training data is overfit, which means that the model predicts the training data very well, but fails for any previously unseen data.

---

4 Code to produce this figure can be found in the online appendix (*https://oreil.ly/YfwRC*).

- For some intermediate value, the validation curve has a maximum. This level of complexity indicates a suitable trade-off between bias and variance.

The means of tuning the model complexity varies from model to model; when we discuss individual models in depth in later chapters, we will see how each model allows for such tuning.

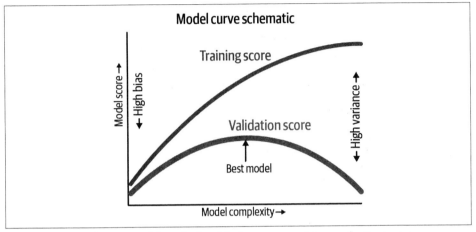

Figure 39-5. A schematic of the relationship between model complexity, training score, and validation score[5]

Validation Curves in Scikit-Learn

Let's look at an example of using cross-validation to compute the validation curve for a class of models. Here we will use a *polynomial regression* model, a generalized linear model in which the degree of the polynomial is a tunable parameter. For example, a degree-1 polynomial fits a straight line to the data; for model parameters *a* and *b*:

$$y = ax + b$$

A degree-3 polynomial fits a cubic curve to the data; for model parameters *a*, *b*, *c*, *d*:

$$y = ax^3 + bx^2 + cx + d$$

---

5 Code to produce this figure can be found in the online appendix (*https://oreil.ly/4AK15*).

We can generalize this to any number of polynomial features. In Scikit-Learn, we can implement this with a linear regression classifier combined with the polynomial preprocessor. We will use a *pipeline* to string these operations together (we will discuss polynomial features and pipelines more fully in Chapter 40):

```
In [10]: from sklearn.preprocessing import PolynomialFeatures
         from sklearn.linear_model import LinearRegression
         from sklearn.pipeline import make_pipeline

         def PolynomialRegression(degree=2, **kwargs):
             return make_pipeline(PolynomialFeatures(degree),
                                  LinearRegression(**kwargs))
```

Now let's create some data to which we will fit our model:

```
In [11]: import numpy as np

         def make_data(N, err=1.0, rseed=1):
             # randomly sample the data
             rng = np.random.RandomState(rseed)
             X = rng.rand(N, 1) ** 2
             y = 10 - 1. / (X.ravel() + 0.1)
             if err > 0:
                 y += err * rng.randn(N)
             return X, y

         X, y = make_data(40)
```

We can now visualize our data, along with polynomial fits of several degrees (see Figure 39-6).

```
In [12]: %matplotlib inline
         import matplotlib.pyplot as plt
         plt.style.use('seaborn-whitegrid')

         X_test = np.linspace(-0.1, 1.1, 500)[:, None]

         plt.scatter(X.ravel(), y, color='black')
         axis = plt.axis()
         for degree in [1, 3, 5]:
             y_test = PolynomialRegression(degree).fit(X, y).predict(X_test)
             plt.plot(X_test.ravel(), y_test, label='degree={0}'.format(degree))
         plt.xlim(-0.1, 1.0)
         plt.ylim(-2, 12)
         plt.legend(loc='best');
```

The knob controlling model complexity in this case is the degree of the polynomial, which can be any nonnegative integer. A useful question to answer is this: what degree of polynomial provides a suitable trade-off between bias (underfitting) and variance (overfitting)?

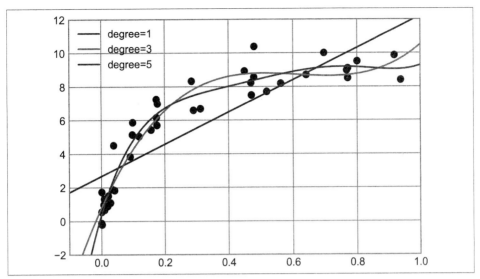

Figure 39-6. Three different polynomial models fit to a dataset[6]

We can make progress in this by visualizing the validation curve for this particular data and model; this can be done straightforwardly using the `validation_curve` convenience routine provided by Scikit-Learn. Given a model, data, parameter name, and a range to explore, this function will automatically compute both the training score and the validation score across the range (see Figure 39-7).

```
In [13]: from sklearn.model_selection import validation_curve
         degree = np.arange(0, 21)
         train_score, val_score = validation_curve(
             PolynomialRegression(), X, y,
             param_name='polynomialfeatures__degree',
             param_range=degree, cv=7)

         plt.plot(degree, np.median(train_score, 1),
                 color='blue', label='training score')
         plt.plot(degree, np.median(val_score, 1),
                 color='red', label='validation score')
         plt.legend(loc='best')
         plt.ylim(0, 1)
         plt.xlabel('degree')
         plt.ylabel('score');
```

---

6 A full-color version of this figure can be found on GitHub (*https://oreil.ly/PDSH_GitHub*).

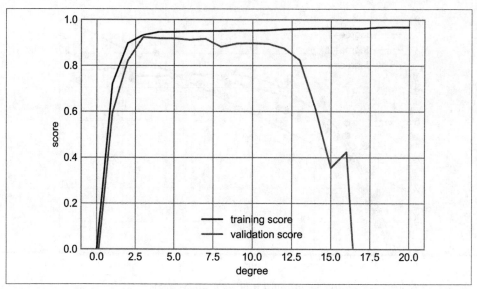

Figure 39-7. The validation curves for the data in Figure 39-9

This shows precisely the qualitative behavior we expect: the training score is everywhere higher than the validation score, the training score is monotonically improving with increased model complexity, and the validation score reaches a maximum before dropping off as the model becomes overfit.

From the validation curve, we can determine that the optimal trade-off between bias and variance is found for a third-order polynomial. We can compute and display this fit over the original data as follows (see Figure 39-8).

```
In [14]: plt.scatter(X.ravel(), y)
         lim = plt.axis()
         y_test = PolynomialRegression(3).fit(X, y).predict(X_test)
         plt.plot(X_test.ravel(), y_test);
         plt.axis(lim);
```

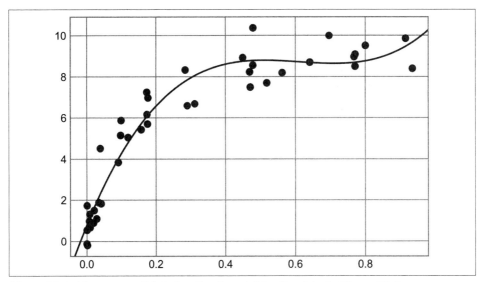

Figure 39-8. The cross-validated optimal model for the data in Figure 39-6

Notice that finding this optimal model did not actually require us to compute the training score, but examining the relationship between the training score and validation score can give us useful insight into the performance of the model.

Learning Curves

One important aspect of model complexity is that the optimal model will generally depend on the size of your training data. For example, let's generate a new dataset with five times as many points (see Figure 39-9).

```
In [15]: X2, y2 = make_data(200)
         plt.scatter(X2.ravel(), y2);
```

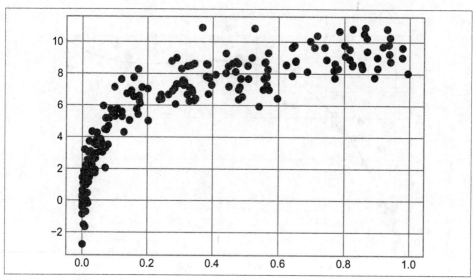

Figure 39-9. Data to demonstrate learning curves

Now let's duplicate the preceding code to plot the validation curve for this larger dataset; for reference, we'll overplot the previous results as well (see Figure 39-10).

```
In [16]: degree = np.arange(21)
         train_score2, val_score2 = validation_curve(
             PolynomialRegression(), X2, y2,
             param_name='polynomialfeatures__degree',
             param_range=degree, cv=7)

         plt.plot(degree, np.median(train_score2, 1),
                  color='blue', label='training score')
         plt.plot(degree, np.median(val_score2, 1),
                  color='red', label='validation score')
         plt.plot(degree, np.median(train_score, 1),
                  color='blue', alpha=0.3, linestyle='dashed')
         plt.plot(degree, np.median(val_score, 1),
                  color='red', alpha=0.3, linestyle='dashed')
         plt.legend(loc='lower center')
         plt.ylim(0, 1)
         plt.xlabel('degree')
         plt.ylabel('score');
```

The solid lines show the new results, while the fainter dashed lines show the results on the previous smaller dataset. It is clear from the validation curve that the larger dataset can support a much more complicated model: the peak here is probably around a degree of 6, but even a degree-20 model isn't seriously overfitting the data— the validation and training scores remain very close.

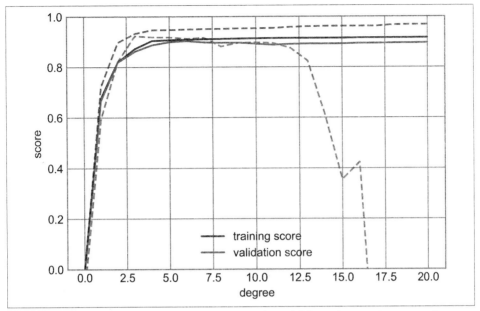

Figure 39-10. Learning curves for the polynomial model fit to data in Figure 39-9[7]

So, the behavior of the validation curve has not one but two important inputs: the model complexity and the number of training points. We can gain further insight by exploring the behavior of the model as a function of the number of training points, which we can do by using increasingly larger subsets of the data to fit our model. A plot of the training/validation score with respect to the size of the training set is sometimes known as a *learning curve*.

The general behavior we would expect from a learning curve is this:

- A model of a given complexity will *overfit* a small dataset: this means the training score will be relatively high, while the validation score will be relatively low.

- A model of a given complexity will *underfit* a large dataset: this means that the training score will decrease, but the validation score will increase.

- A model will never, except by chance, give a better score to the validation set than the training set: this means the curves should keep getting closer together but never cross.

With these features in mind, we would expect a learning curve to look qualitatively like that shown in Figure 39-11.

---

7 A full-color version of this figure can be found on GitHub (*https://oreil.ly/PDSH_GitHub*).

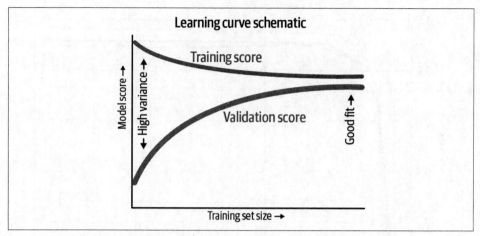

Figure 39-11. Schematic showing the typical interpretation of learning curves[8]

The notable feature of the learning curve is the convergence to a particular score as the number of training samples grows. In particular, once you have enough points that a particular model has converged, *adding more training data will not help you!* The only way to increase model performance in this case is to use another (often more complex) model.

Scikit-Learn offers a convenient utility for computing such learning curves from your models; here we will compute a learning curve for our original dataset with a second-order polynomial model and a ninth-order polynomial (see Figure 39-12).

```
In [17]: from sklearn.model_selection import learning_curve

         fig, ax = plt.subplots(1, 2, figsize=(16, 6))
         fig.subplots_adjust(left=0.0625, right=0.95, wspace=0.1)

         for i, degree in enumerate([2, 9]):
             N, train_lc, val_lc = learning_curve(
                 PolynomialRegression(degree), X, y, cv=7,
                 train_sizes=np.linspace(0.3, 1, 25))

             ax[i].plot(N, np.mean(train_lc, 1),
                     color='blue', label='training score')
             ax[i].plot(N, np.mean(val_lc, 1),
                     color='red', label='validation score')
             ax[i].hlines(np.mean([train_lc[-1], val_lc[-1]]), N[0],
                     N[-1], color='gray', linestyle='dashed')

             ax[i].set_ylim(0, 1)
```

---

8 Code to produce this figure can be found in the online appendix (*https://oreil.ly/omZ1c*).

```
ax[i].set_xlim(N[0], N[-1])
ax[i].set_xlabel('training size')
ax[i].set_ylabel('score')
ax[i].set_title('degree = {0}'.format(degree), size=14)
ax[i].legend(loc='best')
```

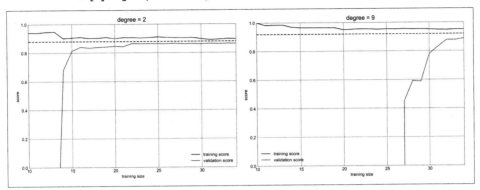

Figure 39-12. Learning curves for a low-complexity model (left) and a high-complexity model (right)[9]

This is a valuable diagnostic, because it gives us a visual depiction of how our model responds to increasing amounts of training data. In particular, when the learning curve has already converged (i.e., when the training and validation curves are already close to each other) *adding more training data will not significantly improve the fit!* This situation is seen in the left panel, with the learning curve for the degree-2 model.

The only way to increase the converged score is to use a different (usually more complicated) model. We see this in the right panel: by moving to a much more complicated model, we increase the score of convergence (indicated by the dashed line), but at the expense of higher model variance (indicated by the difference between the training and validation scores). If we were to add even more data points, the learning curve for the more complicated model would eventually converge.

Plotting a learning curve for your particular choice of model and dataset can help you to make this type of decision about how to move forward in improving your analysis.

The solid lines show the new results, while the fainter dashed lines show the results on the previous smaller dataset. It is clear from the validation curve that the larger dataset can support a much more complicated model: the peak here is probably around a degree of 6, but even a degree-20 model isn't seriously overfitting the data—the validation and training scores remain very close.

---

9 A full-size version of this figure can be found on GitHub (*https://oreil.ly/PDSH_GitHub*).

Validation in Practice: Grid Search

The preceding discussion is meant to give you some intuition into the trade-off between bias and variance, and its dependence on model complexity and training set size. In practice, models generally have more than one knob to turn, meaning plots of validation and learning curves change from lines to multidimensional surfaces. In these cases, such visualizations are difficult, and we would rather simply find the particular model that maximizes the validation score.

Scikit-Learn provides some tools to make this kind of search more convenient: here we'll consider the use of grid search to find the optimal polynomial model. We will explore a two-dimensional grid of model features, namely the polynomial degree and the flag telling us whether to fit the intercept. This can be set up using Scikit-Learn's GridSearchCV meta-estimator:

```
In [18]: from sklearn.model_selection import GridSearchCV

         param_grid = {'polynomialfeatures__degree': np.arange(21),
                       'linearregression__fit_intercept': [True, False]}

         grid = GridSearchCV(PolynomialRegression(), param_grid, cv=7)
```

Notice that like a normal estimator, this has not yet been applied to any data. Calling the fit method will fit the model at each grid point, keeping track of the scores along the way:

```
In [19]: grid.fit(X, y);
```

Now that the model is fit, we can ask for the best parameters as follows:

```
In [20]: grid.best_params_
Out[20]: {'linearregression__fit_intercept': False, 'polynomialfeatures__degree': 4}
```

Finally, if we wish, we can use the best model and show the fit to our data using code from before (see Figure 39-13).

```
In [21]: model = grid.best_estimator_

         plt.scatter(X.ravel(), y)
         lim = plt.axis()
         y_test = model.fit(X, y).predict(X_test)
         plt.plot(X_test.ravel(), y_test);
         plt.axis(lim);
```

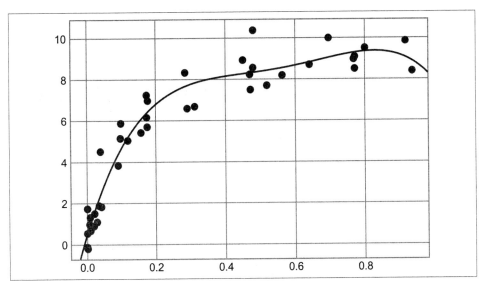

Figure 39-13. The best-fit model determined via an automatic grid search

Other options in `GridSearchCV` include the ability to specify a custom scoring function, to parallelize the computations, to do randomized searches, and more. For more information, see the examples in Chapters 49 and 50, or refer to Scikit-Learn's grid search documentation (*https://oreil.ly/xft8j*).

Summary

In this chapter we began to explore the concept of model validation and hyperparameter optimization, focusing on intuitive aspects of the bias–variance trade-off and how it comes into play when fitting models to data. In particular, we found that the use of a validation set or cross-validation approach is vital when tuning parameters in order to avoid overfitting for more complex/flexible models.

In later chapters, we will discuss the details of particularly useful models, what tuning is available for these models, and how these free parameters affect model complexity. Keep the lessons of this chapter in mind as you read on and learn about these machine learning approaches!

## CHAPTER 40

Feature Engineering

The previous chapters outlined the fundamental ideas of machine learning, but all of the examples so far have assumed that you have numerical data in a tidy, [n_sam ples, n_features] format. In the real world, data rarely comes in such a form. With this in mind, one of the more important steps in using machine learning in practice is *feature engineering*: that is, taking whatever information you have about your problem and turning it into numbers that you can use to build your feature matrix.

In this chapter, we will cover a few common examples of feature engineering tasks: we'll look at features for representing categorical data, text, and images. Additionally, we will discuss derived features for increasing model complexity and imputation of missing data. This process is commonly referred to as vectorization, as it involves converting arbitrary data into well-behaved vectors.

Categorical Features

One common type of nonnumerical data is *categorical* data. For example, imagine you are exploring some data on housing prices, and along with numerical features like "price" and "rooms," you also have "neighborhood" information. For example, your data might look something like this:

```
In [1]: data = [
            {'price': 850000, 'rooms': 4, 'neighborhood': 'Queen Anne'},
            {'price': 700000, 'rooms': 3, 'neighborhood': 'Fremont'},
            {'price': 650000, 'rooms': 3, 'neighborhood': 'Wallingford'},
            {'price': 600000, 'rooms': 2, 'neighborhood': 'Fremont'}
        ]
```

You might be tempted to encode this data with a straightforward numerical mapping:

```
In [2]: {'Queen Anne': 1, 'Fremont': 2, 'Wallingford': 3};
```

But it turns out that this is not generally a useful approach in Scikit-Learn. The package's models make the fundamental assumption that numerical features reflect algebraic quantities, so such a mapping would imply, for example, that *Queen Anne* < *Fremont* < *Wallingford*, or even that *Wallingford–Queen Anne = Fremont*, which (niche demographic jokes aside) does not make much sense.

In this case, one proven technique is to use *one-hot encoding*, which effectively creates extra columns indicating the presence or absence of a category with a value of 1 or 0, respectively. When your data takes the form of a list of dictionaries, Scikit-Learn's `DictVectorizer` will do this for you:

```
In [3]: from sklearn.feature_extraction import DictVectorizer
        vec = DictVectorizer(sparse=False, dtype=int)
        vec.fit_transform(data)
Out[3]: array([[      0,      1,      0, 850000,      4],
               [      1,      0,      0, 700000,      3],
               [      0,      0,      1, 650000,      3],
               [      1,      0,      0, 600000,      2]])
```

Notice that the `neighborhood` column has been expanded into three separate columns representing the three neighborhood labels, and that each row has a 1 in the column associated with its neighborhood. With these categorical features thus encoded, you can proceed as normal with fitting a Scikit-Learn model.

To see the meaning of each column, you can inspect the feature names:

```
In [4]: vec.get_feature_names_out()
Out[4]: array(['neighborhood=Fremont', 'neighborhood=Queen Anne',
               'neighborhood=Wallingford', 'price', 'rooms'], dtype=object)
```

There is one clear disadvantage of this approach: if your category has many possible values, this can *greatly* increase the size of your dataset. However, because the encoded data contains mostly zeros, a sparse output can be a very efficient solution:

```
In [5]: vec = DictVectorizer(sparse=True, dtype=int)
        vec.fit_transform(data)
Out[5]: <4x5 sparse matrix of type '<class 'numpy.int64'>'
                with 12 stored elements in Compressed Sparse Row format>
```

Nearly all of the Scikit-Learn estimators accept such sparse inputs when fitting and evaluating models. Two additional tools that Scikit-Learn includes to support this type of encoding are `sklearn.preprocessing.OneHotEncoder` and `sklearn.feature_extraction.FeatureHasher`.

Text Features

Another common need in feature engineering is to convert text to a set of representative numerical values. For example, most automatic mining of social media data relies on some form of encoding the text as numbers. One of the simplest methods of encoding this type of data is by *word counts*: you take each snippet of text, count the occurrences of each word within it, and put the results in a table.

For example, consider the following set of three phrases:

```
In [6]: sample = ['problem of evil',
                  'evil queen',
                  'horizon problem']
```

For a vectorization of this data based on word count, we could construct individual columns representing the words "problem," "of," "evil," and so on. While doing this by hand would be possible for this simple example, the tedium can be avoided by using Scikit-Learn's CountVectorizer:

```
In [7]: from sklearn.feature_extraction.text import CountVectorizer

        vec = CountVectorizer()
        X = vec.fit_transform(sample)
        X
Out[7]: <3x5 sparse matrix of type '<class 'numpy.int64'>'
            with 7 stored elements in Compressed Sparse Row format>
```

The result is a sparse matrix recording the number of times each word appears; it is easier to inspect if we convert this to a DataFrame with labeled columns:

```
In [8]: import pandas as pd
        pd.DataFrame(X.toarray(), columns=vec.get_feature_names_out())
Out[8]:    evil  horizon  of  problem  queen
        0    1        0   1        1      0
        1    1        0   0        0      1
        2    0        1   0        1      0
```

There are some issues with using a simple raw word count, however: it can lead to features that put too much weight on words that appear very frequently, and this can be suboptimal in some classification algorithms. One approach to fix this is known as *term frequency–inverse document frequency (TF–IDF)*, which weights the word counts by a measure of how often they appear in the documents. The syntax for computing these features is similar to the previous example:

The solid lines show the new results, while the fainter dashed lines show the results on the previous smaller dataset. It is clear from the validation curve that the larger dataset can support a much more complicated model: the peak here is probably around a degree of 6, but even a degree-20 model isn't seriously overfitting the data—the validation and training scores remain very close.

```
In [9]: from sklearn.feature_extraction.text import TfidfVectorizer
        vec = TfidfVectorizer()
        X = vec.fit_transform(sample)
        pd.DataFrame(X.toarray(), columns=vec.get_feature_names_out())
Out[9]:        evil    horizon        of   problem     queen
        0  0.517856  0.000000  0.680919  0.517856  0.000000
        1  0.605349  0.000000  0.000000  0.000000  0.795961
        2  0.000000  0.795961  0.000000  0.605349  0.000000
```

For an example of using TF-IDF in a classification problem, see Chapter 41.

Image Features

Another common need is to suitably encode images for machine learning analysis. The simplest approach is what we used for the digits data in Chapter 38: simply using the pixel values themselves. But depending on the application, such an approach may not be optimal.

A comprehensive summary of feature extraction techniques for images is well beyond the scope of this chapter, but you can find excellent implementations of many of the standard approaches in the Scikit-Image project (*http://scikit-image.org*). For one example of using Scikit-Learn and Scikit-Image together, see Chapter 50.

Derived Features

Another useful type of feature is one that is mathematically derived from some input features. We saw an example of this in Chapter 39 when we constructed *polynomial features* from our input data. We saw that we could convert a linear regression into a polynomial regression not by changing the model, but by transforming the input!

For example, this data clearly cannot be well described by a straight line (see Figure 40-1):

```
In [10]: %matplotlib inline
         import numpy as np
         import matplotlib.pyplot as plt

         x = np.array([1, 2, 3, 4, 5])
         y = np.array([4, 2, 1, 3, 7])
         plt.scatter(x, y);
```

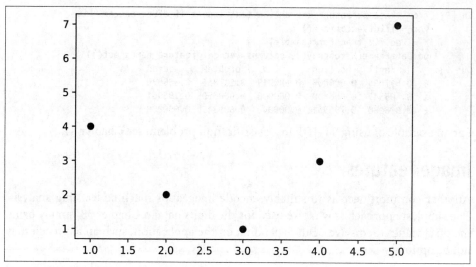

Figure 40-1. Data that is not well described by a straight line

We can still fit a line to the data using `LinearRegression` and get the optimal result, as shown in Figure 40-2:

```
In [11]: from sklearn.linear_model import LinearRegression
         X = x[:, np.newaxis]
         model = LinearRegression().fit(X, y)
         yfit = model.predict(X)
         plt.scatter(x, y)
         plt.plot(x, yfit);
```

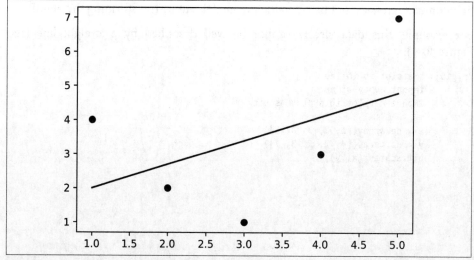

Figure 40-2. A poor straight-line fit

But it's clear that we need a more sophisticated model to describe the relationship between x and y.

One approach to this is to transform the data, adding extra columns of features to drive more flexibility in the model. For example, we can add polynomial features to the data this way:

```
In [12]: from sklearn.preprocessing import PolynomialFeatures
         poly = PolynomialFeatures(degree=3, include_bias=False)
         X2 = poly.fit_transform(X)
         print(X2)
Out[12]: [[  1.    1.    1.]
         [  2.    4.    8.]
         [  3.    9.   27.]
         [  4.   16.   64.]
         [  5.   25.  125.]]
```

The derived feature matrix has one column representing x, a second column representing x^2, and a third column representing x^3. Computing a linear regression on this expanded input gives a much closer fit to our data, as you can see in Figure 40-3:

```
In [13]: model = LinearRegression().fit(X2, y)
         yfit = model.predict(X2)
         plt.scatter(x, y)
         plt.plot(x, yfit);
```

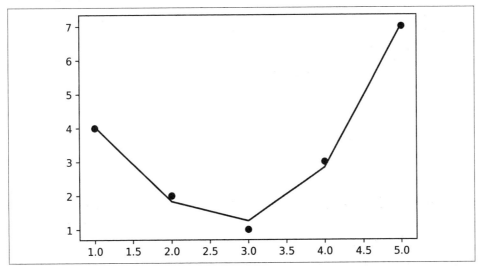

Figure 40-3. A linear fit to polynomial features derived from the data

This idea of improving a model not by changing the model, but by transforming the inputs, is fundamental to many of the more powerful machine learning methods. We'll explore this idea further in Chapter 42 in the context of *basis function regression*.

More generally, this is one motivational path to the powerful set of techniques known as *kernel methods*, which we will explore in Chapter 43.

Imputation of Missing Data

Another common need in feature engineering is handling of missing data. We discussed the handling of missing data in `DataFrame` objects in Chapter 16, and saw that NaN is often is used to mark missing values. For example, we might have a dataset that looks like this:

```
In [14]: from numpy import nan
         X = np.array([[ nan, 0,    3  ],
                       [ 3,    7,    9  ],
                       [ 3,    5,    2  ],
                       [ 4,    nan,  6  ],
                       [ 8,    8,    1  ]])
         y = np.array([14, 16, -1,  8, -5])
```

When applying a typical machine learning model to such data, we will need to first replace the missing values with some appropriate fill value. This is known as *imputation* of missing values, and strategies range from simple (e.g., replacing missing values with the mean of the column) to sophisticated (e.g., using matrix completion or a robust model to handle such data).

The sophisticated approaches tend to be very application-specific, and we won't dive into them here. For a baseline imputation approach using the mean, median, or most frequent value, Scikit-Learn provides the `SimpleImputer` class:

```
In [15]: from sklearn.impute import SimpleImputer
         imp = SimpleImputer(strategy='mean')
         X2 = imp.fit_transform(X)
         X2
Out[15]: array([[4.5, 0. , 3. ],
                [3. , 7. , 9. ],
                [3. , 5. , 2. ],
                [4. , 5. , 6. ],
                [8. , 8. , 1. ]])
```

We see that in the resulting data, the two missing values have been replaced with the mean of the remaining values in the column. This imputed data can then be fed directly into, for example, a `LinearRegression` estimator:

```
In [16]: model = LinearRegression().fit(X2, y)
         model.predict(X2)
Out[16]: array([13.14869292, 14.3784627 , -1.15539732, 10.96606197, -5.33782027])
```

Feature Pipelines

With any of the preceding examples, it can quickly become tedious to do the transformations by hand, especially if you wish to string together multiple steps. For example, we might want a processing pipeline that looks something like this:

1. Impute missing values using the mean.
2. Transform features to quadratic.
3. Fit a linear regression model.

To streamline this type of processing pipeline, Scikit-Learn provides a `Pipeline` object, which can be used as follows:

```
In [17]: from sklearn.pipeline import make_pipeline

         model = make_pipeline(SimpleImputer(strategy='mean'),
                               PolynomialFeatures(degree=2),
                               LinearRegression())
```

This pipeline looks and acts like a standard Scikit-Learn object, and will apply all the specified steps to any input data:

```
In [18]: model.fit(X, y)  # X with missing values, from above
         print(y)
         print(model.predict(X))
Out[18]: [14 16 -1  8 -5]
         [14. 16. -1.  8. -5.]
```

All the steps of the model are applied automatically. Notice that for simplicity, in this demonstration we've applied the model to the data it was trained on; this is why it was able to perfectly predict the result (refer back to Chapter 39 for further discussion).

For some examples of Scikit-Learn pipelines in action, see the following chapter on naive Bayes classification, as well as Chapters 42 and 43.

## CHAPTER 41

In Depth: Naive Bayes Classification

The previous four chapters have given a general overview of the concepts of machine learning. In the rest of Part V, we will be taking a closer look first at four algorithms for supervised learning, and then at four algorithms for unsupervised learning. We start here with our first supervised method, naive Bayes classification.

Naive Bayes models are a group of extremely fast and simple classification algorithms that are often suitable for very high-dimensional datasets. Because they are so fast and have so few tunable parameters, they end up being useful as a quick-and-dirty baseline for a classification problem. This chapter will provide an intuitive explanation of how naive Bayes classifiers work, followed by a few examples of them in action on some datasets.

Bayesian Classification

Naive Bayes classifiers are built on Bayesian classification methods. These rely on Bayes's theorem, which is an equation describing the relationship of conditional probabilities of statistical quantities. In Bayesian classification, we're interested in finding the probability of a label L given some observed features, which we can write as $P(L \mid \text{features})$. Bayes's theorem tells us how to express this in terms of quantities we can compute more directly:

$$P(L \mid \text{features}) = \frac{P(\text{features} \mid L)P(L)}{P(\text{features})}$$

If we are trying to decide between two labels—let's call them L_1 and L_2—then one way to make this decision is to compute the ratio of the posterior probabilities for each label:

$$\frac{P(L_1 \mid \text{features})}{P(L_2 \mid \text{features})} = \frac{P(\text{features} \mid L_1)\,P(L_1)}{P(\text{features} \mid L_2)\,P(L_2)}$$

All we need now is some model by which we can compute $P(\text{features} \mid L_i)$ for each label. Such a model is called a *generative model* because it specifies the hypothetical random process that generates the data. Specifying this generative model for each label is the main piece of the training of such a Bayesian classifier. The general version of such a training step is a very difficult task, but we can make it simpler through the use of some simplifying assumptions about the form of this model.

This is where the "naive" in "naive Bayes" comes in: if we make very naive assumptions about the generative model for each label, we can find a rough approximation of the generative model for each class, and then proceed with the Bayesian classification. Different types of naive Bayes classifiers rest on different naive assumptions about the data, and we will examine a few of these in the following sections.

We begin with the standard imports:

```
In [1]: %matplotlib inline
        import numpy as np
        import matplotlib.pyplot as plt
        import seaborn as sns
        plt.style.use('seaborn-whitegrid')
```

Gaussian Naive Bayes

Perhaps the easiest naive Bayes classifier to understand is Gaussian naive Bayes. With this classifier, the assumption is that *data from each label is drawn from a simple Gaussian distribution*. Imagine that we have the following data, shown in Figure 41-1:

```
In [2]: from sklearn.datasets import make_blobs
        X, y = make_blobs(100, 2, centers=2, random_state=2, cluster_std=1.5)
        plt.scatter(X[:, 0], X[:, 1], c=y, s=50, cmap='RdBu');
```

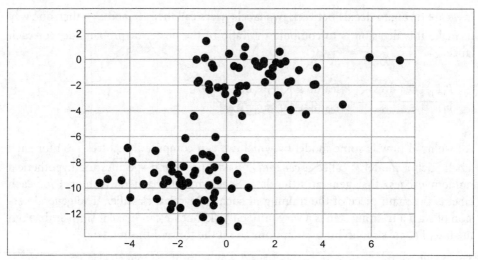

Figure 41-1. Data for Gaussian naive Bayes classification[1]

The simplest Gaussian model is to assume that the data is described by a Gaussian distribution with no covariance between dimensions. This model can be fit by computing the mean and standard deviation of the points within each label, which is all we need to define such a distribution. The result of this naive Gaussian assumption is shown in Figure 41-2.

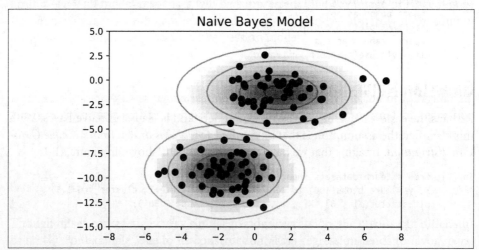

Figure 41-2. Visualization of the Gaussian naive Bayes model[2]

---

1 A full-color version of this figure can be found on GitHub (*https://oreil.ly/PDSH_GitHub*).

2 Code to produce this figure can be found in the online appendix (*https://oreil.ly/o0ENq*).

The ellipses here represent the Gaussian generative model for each label, with larger probability toward the center of the ellipses. With this generative model in place for each class, we have a simple recipe to compute the likelihood $P(\text{features} \mid L_1)$ for any data point, and thus we can quickly compute the posterior ratio and determine which label is the most probable for a given point.

This procedure is implemented in Scikit-Learn's `sklearn.naive_bayes.GaussianNB` estimator:

```
In [3]: from sklearn.naive_bayes import GaussianNB
        model = GaussianNB()
        model.fit(X, y);
```

Let's generate some new data and predict the label:

```
In [4]: rng = np.random.RandomState(0)
        Xnew = [-6, -14] + [14, 18] * rng.rand(2000, 2)
        ynew = model.predict(Xnew)
```

Now we can plot this new data to get an idea of where the decision boundary is (see Figure 41-3).

```
In [5]: plt.scatter(X[:, 0], X[:, 1], c=y, s=50, cmap='RdBu')
        lim = plt.axis()
        plt.scatter(Xnew[:, 0], Xnew[:, 1], c=ynew, s=20, cmap='RdBu', alpha=0.1)
        plt.axis(lim);
```

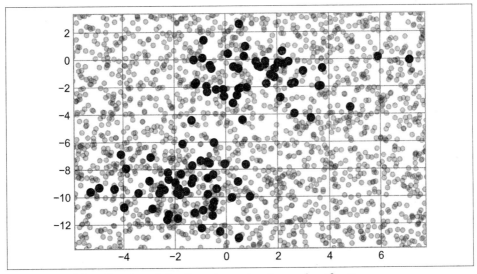

Figure 41-3. Visualization of the Gaussian naive Bayes classification

We see a slightly curved boundary in the classifications—in general, the boundary produced by a Gaussian naive Bayes model will be quadratic.

A nice aspect of this Bayesian formalism is that it naturally allows for probabilistic classification, which we can compute using the `predict_proba` method:

```
In [6]: yprob = model.predict_proba(Xnew)
        yprob[-8:].round(2)
Out[6]: array([[0.89, 0.11],
               [1.  , 0.  ],
               [1.  , 0.  ],
               [1.  , 0.  ],
               [1.  , 0.  ],
               [1.  , 0.  ],
               [0.  , 1.  ],
               [0.15, 0.85]])
```

The columns give the posterior probabilities of the first and second labels, respectively. If you are looking for estimates of uncertainty in your classification, Bayesian approaches like this can be a good place to start.

Of course, the final classification will only be as good as the model assumptions that lead to it, which is why Gaussian naive Bayes often does not produce very good results. Still, in many cases—especially as the number of features becomes large—this assumption is not detrimental enough to prevent Gaussian naive Bayes from being a reliable method.

Multinomial Naive Bayes

The Gaussian assumption just described is by no means the only simple assumption that could be used to specify the generative distribution for each label. Another useful example is multinomial naive Bayes, where the features are assumed to be generated from a simple multinomial distribution. The multinomial distribution describes the probability of observing counts among a number of categories, and thus multinomial naive Bayes is most appropriate for features that represent counts or count rates.

The idea is precisely the same as before, except that instead of modeling the data distribution with the best-fit Gaussian, we model it with a best-fit multinomial distribution.

Example: Classifying Text

One place where multinomial naive Bayes is often used is in text classification, where the features are related to word counts or frequencies within the documents to be classified. We discussed the extraction of such features from text in Chapter 40; here we will use the sparse word count features from the 20 Newsgroups corpus made available through Scikit-Learn to show how we might classify these short documents into categories.

Let's download the data and take a look at the target names:

```
In [7]: from sklearn.datasets import fetch_20newsgroups

        data = fetch_20newsgroups()
        data.target_names
Out[7]: ['alt.atheism',
         'comp.graphics',
         'comp.os.ms-windows.misc',
         'comp.sys.ibm.pc.hardware',
         'comp.sys.mac.hardware',
         'comp.windows.x',
         'misc.forsale',
         'rec.autos',
         'rec.motorcycles',
         'rec.sport.baseball',
         'rec.sport.hockey',
         'sci.crypt',
         'sci.electronics',
         'sci.med',
         'sci.space',
         'soc.religion.christian',
         'talk.politics.guns',
         'talk.politics.mideast',
         'talk.politics.misc',
         'talk.religion.misc']
```

For simplicity here, we will select just a few of these categories and download the training and testing sets:

```
In [8]: categories = ['talk.religion.misc', 'soc.religion.christian',
                       'sci.space', 'comp.graphics']
        train = fetch_20newsgroups(subset='train', categories=categories)
        test = fetch_20newsgroups(subset='test', categories=categories)
```

Here is a representative entry from the data:

```
In [9]: print(train.data[5][48:])
Out[9]: Subject: Federal Hearing
        Originator: dmcgee@uluhe
        Organization: School of Ocean and Earth Science and Technology
        Distribution: usa
        Lines: 10


        Fact or rumor....?  Madalyn Murray O'Hare an atheist who eliminated the
        use of the bible reading and prayer in public schools 15 years ago is now
        going to appear before the FCC with a petition to stop the reading of the
        Gospel on the airways of America.  And she is also campaigning to remove
        Christmas programs, songs, etc from the public schools.  If it is true
        then mail to Federal Communications Commission 1919 H Street Washington DC
        20054 expressing your opposition to her request.  Reference Petition number

        2493.
```

In order to use this data for machine learning, we need to be able to convert the content of each string into a vector of numbers. For this we will use the TF-IDF vectorizer (introduced in Chapter 40), and create a pipeline that attaches it to a multinomial naive Bayes classifier:

```
In [10]: from sklearn.feature_extraction.text import TfidfVectorizer
         from sklearn.naive_bayes import MultinomialNB
         from sklearn.pipeline import make_pipeline

         model = make_pipeline(TfidfVectorizer(), MultinomialNB())
```

With this pipeline, we can apply the model to the training data and predict labels for the test data:

```
In [11]: model.fit(train.data, train.target)
         labels = model.predict(test.data)
```

Now that we have predicted the labels for the test data, we can evaluate them to learn about the performance of the estimator. For example, let's take a look at the confusion matrix between the true and predicted labels for the test data (see Figure 41-4).

```
In [12]: from sklearn.metrics import confusion_matrix
         mat = confusion_matrix(test.target, labels)
         sns.heatmap(mat.T, square=True, annot=True, fmt='d', cbar=False,
                     xticklabels=train.target_names, yticklabels=train.target_names,
                     cmap='Blues')
         plt.xlabel('true label')
         plt.ylabel('predicted label');
```

Evidently, even this very simple classifier can successfully separate space discussions from computer discussions, but it gets confused between discussions about religion and discussions about Christianity. This is perhaps to be expected!

The cool thing here is that we now have the tools to determine the category for *any* string, using the predict method of this pipeline. Here's a utility function that will return the prediction for a single string:

```
In [13]: def predict_category(s, train=train, model=model):
             pred = model.predict([s])
             return train.target_names[pred[0]]
```

Let's try it out:

```
In [14]: predict_category('sending a payload to the ISS')
Out[14]: 'sci.space'
```

```
In [15]: predict_category('discussing the existence of God')
Out[15]: 'soc.religion.christian'
```

```
In [16]: predict_category('determining the screen resolution')
Out[16]: 'comp.graphics'
```

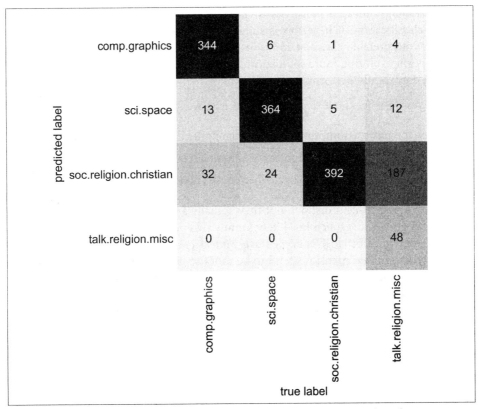

Figure 41-4. Confusion matrix for the multinomial naive Bayes text classifier

Remember that this is nothing more sophisticated than a simple probability model for the (weighted) frequency of each word in the string; nevertheless, the result is striking. Even a very naive algorithm, when used carefully and trained on a large set of high-dimensional data, can be surprisingly effective.

When to Use Naive Bayes

Because naive Bayes classifiers make such stringent assumptions about data, they will generally not perform as well as more complicated models. That said, they have several advantages:

- They are fast for both training and prediction.
- They provide straightforward probabilistic prediction.
- They are often easily interpretable.
- They have few (if any) tunable parameters.

These advantages mean a naive Bayes classifier is often a good choice as an initial baseline classification. If it performs suitably, then congratulations: you have a very fast, very interpretable classifier for your problem. If it does not perform well, then you can begin exploring more sophisticated models, with some baseline knowledge of how well they should perform.

Naive Bayes classifiers tend to perform especially well in the following situations:

- When the naive assumptions actually match the data (very rare in practice)
- For very well-separated categories, when model complexity is less important
- For very high-dimensional data, when model complexity is less important

The last two points seem distinct, but they actually are related: as the dimensionality of a dataset grows, it is much less likely for any two points to be found close together (after all, they must be close in *every single dimension* to be close overall). This means that clusters in high dimensions tend to be more separated, on average, than clusters in low dimensions, assuming the new dimensions actually add information. For this reason, simplistic classifiers like the ones discussed here tend to work as well or better than more complicated classifiers as the dimensionality grows: once you have enough data, even a simple model can be very powerful.

## CHAPTER 42

In Depth: Linear Regression

Just as naive Bayes (discussed in Chapter 41) is a good starting point for classification tasks, linear regression models are a good starting point for regression tasks. Such models are popular because they can be fit quickly and are straightforward to interpret. You are already familiar with the simplest form of linear regression model (i.e., fitting a straight line to two-dimensional data), but such models can be extended to model more complicated data behavior.

In this chapter we will start with a quick walkthrough of the mathematics behind this well-known problem, before moving on to see how linear models can be generalized to account for more complicated patterns in data.

We begin with the standard imports:

```
In [1]: %matplotlib inline
        import matplotlib.pyplot as plt
        plt.style.use('seaborn-whitegrid')
        import numpy as np
```

Simple Linear Regression

We will start with the most familiar linear regression, a straight-line fit to data. A straight-line fit is a model of the form:

$$y = ax + b$$

where a is commonly known as the *slope*, and b is commonly known as the *intercept*.

Consider the following data, which is scattered about a line with a slope of 2 and an intercept of –5 (see Figure 42-1).

```
In [2]: rng = np.random.RandomState(1)
        x = 10 * rng.rand(50)
        y = 2 * x - 5 + rng.randn(50)
        plt.scatter(x, y);
```

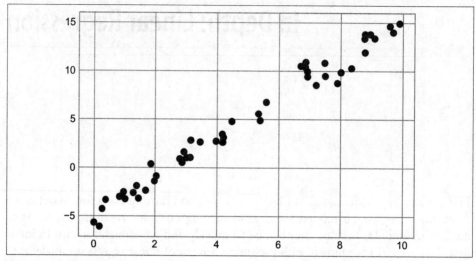

Figure 42-1. Data for linear regression

We can use Scikit-Learn's `LinearRegression` estimator to fit this data and construct the best-fit line, as shown in Figure 42-2.

```
In [3]: from sklearn.linear_model import LinearRegression
        model = LinearRegression(fit_intercept=True)

        model.fit(x[:, np.newaxis], y)

        xfit = np.linspace(0, 10, 1000)
        yfit = model.predict(xfit[:, np.newaxis])

        plt.scatter(x, y)
        plt.plot(xfit, yfit);
```

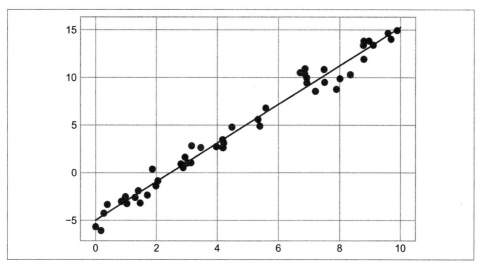

Figure 42-2. A simple linear regression model

The slope and intercept of the data are contained in the model's fit parameters, which in Scikit-Learn are always marked by a trailing underscore. Here the relevant parameters are `coef_` and `intercept_`:

```
In [4]: print("Model slope:    ", model.coef_[0])
        print("Model intercept:", model.intercept_)
Out[4]: Model slope:     2.0272088103606953
        Model intercept: -4.998577085553204
```

We see that the results are very close to the values used to generate the data, as we might hope.

The `LinearRegression` estimator is much more capable than this, however—in addition to simple straight-line fits, it can also handle multidimensional linear models of the form:

$$y = a_0 + a_1 x_1 + a_2 x_2 + \cdots$$

where there are multiple x values. Geometrically, this is akin to fitting a plane to points in three dimensions, or fitting a hyperplane to points in higher dimensions.

The multidimensional nature of such regressions makes them more difficult to visualize, but we can see one of these fits in action by building some example data, using NumPy's matrix multiplication operator:

```
In [5]: rng = np.random.RandomState(1)
        X = 10 * rng.rand(100, 3)
        y = 0.5 + np.dot(X, [1.5, -2., 1.])
```

```
        model.fit(X, y)
        print(model.intercept_)
        print(model.coef_)
Out[5]: 0.50000000000001
        [ 1.5 -2.   1. ]
```

Here the y data is constructed from a linear combination of three random x values, and the linear regression recovers the coefficients used to construct the data.

In this way, we can use the single LinearRegression estimator to fit lines, planes, or hyperplanes to our data. It still appears that this approach would be limited to strictly linear relationships between variables, but it turns out we can relax this as well.

Basis Function Regression

One trick you can use to adapt linear regression to nonlinear relationships between variables is to transform the data according to *basis functions*. We have seen one version of this before, in the PolynomialRegression pipeline used in Chapters 39 and 40. The idea is to take our multidimensional linear model:

$$y = a_0 + a_1 x_1 + a_2 x_2 + a_3 x_3 + \cdots$$

and build the x_1, x_2, x_3, and so on from our single-dimensional input x. That is, we let $x_n = f_n(x)$, where $f_n()$ is some function that transforms our data.

For example, if $f_n(x) = x^n$, our model becomes a polynomial regression:

$$y = a_0 + a_1 x + a_2 x^2 + a_3 x^3 + \cdots$$

Notice that this is *still a linear model*—the linearity refers to the fact that the coefficients a_n never multiply or divide each other. What we have effectively done is taken our one-dimensional x values and projected them into a higher dimension, so that a linear fit can fit more complicated relationships between x and y.

Polynomial Basis Functions

This polynomial projection is useful enough that it is built into Scikit-Learn, using the PolynomialFeatures transformer:

```
In [6]: from sklearn.preprocessing import PolynomialFeatures
        x = np.array([2, 3, 4])
        poly = PolynomialFeatures(3, include_bias=False)
        poly.fit_transform(x[:, None])
Out[6]: array([[ 2.,   4.,   8.],
```

```
       [ 3.,   9.,  27.],
       [ 4.,  16.,  64.]])
```

We see here that the transformer has converted our one-dimensional array into a three-dimensional array, where each column contains the exponentiated value. This new, higher-dimensional data representation can then be plugged into a linear regression.

As we saw in Chapter 40, the cleanest way to accomplish this is to use a pipeline. Let's make a 7th-degree polynomial model in this way:

```
In [7]: from sklearn.pipeline import make_pipeline
        poly_model = make_pipeline(PolynomialFeatures(7),
                                   LinearRegression())
```

With this transform in place, we can use the linear model to fit much more compli-cated relationships between x and y. For example, here is a sine wave with noise (see Figure 42-3).

```
In [8]: rng = np.random.RandomState(1)
        x = 10 * rng.rand(50)
        y = np.sin(x) + 0.1 * rng.randn(50)

        poly_model.fit(x[:, np.newaxis], y)
        yfit = poly_model.predict(xfit[:, np.newaxis])

        plt.scatter(x, y)
        plt.plot(xfit, yfit);
```

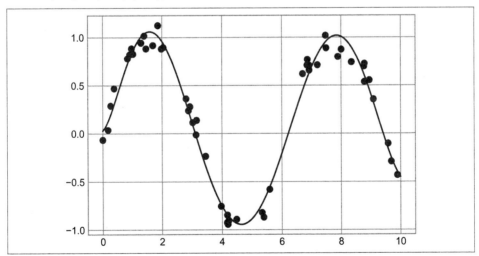

Figure 42-3. A linear polynomial fit to nonlinear training data

Our linear model, through the use of seventh-order polynomial basis functions, can provide an excellent fit to this nonlinear data!

Gaussian Basis Functions

Of course, other basis functions are possible. For example, one useful pattern is to fit a model that is not a sum of polynomial bases, but a sum of Gaussian bases. The result might look something like Figure 42-4.

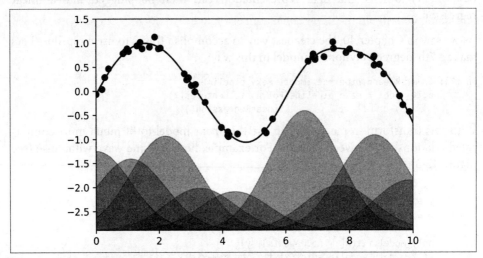

Figure 42-4. A Gaussian basis function fit to nonlinear data[1]

The shaded regions in the plot are the scaled basis functions, and when added together they reproduce the smooth curve through the data. These Gaussian basis functions are not built into Scikit-Learn, but we can write a custom transformer that will create them, as shown here and illustrated in Figure 42-5 (Scikit-Learn transformers are implemented as Python classes; reading Scikit-Learn's source is a good way to see how they can be created):

```
In [9]: from sklearn.base import BaseEstimator, TransformerMixin

        class GaussianFeatures(BaseEstimator, TransformerMixin):
            """Uniformly spaced Gaussian features for one-dimensional input"""

            def __init__(self, N, width_factor=2.0):
                self.N = N
                self.width_factor = width_factor

            @staticmethod
            def _gauss_basis(x, y, width, axis=None):
                arg = (x - y) / width
                return np.exp(-0.5 * np.sum(arg ** 2, axis))
```

---

1 Code to produce this figure can be found in the online appendix (*https://oreil.ly/o1Zya*).

```
    def fit(self, X, y=None):
        # create N centers spread along the data range
        self.centers_ = np.linspace(X.min(), X.max(), self.N)
        self.width_ = self.width_factor*(self.centers_[1]-self.centers_[0])
        return self

    def transform(self, X):
        return self._gauss_basis(X[:, :, np.newaxis], self.centers_,
                                 self.width_, axis=1)

gauss_model = make_pipeline(GaussianFeatures(20),
                            LinearRegression())
gauss_model.fit(x[:, np.newaxis], y)
yfit = gauss_model.predict(xfit[:, np.newaxis])

plt.scatter(x, y)
plt.plot(xfit, yfit)
plt.xlim(0, 10);
```

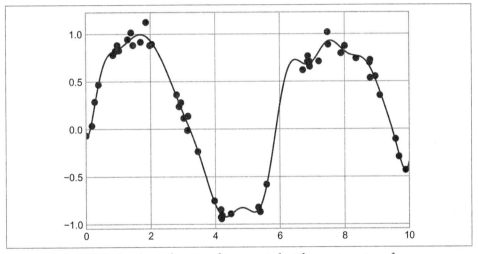

Figure 42-5. A Gaussian basis function fit computed with a custom transformer

I've included this example just to make clear that there is nothing magic about polynomial basis functions: if you have some sort of intuition into the generating process of your data that makes you think one basis or another might be appropriate, you can use that instead.

Regularization

The introduction of basis functions into our linear regression makes the model much more flexible, but it also can very quickly lead to overfitting (refer back to Chapter 39 for a discussion of this). For example, Figure 42-6 shows what happens if we use a large number of Gaussian basis functions:

```
In [10]: model = make_pipeline(GaussianFeatures(30),
                                LinearRegression())
         model.fit(x[:, np.newaxis], y)

         plt.scatter(x, y)
         plt.plot(xfit, model.predict(xfit[:, np.newaxis]))

         plt.xlim(0, 10)
         plt.ylim(-1.5, 1.5);
```

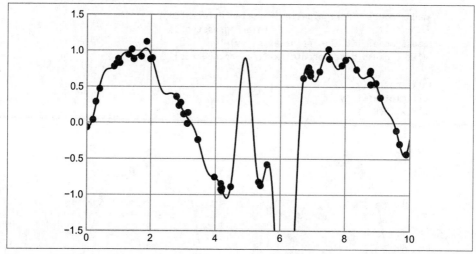

Figure 42-6. An overly complex basis function model that overfits the data

With the data projected to the 30-dimensional basis, the model has far too much flexibility and goes to extreme values between locations where it is constrained by data. We can see the reason for this if we plot the coefficients of the Gaussian bases with respect to their locations, as shown in Figure 42-7.

```
In [11]: def basis_plot(model, title=None):
             fig, ax = plt.subplots(2, sharex=True)
             model.fit(x[:, np.newaxis], y)
             ax[0].scatter(x, y)
             ax[0].plot(xfit, model.predict(xfit[:, np.newaxis]))
             ax[0].set(xlabel='x', ylabel='y', ylim=(-1.5, 1.5))

             if title:
                 ax[0].set_title(title)

             ax[1].plot(model.steps[0][1].centers_,
                        model.steps[1][1].coef_)
             ax[1].set(xlabel='basis location',
                       ylabel='coefficient',
                       xlim=(0, 10))
```

```
model = make_pipeline(GaussianFeatures(30), LinearRegression())
basis_plot(model)
```

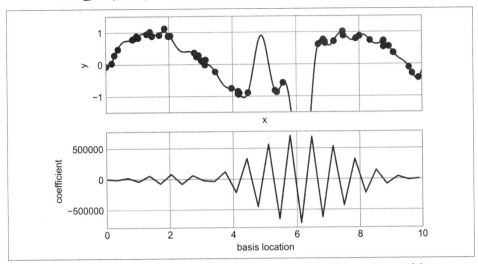

Figure 42-7. The coefficients of the Gaussian bases in the overly complex model

The lower panel of this figure shows the amplitude of the basis function at each location. This is typical overfitting behavior when basis functions overlap: the coefficients of adjacent basis functions blow up and cancel each other out. We know that such behavior is problematic, and it would be nice if we could limit such spikes explicitly in the model by penalizing large values of the model parameters. Such a penalty is known as *regularization*, and comes in several forms.

Ridge Regression (L$_2$ Regularization)

Perhaps the most common form of regularization is known as *ridge regression* or L_2 *regularization* (sometimes also called *Tikhonov regularization*). This proceeds by penalizing the sum of squares (2-norms) of the model coefficients θ_n. In this case, the penalty on the model fit would be:

$$P = \alpha \sum_{n=1}^{N} \theta_n^2$$

where α is a free parameter that controls the strength of the penalty. This type of penalized model is built into Scikit-Learn with the `Ridge` estimator (see Figure 42-8).

```
In [12]: from sklearn.linear_model import Ridge
         model = make_pipeline(GaussianFeatures(30), Ridge(alpha=0.1))
         basis_plot(model, title='Ridge Regression')
```

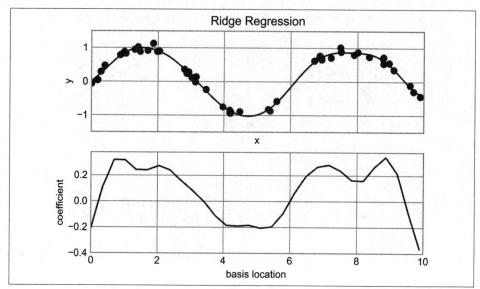

Figure 42-8. Ridge (L₂) regularization applied to the overly complex model (compare to
Figure 42-7)

The α parameter is essentially a knob controlling the complexity of the resulting
model. In the limit $\alpha \rightarrow 0$, we recover the standard linear regression result; in the
limit $\alpha \rightarrow \infty$, all model responses will be suppressed. One advantage of ridge regression in particular is that it can be computed very efficiently—at hardly more computational cost than the original linear regression model.

Lasso Regression (L₁ Regularization)

Another common type of regularization is known as *lasso regression* or *L₁ regularization* and involves penalizing the sum of absolute values (1-norms) of regression
coefficients:

$$P = \alpha \sum_{n=1}^{N} |\theta_n|$$

Though this is conceptually very similar to ridge regression, the results can differ surprisingly. For example, due to its construction, lasso regression tends to favor *sparse
models* where possible: that is, it preferentially sets many model coefficients to exactly
zero.

We can see this behavior if we duplicate the previous example using L_1-normalized
coefficients (see Figure 42-9).

```
In [13]: from sklearn.linear_model import Lasso
         model = make_pipeline(GaussianFeatures(30),
                               Lasso(alpha=0.001, max_iter=2000))
         basis_plot(model, title='Lasso Regression')
```

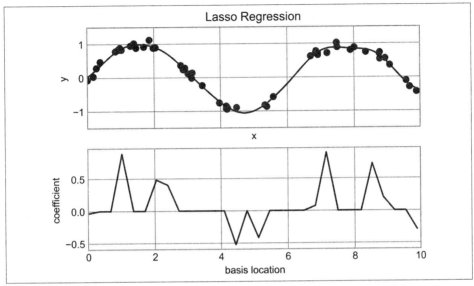

Figure 42-9. Lasso (L_1) regularization applied to the overly complex model (compare to Figure 42-8)

With the lasso regression penalty, the majority of the coefficients are exactly zero, with the functional behavior being modeled by a small subset of the available basis functions. As with ridge regularization, the α parameter tunes the strength of the penalty and should be determined via, for example, cross-validation (refer back to Chapter 39 for a discussion of this).

Example: Predicting Bicycle Traffic

As an example, let's take a look at whether we can predict the number of bicycle trips across Seattle's Fremont Bridge based on weather, season, and other factors. We already saw this data in Chapter 23, but here we will join the bike data with another dataset and try to determine the extent to which weather and seasonal factors—temperature, precipitation, and daylight hours—affect the volume of bicycle traffic through this corridor. Fortunately, the National Oceanic and Atmospheric Administration (NOAA) makes its daily weather station data (*https://oreil.ly/sE5zO*) available —I used station ID USW00024233—and we can easily use Pandas to join the two data sources. We will perform a simple linear regression to relate weather and other information to bicycle counts, in order to estimate how a change in any one of these parameters affects the number of riders on a given day.

In particular, this is an example of how the tools of Scikit-Learn can be used in a statistical modeling framework, in which the parameters of the model are assumed to have interpretable meaning. As discussed previously, this is not a standard approach within machine learning, but such interpretation is possible for some models.

Let's start by loading the two datasets, indexing by date:

```
In [14]: # url = 'https://raw.githubusercontent.com/jakevdp/bicycle-data/main'
         # !curl -O {url}/FremontBridge.csv
         # !curl -O {url}/SeattleWeather.csv
```

```
In [15]: import pandas as pd
         counts = pd.read_csv('FremontBridge.csv',
                              index_col='Date', parse_dates=True)
         weather = pd.read_csv('SeattleWeather.csv',
                               index_col='DATE', parse_dates=True)
```

For simplicity, let's look at data prior to 2020 in order to avoid the effects of the COVID-19 pandemic, which significantly affected commuting patterns in Seattle:

```
In [16]: counts = counts[counts.index < "2020-01-01"]
         weather = weather[weather.index < "2020-01-01"]
```

Next we will compute the total daily bicycle traffic, and put this in its own `DataFrame`:

```
In [17]: daily = counts.resample('d').sum()
         daily['Total'] = daily.sum(axis=1)
         daily = daily[['Total']] # remove other columns
```

We saw previously that the patterns of use generally vary from day to day. Let's account for this in our data by adding binary columns that indicate the day of the week:

```
In [18]: days = ['Mon', 'Tue', 'Wed', 'Thu', 'Fri', 'Sat', 'Sun']
         for i in range(7):
             daily[days[i]] = (daily.index.dayofweek == i).astype(float)
```

Similarly, we might expect riders to behave differently on holidays; let's add an indicator of this as well:

```
In [19]: from pandas.tseries.holiday import USFederalHolidayCalendar
         cal = USFederalHolidayCalendar()
         holidays = cal.holidays('2012', '2020')
         daily = daily.join(pd.Series(1, index=holidays, name='holiday'))
         daily['holiday'].fillna(0, inplace=True)
```

We also might suspect that the hours of daylight would affect how many people ride. Let's use the standard astronomical calculation to add this information (see Figure 42-10).

```
In [20]: def hours_of_daylight(date, axis=23.44, latitude=47.61):
             """Compute the hours of daylight for the given date"""
             days = (date - pd.datetime(2000, 12, 21)).days
             m = (1. - np.tan(np.radians(latitude))
```

```
                * np.tan(np.radians(axis) * np.cos(days * 2 * np.pi / 365.25)))
         return 24. * np.degrees(np.arccos(1 - np.clip(m, 0, 2))) / 180.

         daily['daylight_hrs'] = list(map(hours_of_daylight, daily.index))
         daily[['daylight_hrs']].plot()
         plt.ylim(8, 17)
Out[20]: (8.0, 17.0)
```

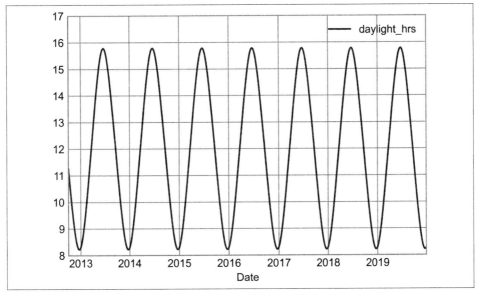

Figure 42-10. Visualization of hours of daylight in Seattle

We can also add the average temperature and total precipitation to the data. In addition to the inches of precipitation, let's add a flag that indicates whether a day is dry (has zero precipitation):

```
In [21]: weather['Temp (F)'] = 0.5 * (weather['TMIN'] + weather['TMAX'])
         weather['Rainfall (in)'] = weather['PRCP']
         weather['dry day'] = (weather['PRCP'] == 0).astype(int)

         daily = daily.join(weather[['Rainfall (in)', 'Temp (F)', 'dry day']])
```

Finally, let's add a counter that increases from day 1, and measures how many years have passed. This will let us measure any observed annual increase or decrease in daily crossings:

```
In [22]: daily['annual'] = (daily.index - daily.index[0]).days / 365.
```

Now that our data is in order, and we can take a look at it:

```
In [23]: daily.head()
Out[23]:             Total  Mon  Tue  Wed  Thu  Fri  Sat  Sun  holiday \
         Date
         2012-10-03  14084.0  0.0  0.0  1.0  0.0  0.0  0.0  0.0     0.0
         2012-10-04  13900.0  0.0  0.0  0.0  1.0  0.0  0.0  0.0     0.0
         2012-10-05  12592.0  0.0  0.0  0.0  0.0  1.0  0.0  0.0     0.0
         2012-10-06   8024.0  0.0  0.0  0.0  0.0  0.0  1.0  0.0     0.0
         2012-10-07   8568.0  0.0  0.0  0.0  0.0  0.0  0.0  1.0     0.0

                     daylight_hrs  Rainfall (in)  Temp (F)  dry day    annual
         Date
         2012-10-03     11.277359            0.0      56.0        1  0.000000
         2012-10-04     11.219142            0.0      56.5        1  0.002740
         2012-10-05     11.161038            0.0      59.5        1  0.005479
         2012-10-06     11.103056            0.0      60.5        1  0.008219
         2012-10-07     11.045208            0.0      60.5        1  0.010959
```

With this in place, we can choose the columns to use, and fit a linear regression model to our data. We will set `fit_intercept=False`, because the daily flags essentially operate as their own day-specific intercepts:

```
In [24]: # Drop any rows with null values
         daily.dropna(axis=0, how='any', inplace=True)

         column_names = ['Mon', 'Tue', 'Wed', 'Thu', 'Fri', 'Sat', 'Sun',
                         'holiday', 'daylight_hrs', 'Rainfall (in)',
                         'dry day', 'Temp (F)', 'annual']
         X = daily[column_names]
         y = daily['Total']

         model = LinearRegression(fit_intercept=False)
         model.fit(X, y)
         daily['predicted'] = model.predict(X)
```

Finally, we can compare the total and predicted bicycle traffic visually (see Figure 42-11).

```
In [25]: daily[['Total', 'predicted']].plot(alpha=0.5);
```

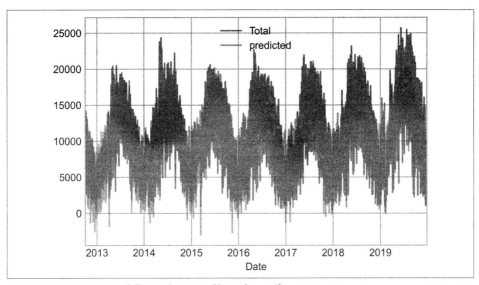

Figure 42-11. Our model's prediction of bicycle traffic

From the fact that the data and model predictions don't line up exactly, it is evident that we have missed some key features. Either our features are not complete (i.e., people decide whether to ride to work based on more than just these features), or there are some nonlinear relationships that we have failed to take into account (e.g., perhaps people ride less at both high and low temperatures). Nevertheless, our rough approximation is enough to give us some insights, and we can take a look at the coefficients of the linear model to estimate how much each feature contributes to the daily bicycle count:

```
In [26]: params = pd.Series(model.coef_, index=X.columns)
         params
Out[26]: Mon              -3309.953439
         Tue              -2860.625060
         Wed              -2962.889892
         Thu              -3480.656444
         Fri              -4836.064503
         Sat             -10436.802843
         Sun             -10795.195718
         holiday          -5006.995232
         daylight_hrs       409.146368
         Rainfall (in)    -2789.860745
         dry day           2111.069565
         Temp (F)           179.026296
         annual             324.437749
         dtype: float64
```

These numbers are difficult to interpret without some measure of their uncertainty. We can compute these uncertainties quickly using bootstrap resamplings of the data:

```
In [27]: from sklearn.utils import resample
         np.random.seed(1)
         err = np.std([model.fit(*resample(X, y)).coef_
                       for i in range(1000)], 0)
```

With these errors estimated, let's again look at the results:

```
In [28]: print(pd.DataFrame({'effect': params.round(0),
                             'uncertainty': err.round(0)}))
Out[28]:              effect  uncertainty
         Mon          -3310.0        265.0
         Tue          -2861.0        274.0
         Wed          -2963.0        268.0
         Thu          -3481.0        268.0
         Fri          -4836.0        261.0
         Sat         -10437.0        259.0
         Sun         -10795.0        267.0
         holiday      -5007.0        401.0
         daylight_hrs    409.0         26.0
         Rainfall (in) -2790.0        186.0
         dry day       2111.0        101.0
         Temp (F)       179.0          7.0
         annual         324.0         22.0
```

The effect column here, roughly speaking, shows how the number of riders is affected by a change of the feature in question. For example, there is a clear divide when it comes to the day of the week: there are thousands fewer riders on weekends than on weekdays. We also see that for each additional hour of daylight, 409 ± 26 more people choose to ride; a temperature increase of one degree Fahrenheit encourages 179 ± 7 people to grab their bicycle; a dry day means an average of $2{,}111 \pm 101$ more riders, and every inch of rainfall leads $2{,}790 \pm 186$ riders to choose another mode of transport. Once all these effects are accounted for, we see a modest increase of 324 ± 22 new daily riders each year.

Our simple model is almost certainly missing some relevant information. For example, as mentioned earlier, nonlinear effects (such as effects of precipitation *and* cold temperature) and nonlinear trends within each variable (such as disinclination to ride at very cold and very hot temperatures) cannot be accounted for in a simple linear model. Additionally, we have thrown away some of the finer-grained information (such as the difference between a rainy morning and a rainy afternoon), and we have ignored correlations between days (such as the possible effect of a rainy Tuesday on Wednesday's numbers, or the effect of an unexpected sunny day after a streak of rainy days). These are all potentially interesting effects, and you now have the tools to begin exploring them if you wish!

## CHAPTER 43

In Depth: Support Vector Machines

Support vector machines (SVMs) are a particularly powerful and flexible class of supervised algorithms for both classification and regression. In this chapter, we will explore the intuition behind SVMs and their use in classification problems.

We begin with the standard imports:

```
In [1]: %matplotlib inline
        import numpy as np
        import matplotlib.pyplot as plt
        plt.style.use('seaborn-whitegrid')
        from scipy import stats
```

 Full-size, full-color figures are available in the supplemental materials on GitHub (*https://oreil.ly/PDSH_GitHub*).

Motivating Support Vector Machines

As part of our discussion of Bayesian classification (see Chapter 41), we learned about a simple kind of model that describes the distribution of each underlying class, and experimented with using it to probabilistically determine labels for new points. That was an example of *generative classification*; here we will consider instead *discriminative classification*. That is, rather than modeling each class, we will simply find a line or curve (in two dimensions) or manifold (in multiple dimensions) that divides the classes from each other.

As an example of this, consider the simple case of a classification task in which the two classes of points are well separated (see Figure 43-1).

```
In [2]: from sklearn.datasets import make_blobs
        X, y = make_blobs(n_samples=50, centers=2,
                          random_state=0, cluster_std=0.60)
        plt.scatter(X[:, 0], X[:, 1], c=y, s=50, cmap='autumn');
```

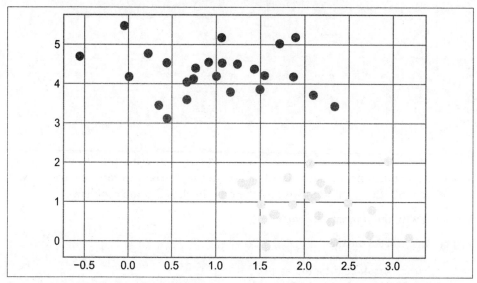

Figure 43-1. Simple data for classification

A linear discriminative classifier would attempt to draw a straight line separating the two sets of data, and thereby create a model for classification. For two-dimensional data like that shown here, this is a task we could do by hand. But immediately we see a problem: there is more than one possible dividing line that can perfectly discriminate between the two classes!

We can draw some of them as follows; Figure 43-2 shows the result:

```
In [3]: xfit = np.linspace(-1, 3.5)
        plt.scatter(X[:, 0], X[:, 1], c=y, s=50, cmap='autumn')
        plt.plot([0.6], [2.1], 'x', color='red', markeredgewidth=2, markersize=10)

        for m, b in [(1, 0.65), (0.5, 1.6), (-0.2, 2.9)]:
            plt.plot(xfit, m * xfit + b, '-k')

        plt.xlim(-1, 3.5);
```

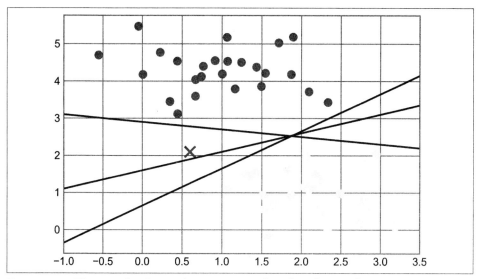

Figure 43-2. Three perfect linear discriminative classifiers for our data

These are three *very* different separators which, nevertheless, perfectly discriminate between these samples. Depending on which you choose, a new data point (e.g., the one marked by the "X" in this plot) will be assigned a different label! Evidently our simple intuition of "drawing a line between classes" is not good enough, and we need to think a bit more deeply.

Support Vector Machines: Maximizing the Margin

Support vector machines offer one way to improve on this. The intuition is this: rather than simply drawing a zero-width line between the classes, we can draw around each line a *margin* of some width, up to the nearest point. Here is an example of how this might look (Figure 43-3).

```
In [4]: xfit = np.linspace(-1, 3.5)
        plt.scatter(X[:, 0], X[:, 1], c=y, s=50, cmap='autumn')

        for m, b, d in [(1, 0.65, 0.33), (0.5, 1.6, 0.55), (-0.2, 2.9, 0.2)]:
            yfit = m * xfit + b
            plt.plot(xfit, yfit, '-k')
            plt.fill_between(xfit, yfit - d, yfit + d, edgecolor='none',
                            color='lightgray', alpha=0.5)

        plt.xlim(-1, 3.5);
```

The line that maximizes this margin is the one we will choose as the optimal model.

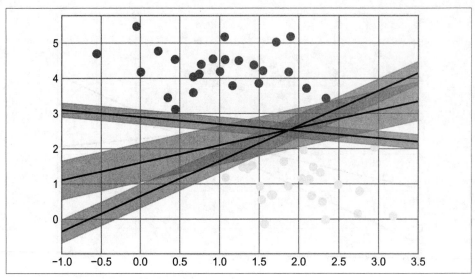

Figure 43-3. Visualization of "margins" within discriminative classifiers

Fitting a Support Vector Machine

Let's see the result of an actual fit to this data: we will use Scikit-Learn's support vector classifier (SVC) to train an SVM model on this data. For the time being, we will use a linear kernel and set the C parameter to a very large number (we'll discuss the meaning of these in more depth momentarily):

```
In [5]: from sklearn.svm import SVC # "Support vector classifier"
        model = SVC(kernel='linear', C=1E10)
        model.fit(X, y)
Out[5]: SVC(C=10000000000.0, kernel='linear')
```

To better visualize what's happening here, let's create a quick convenience function that will plot SVM decision boundaries for us (Figure 43-4).

```
In [6]: def plot_svc_decision_function(model, ax=None, plot_support=True):
            """Plot the decision function for a 2D SVC"""
            if ax is None:
                ax = plt.gca()
            xlim = ax.get_xlim()
            ylim = ax.get_ylim()

            # create grid to evaluate model
            x = np.linspace(xlim[0], xlim[1], 30)
            y = np.linspace(ylim[0], ylim[1], 30)
            Y, X = np.meshgrid(y, x)
            xy = np.vstack([X.ravel(), Y.ravel()]).T
            P = model.decision_function(xy).reshape(X.shape)

            # plot decision boundary and margins
```

```
ax.contour(X, Y, P, colors='k',
            levels=[-1, 0, 1], alpha=0.5,
            linestyles=['--', '-', '--'])

# plot support vectors
if plot_support:
    ax.scatter(model.support_vectors_[:, 0],
                model.support_vectors_[:, 1],
                s=300, linewidth=1, edgecolors='black',
                facecolors='none');
ax.set_xlim(xlim)
ax.set_ylim(ylim)
```
```
In [7]: plt.scatter(X[:, 0], X[:, 1], c=y, s=50, cmap='autumn')
        plot_svc_decision_function(model);
```

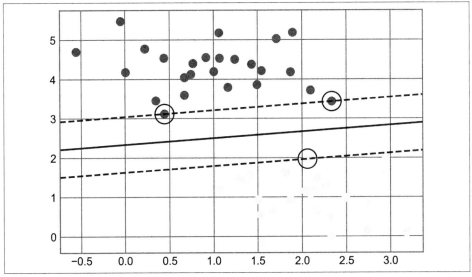

Figure 43-4. A support vector machine classifier fit to the data, with margins (dashed lines) and support vectors (circles) shown

This is the dividing line that maximizes the margin between the two sets of points. Notice that a few of the training points just touch the margin: they are circled in Figure 43-5. These points are the pivotal elements of this fit; they are known as the *support vectors*, and give the algorithm its name. In Scikit-Learn, the identities of these points are stored in the support_vectors_ attribute of the classifier:

```
In [8]: model.support_vectors_
Out[8]: array([[0.44359863, 3.11530945],
               [2.33812285, 3.43116792],
               [2.06156753, 1.96918596]])
```

A key to this classifier's success is that for the fit, only the positions of the support vectors matter; any points further from the margin that are on the correct side do not modify the fit. Technically, this is because these points do not contribute to the loss function used to fit the model, so their position and number do not matter so long as they do not cross the margin.

We can see this, for example, if we plot the model learned from the first 60 points and first 120 points of this dataset (Figure 43-5).

```
In [9]: def plot_svm(N=10, ax=None):
            X, y = make_blobs(n_samples=200, centers=2,
                              random_state=0, cluster_std=0.60)
            X = X[:N]
            y = y[:N]
            model = SVC(kernel='linear', C=1E10)
            model.fit(X, y)

            ax = ax or plt.gca()
            ax.scatter(X[:, 0], X[:, 1], c=y, s=50, cmap='autumn')
            ax.set_xlim(-1, 4)
            ax.set_ylim(-1, 6)
            plot_svc_decision_function(model, ax)

        fig, ax = plt.subplots(1, 2, figsize=(16, 6))
        fig.subplots_adjust(left=0.0625, right=0.95, wspace=0.1)
        for axi, N in zip(ax, [60, 120]):
            plot_svm(N, axi)
            axi.set_title('N = {0}'.format(N))
```

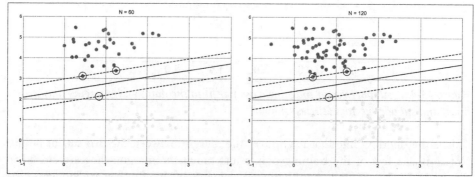

Figure 43-5. The influence of new training points on the SVM model

In the left panel, we see the model and the support vectors for 60 training points. In the right panel, we have doubled the number of training points, but the model has not changed: the three support vectors in the left panel are the same as the support vectors in the right panel. This insensitivity to the exact behavior of distant points is one of the strengths of the SVM model.

If you are running this notebook live, you can use IPython's interactive widgets to view this feature of the SVM model interactively:

```
In [10]: from ipywidgets import interact, fixed
         interact(plot_svm, N=(10, 200), ax=fixed(None));
Out[10]: interactive(children=(IntSlider(value=10, description='N', max=200, min=10),
         > Output()), _dom_classes=('widget-...
```

Beyond Linear Boundaries: Kernel SVM

Where SVM can become quite powerful is when it is combined with *kernels*. We have seen a version of kernels before, in the basis function regressions of Chapter 42. There we projected our data into a higher-dimensional space defined by polynomials and Gaussian basis functions, and thereby were able to fit for nonlinear relationships with a linear classifier.

In SVM models, we can use a version of the same idea. To motivate the need for kernels, let's look at some data that is not linearly separable (Figure 43-6).

```
In [11]: from sklearn.datasets import make_circles
         X, y = make_circles(100, factor=.1, noise=.1)

         clf = SVC(kernel='linear').fit(X, y)

         plt.scatter(X[:, 0], X[:, 1], c=y, s=50, cmap='autumn')
         plot_svc_decision_function(clf, plot_support=False);
```

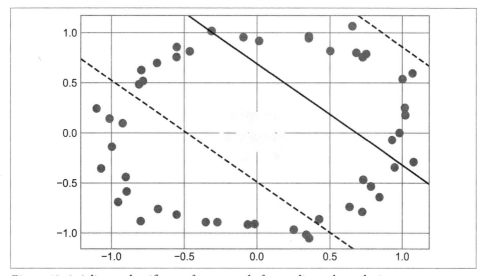

Figure 43-6. A linear classifier performs poorly for nonlinear boundaries

It is clear that no linear discrimination will *ever* be able to separate this data. But we can draw a lesson from the basis function regressions in Chapter 42, and think about how we might project the data into a higher dimension such that a linear separator *would* be sufficient. For example, one simple projection we could use would be to compute a *radial basis function* (RBF) centered on the middle clump:

```
In [12]: r = np.exp(-(X ** 2).sum(1))
```

We can visualize this extra data dimension using a three-dimensional plot, as seen in Figure 43-7.

```
In [13]: from mpl_toolkits import mplot3d

         ax = plt.subplot(projection='3d')
         ax.scatter3D(X[:, 0], X[:, 1], r, c=y, s=50, cmap='autumn')
         ax.view_init(elev=20, azim=30)
         ax.set_xlabel('x')
         ax.set_ylabel('y')
         ax.set_zlabel('r');
```

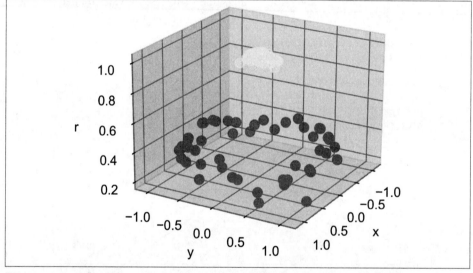

Figure 43-7. A third dimension added to the data allows for linear separation

We can see that with this additional dimension, the data becomes trivially linearly separable, by drawing a separating plane at, say, r=0.7.

In this case we had to choose and carefully tune our projection: if we had not centered our radial basis function in the right location, we would not have seen such clean, linearly separable results. In general, the need to make such a choice is a problem: we would like to somehow automatically find the best basis functions to use.

One strategy to this end is to compute a basis function centered at *every* point in the dataset, and let the SVM algorithm sift through the results. This type of basis function transformation is known as a *kernel transformation*, as it is based on a similarity relationship (or kernel) between each pair of points.

A potential problem with this strategy—projecting N points into N dimensions—is that it might become very computationally intensive as N grows large. However, because of a neat little procedure known as the *kernel trick* (*https://oreil.ly/h7PBj*), a fit on kernel-transformed data can be done implicitly—that is, without ever building the full N-dimensional representation of the kernel projection. This kernel trick is built into the SVM, and is one of the reasons the method is so powerful.

In Scikit-Learn, we can apply kernelized SVM simply by changing our linear kernel to an RBF kernel, using the `kernel` model hyperparameter:

```
In [14]: clf = SVC(kernel='rbf', C=1E6)
         clf.fit(X, y)
Out[14]: SVC(C=1000000.0)
```

Let's use our previously defined function to visualize the fit and identify the support vectors (Figure 43-8).

```
In [15]: plt.scatter(X[:, 0], X[:, 1], c=y, s=50, cmap='autumn')
         plot_svc_decision_function(clf)
         plt.scatter(clf.support_vectors_[:, 0], clf.support_vectors_[:, 1],
                     s=300, lw=1, facecolors='none');
```

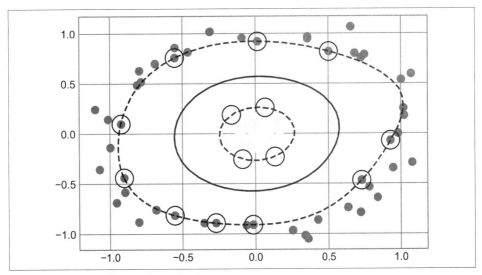

Figure 43-8. Kernel SVM fit to the data

Using this kernelized support vector machine, we learn a suitable nonlinear decision boundary. This kernel transformation strategy is used often in machine learning to turn fast linear methods into fast nonlinear methods, especially for models in which the kernel trick can be used.

Tuning the SVM: Softening Margins

Our discussion thus far has centered around very clean datasets, in which a perfect decision boundary exists. But what if your data has some amount of overlap? For example, you may have data like this (see Figure 43-9).

```
In [16]: X, y = make_blobs(n_samples=100, centers=2,
                           random_state=0, cluster_std=1.2)
         plt.scatter(X[:, 0], X[:, 1], c=y, s=50, cmap='autumn');
```

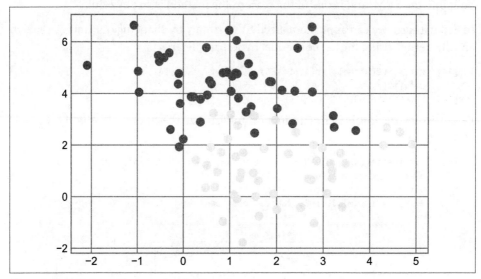

Figure 43-9. Data with some level of overlap

To handle this case, the SVM implementation has a bit of a fudge factor that "softens" the margin: that is, it allows some of the points to creep into the margin if that allows a better fit. The hardness of the margin is controlled by a tuning parameter, most often known as C. For a very large C, the margin is hard, and points cannot lie in it. For a smaller C, the margin is softer and can grow to encompass some points.

The plot shown in Figure 43-10 gives a visual picture of how a changing C affects the final fit via the softening of the margin:

```
In [17]: X, y = make_blobs(n_samples=100, centers=2,
                           random_state=0, cluster_std=0.8)

         fig, ax = plt.subplots(1, 2, figsize=(16, 6))
         fig.subplots_adjust(left=0.0625, right=0.95, wspace=0.1)

         for axi, C in zip(ax, [10.0, 0.1]):
             model = SVC(kernel='linear', C=C).fit(X, y)
             axi.scatter(X[:, 0], X[:, 1], c=y, s=50, cmap='autumn')
             plot_svc_decision_function(model, axi)
```

```
axi.scatter(model.support_vectors_[:, 0],
            model.support_vectors_[:, 1],
            s=300, lw=1, facecolors='none');
axi.set_title('C = {0:.1f}'.format(C), size=14)
```

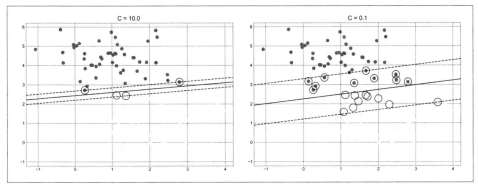

Figure 43-10. The effect of the C parameter on the support vector fit

The optimal value of C will depend on your dataset, and you should tune this parameter using cross-validation or a similar procedure (refer back to Chapter 39).

Example: Face Recognition

As an example of support vector machines in action, let's take a look at the facial recognition problem. We will use the Labeled Faces in the Wild dataset, which consists of several thousand collated photos of various public figures. A fetcher for the dataset is built into Scikit-Learn:

```
In [18]: from sklearn.datasets import fetch_lfw_people
         faces = fetch_lfw_people(min_faces_per_person=60)
         print(faces.target_names)
         print(faces.images.shape)
Out[18]: ['Ariel Sharon' 'Colin Powell' 'Donald Rumsfeld' 'George W Bush'
          'Gerhard Schroeder' 'Hugo Chavez' 'Junichiro Koizumi' 'Tony Blair']
         (1348, 62, 47)
```

Let's plot a few of these faces to see what we're working with (see Figure 43-11).

```
In [19]: fig, ax = plt.subplots(3, 5, figsize=(8, 6))
         for i, axi in enumerate(ax.flat):
             axi.imshow(faces.images[i], cmap='bone')
             axi.set(xticks=[], yticks=[],
                     xlabel=faces.target_names[faces.target[i]])
```

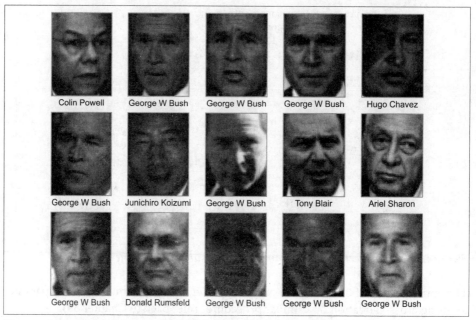

Figure 43-11. Examples from the Labeled Faces in the Wild dataset

Each image contains 62 × 47, or around 3,000, pixels. We could proceed by simply using each pixel value as a feature, but often it is more effective to use some sort of preprocessor to extract more meaningful features; here we will use principal component analysis (see Chapter 45) to extract 150 fundamental components to feed into our support vector machine classifier. We can do this most straightforwardly by packaging the preprocessor and the classifier into a single pipeline:

```
In [20]: from sklearn.svm import SVC
         from sklearn.decomposition import PCA
         from sklearn.pipeline import make_pipeline

         pca = PCA(n_components=150, whiten=True,
                   svd_solver='randomized', random_state=42)
         svc = SVC(kernel='rbf', class_weight='balanced')
         model = make_pipeline(pca, svc)
```

For the sake of testing our classifier output, we will split the data into a training set and a testing set:

```
In [21]: from sklearn.model_selection import train_test_split
         Xtrain, Xtest, ytrain, ytest = train_test_split(faces.data, faces.target,
                                                         random_state=42)
```

Finally, we can use grid search cross-validation to explore combinations of parameters. Here we will adjust C (which controls the margin hardness) and gamma (which controls the size of the radial basis function kernel), and determine the best model:

```
In [22]: from sklearn.model_selection import GridSearchCV
         param_grid = {'svc__C': [1, 5, 10, 50],
                       'svc__gamma': [0.0001, 0.0005, 0.001, 0.005]}
         grid = GridSearchCV(model, param_grid)

         %time grid.fit(Xtrain, ytrain)
         print(grid.best_params_)
Out[22]: CPU times: user 1min 19s, sys: 8.56 s, total: 1min 27s
         Wall time: 36.2 s
         {'svc__C': 10, 'svc__gamma': 0.001}
```

The optimal values fall toward the middle of our grid; if they fell at the edges, we would want to expand the grid to make sure we have found the true optimum.

Now with this cross-validated model we can predict the labels for the test data, which the model has not yet seen:

```
In [23]: model = grid.best_estimator_
         yfit = model.predict(Xtest)
```

Let's take a look at a few of the test images along with their predicted values (see Figure 43-12).

```
In [24]: fig, ax = plt.subplots(4, 6)
         for i, axi in enumerate(ax.flat):
             axi.imshow(Xtest[i].reshape(62, 47), cmap='bone')
             axi.set(xticks=[], yticks=[])
             axi.set_ylabel(faces.target_names[yfit[i]].split()[-1],
                            color='black' if yfit[i] == ytest[i] else 'red')
         fig.suptitle('Predicted Names; Incorrect Labels in Red', size=14);
```

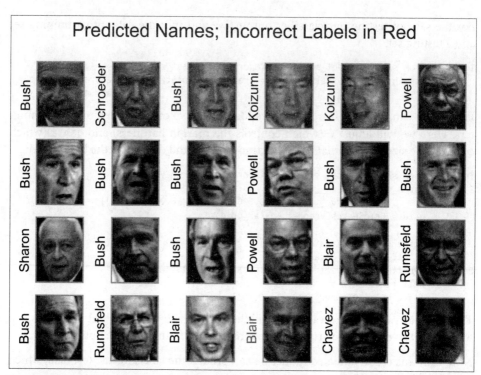

Figure 43-12. Labels predicted by our model

Out of this small sample, our optimal estimator mislabeled only a single face (Bush's face in the bottom row was mislabeled as Blair). We can get a better sense of our estimator's performance using the classification report, which lists recovery statistics label by label:

```
In [25]: from sklearn.metrics import classification_report
         print(classification_report(ytest, yfit,
                              target_names=faces.target_names))
Out[25]:            precision    recall  f1-score   support

     Ariel Sharon       0.65      0.73      0.69        15
     Colin Powell       0.80      0.87      0.83        68
  Donald Rumsfeld       0.74      0.84      0.79        31
    George W Bush       0.92      0.83      0.88       126
 Gerhard Schroeder      0.86      0.83      0.84        23
      Hugo Chavez       0.93      0.70      0.80        20
 Junichiro Koizumi      0.92      1.00      0.96        12
       Tony Blair       0.85      0.95      0.90        42

         accuracy                          0.85       337
        macro avg       0.83      0.84      0.84       337
     weighted avg       0.86      0.85      0.85       337
```

We might also display the confusion matrix between these classes (see Figure 43-13).

```
In [26]: from sklearn.metrics import confusion_matrix
         import seaborn as sns
         mat = confusion_matrix(ytest, yfit)
         sns.heatmap(mat.T, square=True, annot=True, fmt='d',
                 cbar=False, cmap='Blues',
                 xticklabels=faces.target_names,
                 yticklabels=faces.target_names)
         plt.xlabel('true label')
         plt.ylabel('predicted label');
```

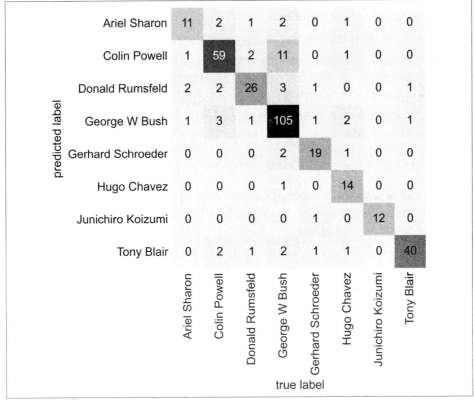

Figure 43-13. Confusion matrix for the faces data

This helps us get a sense of which labels are likely to be confused by the estimator.

For a real-world facial recognition task, in which the photos do not come pre-cropped into nice grids, the only difference in the facial classification scheme is the feature selection: you would need to use a more sophisticated algorithm to find the faces, and extract features that are independent of the pixellation. For this kind of application, one good option is to make use of OpenCV (*http://opencv.org*), which, among other things, includes pretrained implementations of state-of-the-art feature extraction tools for images in general and faces in particular.

Summary

This has been a brief intuitive introduction to the principles behind support vector machines. These models are a powerful classification method for a number of reasons:

- Their dependence on relatively few support vectors means that they are compact and take up very little memory.
- Once the model is trained, the prediction phase is very fast.
- Because they are affected only by points near the margin, they work well with high-dimensional data—even data with more dimensions than samples, which is challenging for other algorithms.
- Their integration with kernel methods makes them very versatile, able to adapt to many types of data.

However, SVMs have several disadvantages as well:

- The scaling with the number of samples N is $\mathcal{O}[N^3]$ at worst, or $\mathcal{O}[N^2]$ for efficient implementations. For large numbers of training samples, this computational cost can be prohibitive.
- The results are strongly dependent on a suitable choice for the softening parameter C. This must be carefully chosen via cross-validation, which can be expensive as datasets grow in size.
- The results do not have a direct probabilistic interpretation. This can be estimated via an internal cross-validation (see the `probability` parameter of SVC), but this extra estimation is costly.

With those traits in mind, I generally only turn to SVMs once other simpler, faster, and less tuning-intensive methods have been shown to be insufficient for my needs. Nevertheless, if you have the CPU cycles to commit to training and cross-validating an SVM on your data, the method can lead to excellent results.

## CHAPTER 44

In Depth: Decision Trees and Random Forests

Previously we have looked in depth at a simple generative classifier (naive Bayes; see Chapter 41) and a powerful discriminative classifier (support vector machines; see Chapter 43). Here we'll take a look at another powerful algorithm: a nonparametric algorithm called *random forests*. Random forests are an example of an *ensemble* method, meaning one that relies on aggregating the results of a set of simpler estimators. The somewhat surprising result with such ensemble methods is that the sum can be greater than the parts: that is, the predictive accuracy of a majority vote among a number of estimators can end up being better than that of any of the individual estimators doing the voting! We will see examples of this in the following sections.

We begin with the standard imports:

```
In [1]: %matplotlib inline
        import numpy as np
        import matplotlib.pyplot as plt
        plt.style.use('seaborn-whitegrid')
```

Motivating Random Forests: Decision Trees

Random forests are an example of an ensemble learner built on decision trees. For this reason, we'll start by discussing decision trees themselves.

Decision trees are extremely intuitive ways to classify or label objects: you simply ask a series of questions designed to zero in on the classification. For example, if you wanted to build a decision tree to classify animals you come across while on a hike, you might construct the one shown in Figure 44-1.

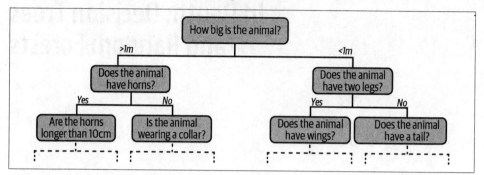

Figure 44-1. An example of a binary decision tree[1]

The binary splitting makes this extremely efficient: in a well-constructed tree, each question will cut the number of options by approximately half, very quickly narrowing the options even among a large number of classes. The trick, of course, comes in deciding which questions to ask at each step. In machine learning implementations of decision trees, the questions generally take the form of axis-aligned splits in the data: that is, each node in the tree splits the data into two groups using a cutoff value within one of the features. Let's now look at an example of this.

Creating a Decision Tree

Consider the following two-dimensional data, which has one of four class labels (see Figure 44-2).

```
In [2]: from sklearn.datasets import make_blobs

        X, y = make_blobs(n_samples=300, centers=4,
                          random_state=0, cluster_std=1.0)
        plt.scatter(X[:, 0], X[:, 1], c=y, s=50, cmap='rainbow');
```

---

1 Code to produce this figure can be found in the online appendix (*https://oreil.ly/xP9ZI*).

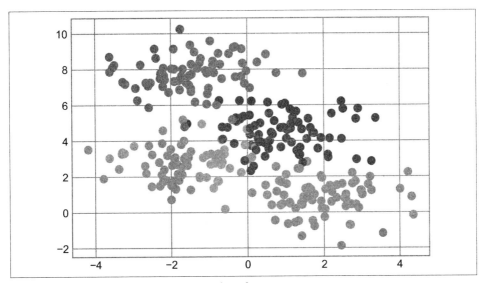

Figure 44-2. Data for the decision tree classifier

A simple decision tree built on this data will iteratively split the data along one or the other axis according to some quantitative criterion, and at each level assign the label of the new region according to a majority vote of points within it. Figure 44-3 presents a visualization of the first four levels of a decision tree classifier for this data.

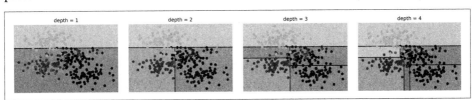

Figure 44-3. Visualization of how the decision tree splits the data[2]

Notice that after the first split, every point in the upper branch remains unchanged, so there is no need to further subdivide this branch. Except for nodes that contain all of one color, at each level *every* region is again split along one of the two features.

This process of fitting a decision tree to our data can be done in Scikit-Learn with the `DecisionTreeClassifier` estimator:

```
In [3]: from sklearn.tree import DecisionTreeClassifier
        tree = DecisionTreeClassifier().fit(X, y)
```

2 Code to produce this figure can be found in the online appendix (*https://oreil.ly/H4WFg*).

Let's write a utility function to help us visualize the output of the classifier:

```
In [4]: def visualize_classifier(model, X, y, ax=None, cmap='rainbow'):
            ax = ax or plt.gca()

            # Plot the training points
            ax.scatter(X[:, 0], X[:, 1], c=y, s=30, cmap=cmap,
                       clim=(y.min(), y.max()), zorder=3)
            ax.axis('tight')
            ax.axis('off')
            xlim = ax.get_xlim()
            ylim = ax.get_ylim()

            # fit the estimator
            model.fit(X, y)
            xx, yy = np.meshgrid(np.linspace(*xlim, num=200),
                                 np.linspace(*ylim, num=200))
            Z = model.predict(np.c_[xx.ravel(), yy.ravel()]).reshape(xx.shape)

            # Create a color plot with the results
            n_classes = len(np.unique(y))
            contours = ax.contourf(xx, yy, Z, alpha=0.3,
                                   levels=np.arange(n_classes + 1) - 0.5,
                                   cmap=cmap, zorder=1)

            ax.set(xlim=xlim, ylim=ylim)
```

Now we can examine what the decision tree classification looks like (see Figure 44-4).

```
In [5]: visualize_classifier(DecisionTreeClassifier(), X, y)
```

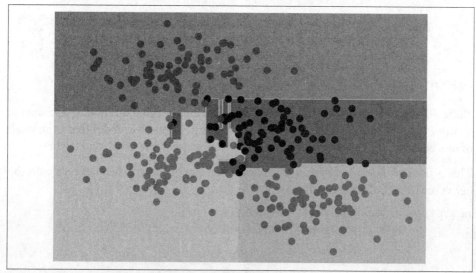

Figure 44-4. Visualization of a decision tree classification

If you're running this notebook live, you can use the helper script included in the online appendix (*https://oreil.ly/etDrN*) to bring up an interactive visualization of the decision tree building process:

```
In [6]: # helpers_05_08 is found in the online appendix
        import helpers_05_08
        helpers_05_08.plot_tree_interactive(X, y);
Out[6]: interactive(children=(Dropdown(description='depth', index=1, options=(1, 5),
          > value=5), Output()), _dom_classes...
```

Notice that as the depth increases, we tend to get very strangely shaped classification regions; for example, at a depth of five, there is a tall and skinny purple region between the yellow and blue regions. It's clear that this is less a result of the true, intrinsic data distribution, and more a result of the particular sampling or noise properties of the data. That is, this decision tree, even at only five levels deep, is clearly overfitting our data.

Decision Trees and Overfitting

Such overfitting turns out to be a general property of decision trees: it is very easy to go too deep in the tree, and thus to fit details of the particular data rather than the overall properties of the distributions it is drawn from. Another way to see this overfitting is to look at models trained on different subsets of the data—for example, in Figure 44-5 we train two different trees, each on half of the original data.

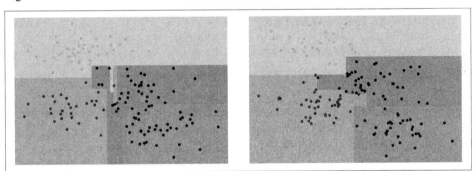

Figure 44-5. An example of two randomized decision trees[3]

It is clear that in some places the two trees produce consistent results (e.g., in the four corners), while in other places the two trees give very different classifications (e.g., in the regions between any two clusters). The key observation is that the inconsistencies tend to happen where the classification is less certain, and thus by using information from *both* of these trees, we might come up with a better result!

---

3 Code to produce this figure can be found in the online appendix (*https://oreil.ly/PessV*).

If you are running this notebook live, the following function will allow you to interactively display the fits of trees trained on a random subset of the data:

```
In [7]: # helpers_05_08 is found in the online appendix
        import helpers_05_08
        helpers_05_08.randomized_tree_interactive(X, y)
Out[7]: interactive(children=(Dropdown(description='random_state', options=(0, 100),
        > value=0), Output()), _dom_classes...
```

Just as using information from two trees improves our results, we might expect that using information from many trees would improve our results even further.

Ensembles of Estimators: Random Forests

This notion—that multiple overfitting estimators can be combined to reduce the effect of this overfitting—is what underlies an ensemble method called *bagging*. Bagging makes use of an ensemble (a grab bag, perhaps) of parallel estimators, each of which overfits the data, and averages the results to find a better classification. An ensemble of randomized decision trees is known as a *random forest*.

This type of bagging classification can be done manually using Scikit-Learn's `Bagging Classifier` meta-estimator, as shown here (see Figure 44-6).

```
In [8]: from sklearn.tree import DecisionTreeClassifier
        from sklearn.ensemble import BaggingClassifier

        tree = DecisionTreeClassifier()
        bag = BaggingClassifier(tree, n_estimators=100, max_samples=0.8,
                                random_state=1)

        bag.fit(X, y)
        visualize_classifier(bag, X, y)
```

In this example, we have randomized the data by fitting each estimator with a random subset of 80% of the training points. In practice, decision trees are more effectively randomized by injecting some stochasticity in how the splits are chosen: this way all the data contributes to the fit each time, but the results of the fit still have the desired randomness. For example, when determining which feature to split on, the randomized tree might select from among the top several features. You can read more technical details about these randomization strategies in the Scikit-Learn documentation (*https://oreil.ly/4jrv4*) and references within.

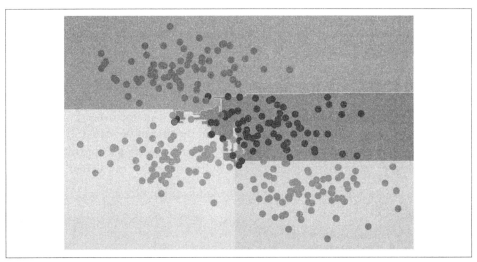

Figure 44-6. Decision boundaries for an ensemble of random decision trees

In Scikit-Learn, such an optimized ensemble of randomized decision trees is implemented in the `RandomForestClassifier` estimator, which takes care of all the randomization automatically. All you need to do is select a number of estimators, and it will very quickly—in parallel, if desired—fit the ensemble of trees (see Figure 44-7).

```
In [9]: from sklearn.ensemble import RandomForestClassifier

        model = RandomForestClassifier(n_estimators=100, random_state=0)
        visualize_classifier(model, X, y);
```

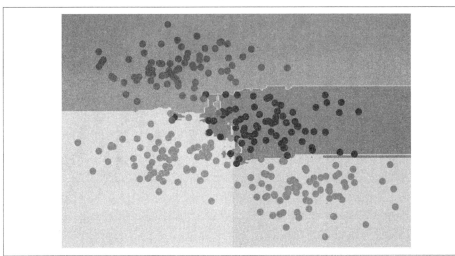

Figure 44-7. Decision boundaries for a random forest, which is an optimized ensemble of decision trees

We see that by averaging over one hundred randomly perturbed models, we end up with an overall model that is much closer to our intuition about how the parameter space should be split.

Random Forest Regression

In the previous section we considered random forests within the context of classification. Random forests can also be made to work in the case of regression (that is, with continuous rather than categorical variables). The estimator to use for this is the Ran domForestRegressor, and the syntax is very similar to what we saw earlier.

Consider the following data, drawn from the combination of a fast and slow oscillation (see Figure 44-8).

```
In [10]: rng = np.random.RandomState(42)
         x = 10 * rng.rand(200)

         def model(x, sigma=0.3):
             fast_oscillation = np.sin(5 * x)
             slow_oscillation = np.sin(0.5 * x)
             noise = sigma * rng.randn(len(x))

             return slow_oscillation + fast_oscillation + noise

         y = model(x)
         plt.errorbar(x, y, 0.3, fmt='o');
```

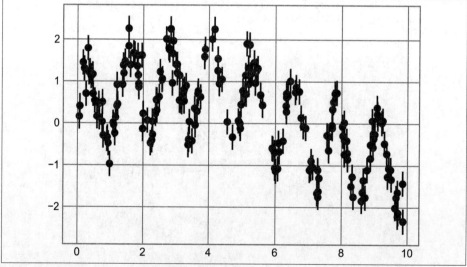

Figure 44-8. Data for random forest regression

Using the random forest regressor, we can find the best-fit curve (Figure 44-9).

```
In [11]: from sklearn.ensemble import RandomForestRegressor
         forest = RandomForestRegressor(200)
         forest.fit(x[:, None], y)

         xfit = np.linspace(0, 10, 1000)
         yfit = forest.predict(xfit[:, None])
         ytrue = model(xfit, sigma=0)

         plt.errorbar(x, y, 0.3, fmt='o', alpha=0.5)
         plt.plot(xfit, yfit, '-r');
         plt.plot(xfit, ytrue, '-k', alpha=0.5);
```

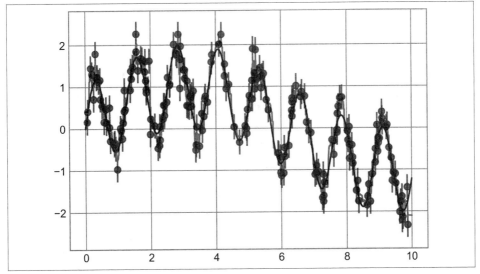

Figure 44-9. Random forest model fit to the data

Here the true model is shown in the smooth gray curve, while the random forest model is shown by the jagged red curve. The nonparametric random forest model is flexible enough to fit the multiperiod data, without us needing to specifying a multiperiod model!

Example: Random Forest for Classifying Digits

In Chapter 38, we worked through an example using the digits dataset included with Scikit-Learn. Let's use that again here to see how the random forest classifier can be applied in this context:

```
In [12]: from sklearn.datasets import load_digits
         digits = load_digits()
         digits.keys()
Out[12]: dict_keys(['data', 'target', 'frame', 'feature_names', 'target_names',
          > 'images', 'DESCR'])
```

To remind us what we're looking at, we'll visualize the first few data points (see Figure 44-10).

```
In [13]: # set up the figure
         fig = plt.figure(figsize=(6, 6))  # figure size in inches
         fig.subplots_adjust(left=0, right=1, bottom=0, top=1,
                             hspace=0.05, wspace=0.05)

         # plot the digits: each image is 8x8 pixels
         for i in range(64):
             ax = fig.add_subplot(8, 8, i + 1, xticks=[], yticks=[])
             ax.imshow(digits.images[i], cmap=plt.cm.binary, interpolation='nearest')

             # label the image with the target value
             ax.text(0, 7, str(digits.target[i]))
```

Figure 44-10. Representation of the digits data

We can classify the digits using a random forest as follows:

```
In [14]: from sklearn.model_selection import train_test_split

         Xtrain, Xtest, ytrain, ytest = train_test_split(digits.data, digits.target,
                                                          random_state=0)
         model = RandomForestClassifier(n_estimators=1000)
         model.fit(Xtrain, ytrain)
         ypred = model.predict(Xtest)
```

Let's look at the classification report for this classifier:

```
In [15]: from sklearn import metrics
         print(metrics.classification_report(ypred, ytest))
Out[15]:            precision    recall  f1-score   support

              0       1.00       0.97      0.99        38
              1       0.98       0.98      0.98        43
              2       0.95       1.00      0.98        42
              3       0.98       0.96      0.97        46
              4       0.97       1.00      0.99        37
              5       0.98       0.96      0.97        49
              6       1.00       1.00      1.00        52
              7       1.00       0.96      0.98        50
              8       0.94       0.98      0.96        46
              9       0.98       0.98      0.98        47

       accuracy                           0.98       450
      macro avg       0.98       0.98      0.98       450
   weighted avg       0.98       0.98      0.98       450
```

And for good measure, plot the confusion matrix (see Figure 44-11).

```
In [16]: from sklearn.metrics import confusion_matrix
         import seaborn as sns
         mat = confusion_matrix(ytest, ypred)
         sns.heatmap(mat.T, square=True, annot=True, fmt='d',
                     cbar=False, cmap='Blues')
         plt.xlabel('true label')
         plt.ylabel('predicted label');
```

We find that a simple, untuned random forest results in a quite accurate classification of the digits data.

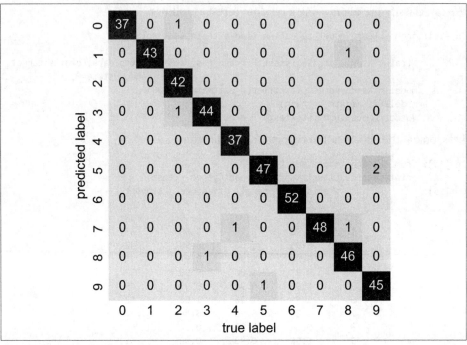

Figure 44-11. Confusion matrix for digit classification with random forests

Summary

This chapter provided a brief introduction to the concept of ensemble estimators, and in particular the random forest, an ensemble of randomized decision trees. Random forests are a powerful method with several advantages:

- Both training and prediction are very fast, because of the simplicity of the underlying decision trees. In addition, both tasks can be straightforwardly parallelized, because the individual trees are entirely independent entities.

- The multiple trees allow for a probabilistic classification: a majority vote among estimators gives an estimate of the probability (accessed in Scikit-Learn with the `predict_proba` method).

- The nonparametric model is extremely flexible and can thus perform well on tasks that are underfit by other estimators.

A primary disadvantage of random forests is that the results are not easily interpretable: that is, if you would like to draw conclusions about the *meaning* of the classification model, random forests may not be the best choice.

## CHAPTER 45

In Depth: Principal Component Analysis

Up until now, we have been looking in depth at supervised learning estimators: those estimators that predict labels based on labeled training data. Here we begin looking at several unsupervised estimators, which can highlight interesting aspects of the data without reference to any known labels.

In this chapter we will explore what is perhaps one of the most broadly used unsupervised algorithms, principal component analysis (PCA). PCA is fundamentally a dimensionality reduction algorithm, but it can also be useful as a tool for visualization, noise filtering, feature extraction and engineering, and much more. After a brief conceptual discussion of the PCA algorithm, we will explore a couple examples of these further applications.

We begin with the standard imports:

```
In [1]: %matplotlib inline
        import numpy as np
        import matplotlib.pyplot as plt
        plt.style.use('seaborn-whitegrid')
```

Introducing Principal Component Analysis

Principal component analysis is a fast and flexible unsupervised method for dimensionality reduction in data, which we saw briefly in Chapter 38. Its behavior is easiest to visualize by looking at a two-dimensional dataset. Consider these 200 points (see Figure 45-1).

```
In [2]: rng = np.random.RandomState(1)
        X = np.dot(rng.rand(2, 2), rng.randn(2, 200)).T
        plt.scatter(X[:, 0], X[:, 1])
        plt.axis('equal');
```

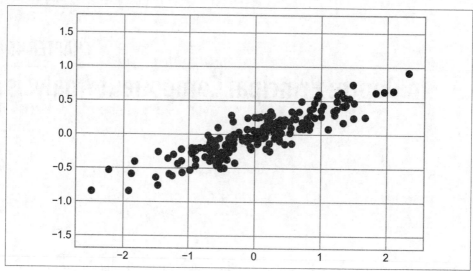

Figure 45-1. Data for demonstration of PCA

By eye, it is clear that there is a nearly linear relationship between the *x* and *y* variables. This is reminiscent of the linear regression data we explored in Chapter 42, but the problem setting here is slightly different: rather than attempting to *predict* the *y* values from the *x* values, the unsupervised learning problem attempts to learn about the *relationship* between the *x* and *y* values.

In principal component analysis, this relationship is quantified by finding a list of the *principal axes* in the data, and using those axes to describe the dataset. Using Scikit-Learn's PCA estimator, we can compute this as follows:

```
In [3]: from sklearn.decomposition import PCA
        pca = PCA(n_components=2)
        pca.fit(X)
Out[3]: PCA(n_components=2)
```

The fit learns some quantities from the data, most importantly the components and explained variance:

```
In [4]: print(pca.components_)
Out[4]: [[-0.94446029 -0.32862557]
         [-0.32862557  0.94446029]]

In [5]: print(pca.explained_variance_)
Out[5]: [0.7625315 0.0184779]
```

To see what these numbers mean, let's visualize them as vectors over the input data, using the components to define the direction of the vector and the explained variance to define the squared length of the vector (see Figure 45-2).

```
In [6]: def draw_vector(v0, v1, ax=None):
            ax = ax or plt.gca()
            arrowprops=dict(arrowstyle='->', linewidth=2,
                            shrinkA=0, shrinkB=0)
            ax.annotate('', v1, v0, arrowprops=arrowprops)

        # plot data
        plt.scatter(X[:, 0], X[:, 1], alpha=0.2)
        for length, vector in zip(pca.explained_variance_, pca.components_):
            v = vector * 3 * np.sqrt(length)
            draw_vector(pca.mean_, pca.mean_ + v)
        plt.axis('equal');
```

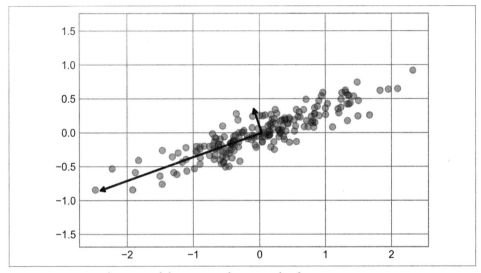

Figure 45-2. Visualization of the principal axes in the data

These vectors represent the principal axes of the data, and the length of each vector is an indication of how "important" that axis is in describing the distribution of the data —more precisely, it is a measure of the variance of the data when projected onto that axis. The projection of each data point onto the principal axes are the principal components of the data.

If we plot these principal components beside the original data, we see the plots shown in Figure 45-3.

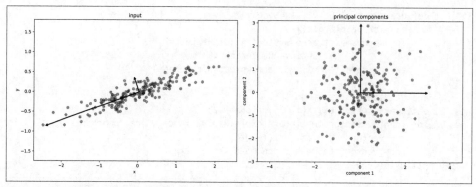

Figure 45-3. Transformed principal axes in the data[1]

This transformation from data axes to principal axes is an *affine transformation*, which means it is composed of a translation, rotation, and uniform scaling.

While this algorithm to find principal components may seem like just a mathematical curiosity, it turns out to have very far-reaching applications in the world of machine learning and data exploration.

PCA as Dimensionality Reduction

Using PCA for dimensionality reduction involves zeroing out one or more of the smallest principal components, resulting in a lower-dimensional projection of the data that preserves the maximal data variance.

Here is an example of using PCA as a dimensionality reduction transform:

```
In [7]: pca = PCA(n_components=1)
        pca.fit(X)
        X_pca = pca.transform(X)
        print("original shape:   ", X.shape)
        print("transformed shape:", X_pca.shape)
Out[7]: original shape:    (200, 2)
        transformed shape: (200, 1)
```

The transformed data has been reduced to a single dimension. To understand the effect of this dimensionality reduction, we can perform the inverse transform of this reduced data and plot it along with the original data (see Figure 45-4).

```
In [8]: X_new = pca.inverse_transform(X_pca)
        plt.scatter(X[:, 0], X[:, 1], alpha=0.2)
        plt.scatter(X_new[:, 0], X_new[:, 1], alpha=0.8)
        plt.axis('equal');
```

---

1 Code to produce this figure can be found in the online appendix (*https://oreil.ly/VmpjC*).

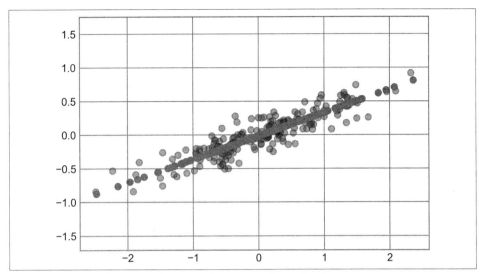

Figure 45-4. Visualization of PCA as dimensionality reduction

The light points are the original data, while the dark points are the projected version. This makes clear what a PCA dimensionality reduction means: the information along the least important principal axis or axes is removed, leaving only the component(s) of the data with the highest variance. The fraction of variance that is cut out (proportional to the spread of points about the line formed in the preceding figure) is roughly a measure of how much "information" is discarded in this reduction of dimensionality.

This reduced-dimension dataset is in some senses "good enough" to encode the most important relationships between the points: despite reducing the number of data features by 50%, the overall relationships between the data points are mostly preserved.

PCA for Visualization: Handwritten Digits

The usefulness of dimensionality reduction may not be entirely apparent in only two dimensions, but it becomes clear when looking at high-dimensional data. To see this, let's take a quick look at the application of PCA to the digits dataset we worked with in Chapter 44.

We'll start by loading the data:

```
In [9]: from sklearn.datasets import load_digits
        digits = load_digits()
        digits.data.shape
Out[9]: (1797, 64)
```

Recall that the digits dataset consists of 8 × 8–pixel images, meaning that they are 64-dimensional. To gain some intuition into the relationships between these points, we can use PCA to project them into a more manageable number of dimensions, say two:

```
In [10]: pca = PCA(2)  # project from 64 to 2 dimensions
         projected = pca.fit_transform(digits.data)
         print(digits.data.shape)
         print(projected.shape)
Out[10]: (1797, 64)
         (1797, 2)
```

We can now plot the first two principal components of each point to learn about the data, as seen in Figure 45-5.

```
In [11]: plt.scatter(projected[:, 0], projected[:, 1],
                      c=digits.target, edgecolor='none', alpha=0.5,
                      cmap=plt.cm.get_cmap('rainbow', 10))
         plt.xlabel('component 1')
         plt.ylabel('component 2')
         plt.colorbar();
```

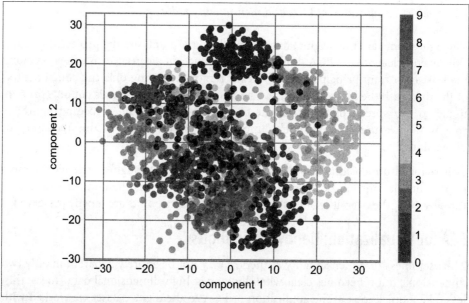

Figure 45-5. PCA applied to the handwritten digits data

Recall what these components mean: the full data is a 64-dimensional point cloud, and these points are the projection of each data point along the directions with the largest variance. Essentially, we have found the optimal stretch and rotation in 64-dimensional space that allows us to see the layout of the data in two dimensions, and we have done this in an unsupervised manner—that is, without reference to the labels.

What Do the Components Mean?

We can go a bit further here, and begin to ask what the reduced dimensions *mean*. This meaning can be understood in terms of combinations of basis vectors. For example, each image in the training set is defined by a collection of 64 pixel values, which we will call the vector x:

$$x = [x_1, x_2, x_3 \cdots x_{64}]$$

One way we can think about this is in terms of a pixel basis. That is, to construct the image, we multiply each element of the vector by the pixel it describes, and then add the results together to build the image:

$$\text{image}(x) = x_1 \cdot (\text{pixel 1}) + x_2 \cdot (\text{pixel 2}) + x_3 \cdot (\text{pixel 3}) \cdots x_{64} \cdot (\text{pixel 64})$$

One way we might imagine reducing the dimensionality of this data is to zero out all but a few of these basis vectors. For example, if we use only the first eight pixels, we get an eight-dimensional projection of the data (Figure 45-6). However, it is not very reflective of the whole image: we've thrown out nearly 90% of the pixels!

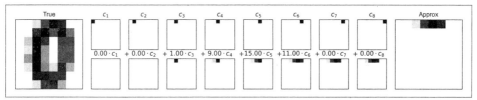

Figure 45-6. A naive dimensionality reduction achieved by discarding pixels[2]

The upper row of panels shows the individual pixels, and the lower row shows the cumulative contribution of these pixels to the construction of the image. Using only eight of the pixel-basis components, we can only construct a small portion of the 64-pixel image. Were we to continue this sequence and use all 64 pixels, we would recover the original image.

But the pixel-wise representation is not the only choice of basis. We can also use other basis functions, which each contain some predefined contribution from each pixel, and write something like:

$$image(x) = \text{mean} + x_1 \cdot (\text{basis 1}) + x_2 \cdot (\text{basis 2}) + x_3 \cdot (\text{basis 3}) \cdots$$

---

2 Code to produce this figure can be found in the online appendix (*https://oreil.ly/ixfc1*).

PCA can be thought of as a process of choosing optimal basis functions, such that adding together just the first few of them is enough to suitably reconstruct the bulk of the elements in the dataset. The principal components, which act as the low-dimensional representation of our data, are simply the coefficients that multiply each of the elements in this series. Figure 45-7 shows a similar depiction of reconstructing the same digit using the mean plus the first eight PCA basis functions.

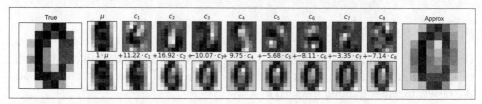

Figure 45-7. A more sophisticated dimensionality reduction achieved by discarding the least important principal components (compare to Figure 45-6)[3]

Unlike the pixel basis, the PCA basis allows us to recover the salient features of the input image with just a mean, plus eight components! The amount of each pixel in each component is the corollary of the orientation of the vector in our two-dimensional example. This is the sense in which PCA provides a low-dimensional representation of the data: it discovers a set of basis functions that are more efficient than the native pixel basis of the input data.

Choosing the Number of Components

A vital part of using PCA in practice is the ability to estimate how many components are needed to describe the data. This can be determined by looking at the cumulative *explained variance ratio* as a function of the number of components (see Figure 45-8).

```
In [12]: pca = PCA().fit(digits.data)
         plt.plot(np.cumsum(pca.explained_variance_ratio_))
         plt.xlabel('number of components')
         plt.ylabel('cumulative explained variance');
```

This curve quantifies how much of the total, 64-dimensional variance is contained within the first N components. For example, we see that with the digits data the first 10 components contain approximately 75% of the variance, while you need around 50 components to describe close to 100% of the variance.

---

3 Code to produce this figure can be found in the online appendix (*https://oreil.ly/WSeOT*).

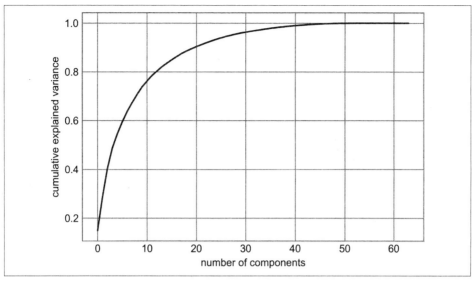

Figure 45-8. The cumulative explained variance, which measures how well PCA preserves the content of the data

This tells us that our two-dimensional projection loses a lot of information (as measured by the explained variance) and that we'd need about 20 components to retain 90% of the variance. Looking at this plot for a high-dimensional dataset can help you understand the level of redundancy present in its features.

PCA as Noise Filtering

PCA can also be used as a filtering approach for noisy data. The idea is this: any components with variance much larger than the effect of the noise should be relatively unaffected by the noise. So, if you reconstruct the data using just the largest subset of principal components, you should be preferentially keeping the signal and throwing out the noise.

Let's see how this looks with the digits data. First we will plot several of the input noise-free input samples (Figure 45-9).

```
In [13]: def plot_digits(data):
             fig, axes = plt.subplots(4, 10, figsize=(10, 4),
                                 subplot_kw={'xticks':[], 'yticks':[]},
                                 gridspec_kw=dict(hspace=0.1, wspace=0.1))
             for i, ax in enumerate(axes.flat):
                 ax.imshow(data[i].reshape(8, 8),
                         cmap='binary', interpolation='nearest',
                         clim=(0, 16))
         plot_digits(digits.data)
```

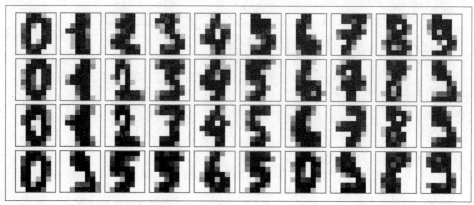

Figure 45-9. Digits without noise

Now let's add some random noise to create a noisy dataset, and replot it (Figure 45-10).

```
In [14]: rng = np.random.default_rng(42)
         rng.normal(10, 2)
Out[14]: 10.609434159508863
```

```
In [15]: rng = np.random.default_rng(42)
         noisy = rng.normal(digits.data, 4)
         plot_digits(noisy)
```

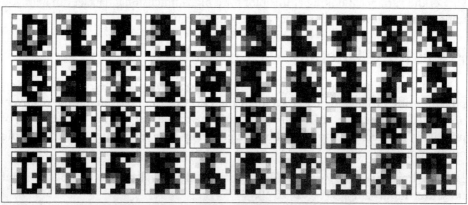

Figure 45-10. Digits with Gaussian random noise added

The visualization makes the presence of this random noise clear. Let's train a PCA model on the noisy data, requesting that the projection preserve 50% of the variance:

```
In [16]: pca = PCA(0.50).fit(noisy)
         pca.n_components_
Out[16]: 12
```

Here 50% of the variance amounts to 12 principal components, out of the 64 original features. Now we compute these components, and then use the inverse of the transform to reconstruct the filtered digits; Figure 45-11 shows the result.

```
In [17]: components = pca.transform(noisy)
         filtered = pca.inverse_transform(components)
         plot_digits(filtered)
```

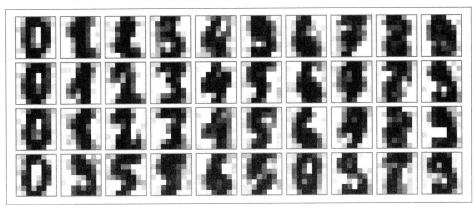

Figure 45-11. Digits "denoised" using PCA

This signal preserving/noise filtering property makes PCA a very useful feature selection routine—for example, rather than training a classifier on very high-dimensional data, you might instead train the classifier on the lower-dimensional principal component representation, which will automatically serve to filter out random noise in the inputs.

Example: Eigenfaces

Earlier we explored an example of using a PCA projection as a feature selector for facial recognition with a support vector machine (see Chapter 43). Here we will take a look back and explore a bit more of what went into that. Recall that we were using the Labeled Faces in the Wild (LFW) dataset made available through Scikit-Learn:

```
In [18]: from sklearn.datasets import fetch_lfw_people
         faces = fetch_lfw_people(min_faces_per_person=60)
         print(faces.target_names)
         print(faces.images.shape)
Out[18]: ['Ariel Sharon' 'Colin Powell' 'Donald Rumsfeld' 'George W Bush'
          'Gerhard Schroeder' 'Hugo Chavez' 'Junichiro Koizumi' 'Tony Blair']
         (1348, 62, 47)
```

Let's take a look at the principal axes that span this dataset. Because this is a large dataset, we will use the "random" eigensolver in the PCA estimator: it uses a randomized method to approximate the first N principal components more quickly than the standard approach, at the expense of some accuracy. This trade-off can be useful for high-dimensional data (here, a dimensionality of nearly 3,000). We will take a look at the first 150 components:

```
In [19]: pca = PCA(150, svd_solver='randomized', random_state=42)
         pca.fit(faces.data)
Out[19]: PCA(n_components=150, random_state=42, svd_solver='randomized')
```

In this case, it can be interesting to visualize the images associated with the first several principal components (these components are technically known as *eigenvectors*, so these types of images are often called *eigenfaces*; as you can see in Figure 45-12, they are as creepy as they sound):

```
In [20]: fig, axes = plt.subplots(3, 8, figsize=(9, 4),
                        subplot_kw={'xticks':[], 'yticks':[]},
                        gridspec_kw=dict(hspace=0.1, wspace=0.1))
         for i, ax in enumerate(axes.flat):
             ax.imshow(pca.components_[i].reshape(62, 47), cmap='bone')
```

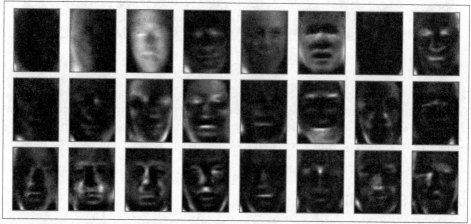

Figure 45-12. A visualization of eigenfaces learned from the LFW dataset

The results are very interesting, and give us insight into how the images vary: for example, the first few eigenfaces (from the top left) seem to be associated with the angle of lighting on the face, and later principal vectors seem to be picking out certain features, such as eyes, noses, and lips. Let's take a look at the cumulative variance of these components to see how much of the data information the projection is preserving (see Figure 45-13).

```
In [21]: plt.plot(np.cumsum(pca.explained_variance_ratio_))
         plt.xlabel('number of components')
         plt.ylabel('cumulative explained variance');
```

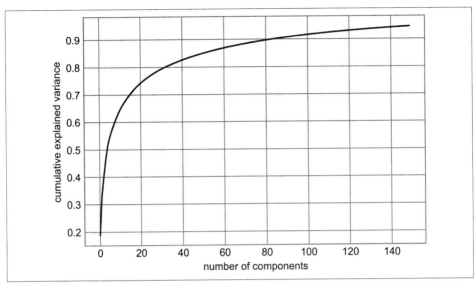

Figure 45-13. Cumulative explained variance for the LFW data

The 150 components we have chosen account for just over 90% of the variance. That would lead us to believe that using these 150 components, we would recover most of the essential characteristics of the data. To make this more concrete, we can compare the input images with the images reconstructed from these 150 components (see Figure 45-14).

```
In [22]: # Compute the components and projected faces
         pca = pca.fit(faces.data)
         components = pca.transform(faces.data)
         projected = pca.inverse_transform(components)
```

```
In [23]: # Plot the results
         fig, ax = plt.subplots(2, 10, figsize=(10, 2.5),
                                subplot_kw={'xticks':[], 'yticks':[]},
                                gridspec_kw=dict(hspace=0.1, wspace=0.1))
         for i in range(10):
             ax[0, i].imshow(faces.data[i].reshape(62, 47), cmap='binary_r')
             ax[1, i].imshow(projected[i].reshape(62, 47), cmap='binary_r')

         ax[0, 0].set_ylabel('full-dim\ninput')
         ax[1, 0].set_ylabel('150-dim\nreconstruction');
```

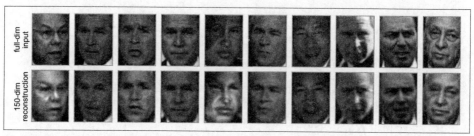

Figure 45-14. 150-dimensional PCA reconstruction of the LFW data

The top row here shows the input images, while the bottom row shows the reconstruction of the images from just 150 of the ~3,000 initial features. This visualization makes clear why the PCA feature selection used in Chapter 43 was so successful: although it reduces the dimensionality of the data by nearly a factor of 20, the projected images contain enough information that we might, by eye, recognize the individuals in each image. This means our classification algorithm only needs to be trained on 150-dimensional data rather than 3,000-dimensional data, which, depending on the particular algorithm we choose, can lead to much more efficient classification.

Summary

In this chapter we explored the use of principal component analysis for dimensionality reduction, visualization of high-dimensional data, noise filtering, and feature selection within high-dimensional data. Because of its versatility and interpretability, PCA has been shown to be effective in a wide variety of contexts and disciplines. Given any high-dimensional dataset, I tend to start with PCA in order to visualize the relationships between points (as we did with the digits data), to understand the main variance in the data (as we did with the eigenfaces), and to understand the intrinsic dimensionality (by plotting the explained variance ratio). Certainly PCA is not useful for every high-dimensional dataset, but it offers a straightforward and efficient path to gaining insight into high-dimensional data.

PCA's main weakness is that it tends to be highly affected by outliers in the data. For this reason, several robust variants of PCA have been developed, many of which act to iteratively discard data points that are poorly described by the initial components. Scikit-Learn includes a number of interesting variants on PCA in the `sklearn` `.decomposition` submodule; one example is `SparsePCA`, which introduces a regularization term (see Chapter 42) that serves to enforce sparsity of the components.

In the following chapters, we will look at other unsupervised learning methods that build on some of the ideas of PCA.

## CHAPTER 46

In Depth: Manifold Learning

In the previous chapter we saw how PCA can be used for dimensionality reduction, reducing the number of features of a dataset while maintaining the essential relationships between the points. While PCA is flexible, fast, and easily interpretable, it does not perform so well when there are *nonlinear* relationships within the data, some examples of which we will see shortly.

To address this deficiency, we can turn to *manifold learning algorithms*—a class of unsupervised estimators that seek to describe datasets as low-dimensional manifolds embedded in high-dimensional spaces. When you think of a manifold, I'd suggest imagining a sheet of paper: this is a two-dimensional object that lives in our familiar three-dimensional world.

In the parlance of manifold learning, you can think of this sheet as a two-dimensional manifold embedded in three-dimensional space. Rotating, reorienting, or stretching the piece of paper in three-dimensional space doesn't change its flat geometry: such operations are akin to linear embeddings. If you bend, curl, or crumple the paper, it is still a two-dimensional manifold, but the embedding into the three-dimensional space is no longer linear. Manifold learning algorithms seek to learn about the fundamental two-dimensional nature of the paper, even as it is contorted to fill the three-dimensional space.

Here we will examine a number of manifold methods, going most deeply into a subset of these techniques: multidimensional scaling (MDS), locally linear embedding (LLE), and isometric mapping (Isomap).

We begin with the standard imports:

```
In [1]: %matplotlib inline
        import matplotlib.pyplot as plt
        plt.style.use('seaborn-whitegrid')
        import numpy as np
```

Manifold Learning: "HELLO"

To make these concepts more clear, let's start by generating some two-dimensional data that we can use to define a manifold. Here is a function that will create data in the shape of the word "HELLO":

```
In [2]: def make_hello(N=1000, rseed=42):
            # Make a plot with "HELLO" text; save as PNG
            fig, ax = plt.subplots(figsize=(4, 1))
            fig.subplots_adjust(left=0, right=1, bottom=0, top=1)
            ax.axis('off')
            ax.text(0.5, 0.4, 'HELLO', va='center', ha='center',
                    weight='bold', size=85)
            fig.savefig('hello.png')
            plt.close(fig)

            # Open this PNG and draw random points from it
            from matplotlib.image import imread
            data = imread('hello.png')[::-1, :, 0].T
            rng = np.random.RandomState(rseed)
            X = rng.rand(4 * N, 2)
            i, j = (X * data.shape).astype(int).T
            mask = (data[i, j] < 1)
            X = X[mask]
            X[:, 0] *= (data.shape[0] / data.shape[1])
            X = X[:N]
            return X[np.argsort(X[:, 0])]
```

Let's call the function and visualize the resulting data (Figure 46-1).

```
In [3]: X = make_hello(1000)
        colorize = dict(c=X[:, 0], cmap=plt.cm.get_cmap('rainbow', 5))
        plt.scatter(X[:, 0], X[:, 1], **colorize)
        plt.axis('equal');
```

The output is two dimensional, and consists of points drawn in the shape of the word "HELLO". This data form will help us to see visually what these algorithms are doing.

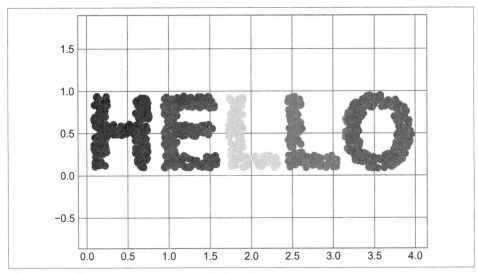

Figure 46-1. Data for use with manifold learning

Multidimensional Scaling

Looking at data like this, we can see that the particular choices of x and y values of the dataset are not the most fundamental description of the data: we can scale, shrink, or rotate the data, and the "HELLO" will still be apparent. For example, if we use a rotation matrix to rotate the data, the x and y values change, but the data is still fundamentally the same (see Figure 46-2).

```
In [4]: def rotate(X, angle):
            theta = np.deg2rad(angle)
            R = [[np.cos(theta), np.sin(theta)],
                 [-np.sin(theta), np.cos(theta)]]
            return np.dot(X, R)

        X2 = rotate(X, 20) + 5
        plt.scatter(X2[:, 0], X2[:, 1], **colorize)
        plt.axis('equal');
```

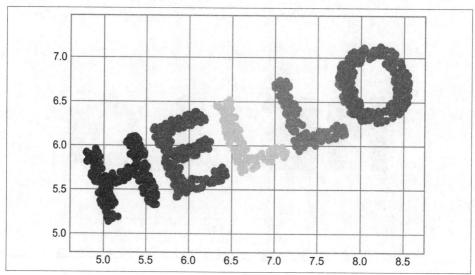

Figure 46-2. Rotated dataset

This confirms that the *x* and *y* values are not necessarily fundamental to the relationships in the data. What *is* fundamental, in this case, is the *distance* between each point within the dataset. A common way to represent this is to use a distance matrix: for N points, we construct an $N \times N$ array such that entry (i, j) contains the distance between point i and point j. Let's use Scikit-Learn's efficient `pairwise_distances` function to do this for our original data:

```
In [5]: from sklearn.metrics import pairwise_distances
        D = pairwise_distances(X)
        D.shape
Out[5]: (1000, 1000)
```

As promised, for our N=1,000 points, we obtain a 1,000 × 1,000 matrix, which can be visualized as shown here (see Figure 46-3).

```
In [6]: plt.imshow(D, zorder=2, cmap='viridis', interpolation='nearest')
        plt.colorbar();
```

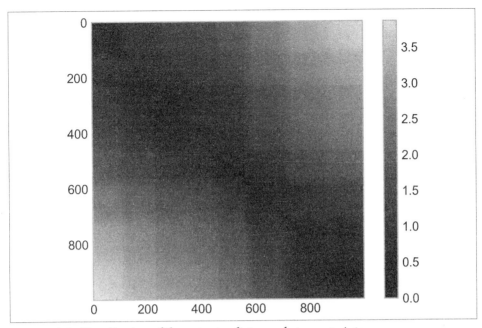

Figure 46-3. Visualization of the pairwise distances between points

If we similarly construct a distance matrix for our rotated and translated data, we see that it is the same:

```
In [7]: D2 = pairwise_distances(X2)
        np.allclose(D, D2)
Out[7]: True
```

This distance matrix gives us a representation of our data that is invariant to rotations and translations, but the visualization of the matrix in Figure 46-3 is not entirely intuitive. In the representation shown there, we have lost any visible sign of the interesting structure in the data: the "HELLO" that we saw before.

Further, while computing this distance matrix from the (x, y) coordinates is straightforward, transforming the distances back into x and y coordinates is rather difficult. This is exactly what the multidimensional scaling algorithm aims to do: given a distance matrix between points, it recovers a D-dimensional coordinate representation of the data. Let's see how it works for our distance matrix, using the `precomputed` dissimilarity to specify that we are passing a distance matrix (Figure 46-4).

```
In [8]: from sklearn.manifold import MDS
        model = MDS(n_components=2, dissimilarity='precomputed', random_state=1701)
        out = model.fit_transform(D)
        plt.scatter(out[:, 0], out[:, 1], **colorize)
        plt.axis('equal');
```

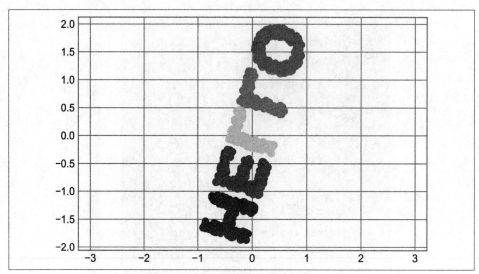

Figure 46-4. An MDS embedding computed from the pairwise distances

The MDS algorithm recovers one of the possible two-dimensional coordinate representations of our data, using *only* the $N \times N$ distance matrix describing the relationship between the data points.

MDS as Manifold Learning

The usefulness of this becomes more apparent when we consider the fact that distance matrices can be computed from data in *any* dimension. So, for example, instead of simply rotating the data in the two-dimensional plane, we can project it into three dimensions using the following function (essentially a three-dimensional generalization of the rotation matrix used earlier):

```
In [9]: def random_projection(X, dimension=3, rseed=42):
            assert dimension >= X.shape[1]
            rng = np.random.RandomState(rseed)
            C = rng.randn(dimension, dimension)
            e, V = np.linalg.eigh(np.dot(C, C.T))
            return np.dot(X, V[:X.shape[1]])

        X3 = random_projection(X, 3)
        X3.shape
Out[9]: (1000, 3)
```

Let's visualize these points to see what we're working with (Figure 46-5).

```
In [10]: from mpl_toolkits import mplot3d
         ax = plt.axes(projection='3d')
         ax.scatter3D(X3[:, 0], X3[:, 1], X3[:, 2],
                      **colorize);
```

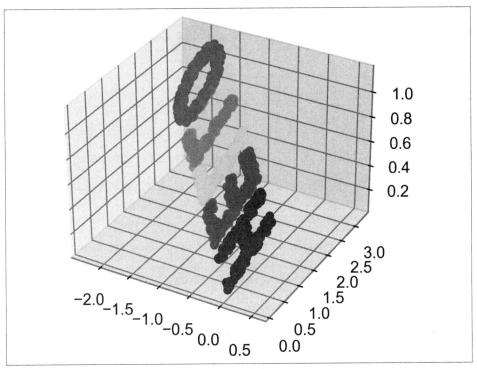

Figure 46-5. Data embedded linearly into three dimensions

We can now ask the MDS estimator to input this three-dimensional data, compute the distance matrix, and then determine the optimal two-dimensional embedding for this distance matrix. The result recovers a representation of the original data, as shown in Figure 46-6.

```
In [11]: model = MDS(n_components=2, random_state=1701)
         out3 = model.fit_transform(X3)
         plt.scatter(out3[:, 0], out3[:, 1], **colorize)
         plt.axis('equal');
```

This is essentially the goal of a manifold learning estimator: given high-dimensional embedded data, it seeks a low-dimensional representation of the data that preserves certain relationships within the data. In the case of MDS, the quantity preserved is the distance between every pair of points.

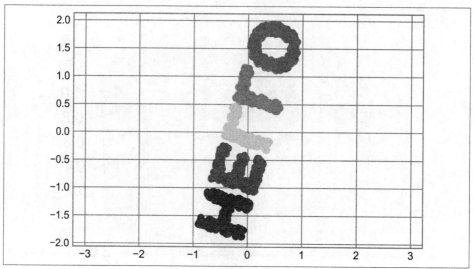

Figure 46-6. The MDS embedding of the three-dimensional data recovers the input up to a rotation and reflection

Nonlinear Embeddings: Where MDS Fails

Our discussion thus far has considered *linear* embeddings, which essentially consist of rotations, translations, and scalings of data into higher-dimensional spaces. Where MDS breaks down is when the embedding is nonlinear—that is, when it goes beyond this simple set of operations. Consider the following embedding, which takes the input and contorts it into an "S" shape in three dimensions:

```
In [12]: def make_hello_s_curve(X):
             t = (X[:, 0] - 2) * 0.75 * np.pi
             x = np.sin(t)
             y = X[:, 1]
             z = np.sign(t) * (np.cos(t) - 1)
             return np.vstack((x, y, z)).T

         XS = make_hello_s_curve(X)
```

This is again three-dimensional data, but as we can see in Figure 46-7 the embedding is much more complicated.

```
In [13]: from mpl_toolkits import mplot3d
         ax = plt.axes(projection='3d')
         ax.scatter3D(XS[:, 0], XS[:, 1], XS[:, 2],
                      **colorize);
```

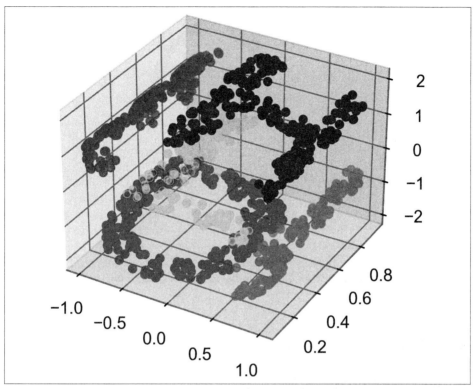

Figure 46-7. Data embedded nonlinearly into three dimensions

The fundamental relationships between the data points are still there, but this time the data has been transformed in a nonlinear way: it has been wrapped up into the shape of an "S."

If we try a simple MDS algorithm on this data, it is not able to "unwrap" this nonlinear embedding, and we lose track of the fundamental relationships in the embedded manifold (see Figure 46-8).

```
In [14]: from sklearn.manifold import MDS
         model = MDS(n_components=2, random_state=2)
         outS = model.fit_transform(XS)
         plt.scatter(outS[:, 0], outS[:, 1], **colorize)
         plt.axis('equal');
```

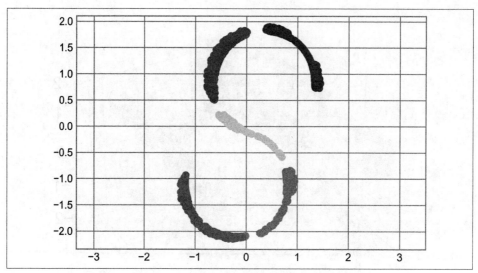

Figure 46-8. The MDS algorithm applied to the nonlinear data; it fails to recover the underlying structure

The best two-dimensional *linear* embedding does not unwrap the S-curve, but instead discards the original y-axis.

Nonlinear Manifolds: Locally Linear Embedding

How can we move forward here? Stepping back, we can see that the source of the problem is that MDS tries to preserve distances between faraway points when constructing the embedding. But what if we instead modified the algorithm such that it only preserves distances between nearby points? The resulting embedding would be closer to what we want.

Visually, we can think of it as illustrated Figure 46-9.

Here each faint line represents a distance that should be preserved in the embedding. On the left is a representation of the model used by MDS: it tries to preserve the distances between each pair of points in the dataset. On the right is a representation of the model used by a manifold learning algorithm called *locally linear embedding*: rather than preserving *all* distances, it instead tries to preserve only the distances between *neighboring points* (in this case, the nearest 100 neighbors of each point).

Thinking about the left panel, we can see why MDS fails: there is no way to unroll this data while adequately preserving the length of every line drawn between the two points. For the right panel, on the other hand, things look a bit more optimistic. We could imagine unrolling the data in a way that keeps the lengths of the lines approximately the same. This is precisely what LLE does, through a global optimization of a cost function reflecting this logic.

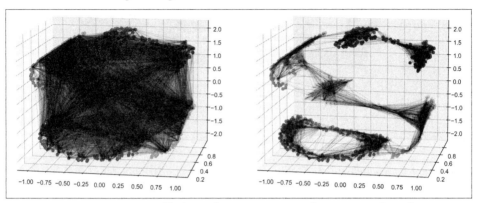

Figure 46-9. Representation of linkages between points within MDS and LLE[1]

LLE comes in a number of flavors; here we will use the *modified LLE* algorithm to recover the embedded two-dimensional manifold. In general, modified LLE does better than other flavors of the algorithm at recovering well-defined manifolds with very little distortion (see Figure 46-10).

```
In [15]: from sklearn.manifold import LocallyLinearEmbedding
         model = LocallyLinearEmbedding(
             n_neighbors=100, n_components=2,
             method='modified', eigen_solver='dense')
         out = model.fit_transform(XS)

         fig, ax = plt.subplots()
         ax.scatter(out[:, 0], out[:, 1], **colorize)
         ax.set_ylim(0.15, -0.15);
```

The result remains somewhat distorted compared to our original manifold, but captures the essential relationships in the data!

---

1 Code to produce this figure can be found in the online appendix (*https://oreil.ly/gu4iE*).

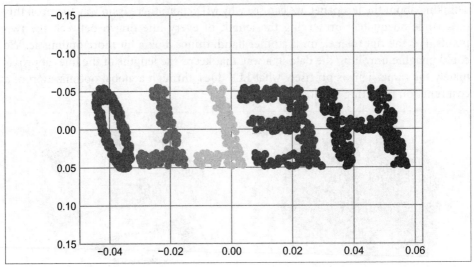

Figure 46-10. Locally linear embedding can recover the underlying data from a nonlinearly embedded input

Some Thoughts on Manifold Methods

Compelling as these examples may be, in practice manifold learning techniques tend to be finicky enough that they are rarely used for anything more than simple qualitative visualization of high-dimensional data.

The following are some of the particular challenges of manifold learning, which all contrast poorly with PCA:

- In manifold learning, there is no good framework for handling missing data. In contrast, there are straightforward iterative approaches for dealing with missing data in PCA.

- In manifold learning, the presence of noise in the data can "short-circuit" the manifold and drastically change the embedding. In contrast, PCA naturally filters noise from the most important components.

- The manifold embedding result is generally highly dependent on the number of neighbors chosen, and there is generally no solid quantitative way to choose an optimal number of neighbors. In contrast, PCA does not involve such a choice.

- In manifold learning, the globally optimal number of output dimensions is difficult to determine. In contrast, PCA lets you find the number of output dimensions based on the explained variance.

- In manifold learning, the meaning of the embedded dimensions is not always clear. In PCA, the principal components have a very clear meaning.

- In manifold learning, the computational expense of manifold methods scales as $O[N^2]$ or $O[N^3]$. For PCA, there exist randomized approaches that are generally much faster (though see the *megaman* package (*https://oreil.ly/VLBly*) for some more scalable implementations of manifold learning).

With all that on the table, the only clear advantage of manifold learning methods over PCA is their ability to preserve nonlinear relationships in the data; for that reason I tend to explore data with manifold methods only after first exploring it with PCA.

Scikit-Learn implements several common variants of manifold learning beyond LLE and Isomap (which we've used in a few of the previous chapters and will look at in the next section): the Scikit-Learn documentation has a nice discussion and comparison of them (*https://oreil.ly/tFzS5*). Based on my own experience, I would give the following recommendations:

- For toy problems such as the S-curve we saw before, LLE and its variants (especially modified LLE) perform very well. This is implemented in `sklearn.mani fold.LocallyLinearEmbedding`.

- For high-dimensional data from real-world sources, LLE often produces poor results, and Isomap seems to generally lead to more meaningful embeddings. This is implemented in `sklearn.manifold.Isomap`.

- For data that is highly clustered, *t-distributed stochastic neighbor embedding* (t-SNE) seems to work very well, though it can be very slow compared to other methods. This is implemented in `sklearn.manifold.TSNE`.

If you're interested in getting a feel for how these work, I'd suggest running each of the methods on the data in this section.

Example: Isomap on Faces

One place manifold learning is often used is in understanding the relationship between high-dimensional data points. A common case of high-dimensional data is images: for example, a set of images with 1,000 pixels each can be thought of as a collection of points in 1,000 dimensions, with the brightness of each pixel in each image defining the coordinate in that dimension.

To illustrate, let's apply Isomap on some data from the Labeled Faces in the Wild dataset, which we previously saw in Chapters 43 and 45. Running this command will download the dataset and cache it in your home directory for later use:

```
In [16]: from sklearn.datasets import fetch_lfw_people
         faces = fetch_lfw_people(min_faces_per_person=30)
         faces.data.shape
Out[16]: (2370, 2914)
```

We have 2,370 images, each with 2,914 pixels. In other words, the images can be thought of as data points in a 2,914-dimensional space!

Let's display several of these images to remind us what we're working with (see Figure 46-11).

```
In [17]: fig, ax = plt.subplots(4, 8, subplot_kw=dict(xticks=[], yticks=[]))
         for i, axi in enumerate(ax.flat):
             axi.imshow(faces.images[i], cmap='gray')
```

Figure 46-11. Examples of the input faces

When we encountered this data in Chapter 45, our goal was essentially compression: to use the components to reconstruct the inputs from the lower-dimensional representation.

PCA is versatile enough that we can also use it in this context, where we would like to plot a low-dimensional embedding of the 2,914-dimensional data to learn the fundamental relationships between the images. Let's again look at the explained variance ratio, which will give us an idea of how many linear features are required to describe the data (see Figure 46-12).

```
In [18]: from sklearn.decomposition import PCA
         model = PCA(100, svd_solver='randomized').fit(faces.data)
         plt.plot(np.cumsum(model.explained_variance_ratio_))
         plt.xlabel('n components')
         plt.ylabel('cumulative variance');
```

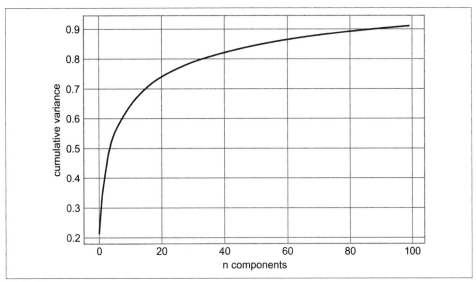

Figure 46-12. Cumulative variance from the PCA projection

We see that for this data, nearly 100 components are required to preserve 90% of the variance. This tells us that the data is intrinsically very high-dimensional—it can't be described linearly with just a few components.

When this is the case, nonlinear manifold embeddings like LLE and Isomap may be helpful. We can compute an Isomap embedding on these faces using the same pattern shown before:

```
In [19]: from sklearn.manifold import Isomap
         model = Isomap(n_components=2)
         proj = model.fit_transform(faces.data)
         proj.shape
Out[19]: (2370, 2)
```

The output is a two-dimensional projection of all the input images. To get a better idea of what the projection tells us, let's define a function that will output image thumbnails at the locations of the projections:

```
In [20]: from matplotlib import offsetbox

         def plot_components(data, model, images=None, ax=None,
                             thumb_frac=0.05, cmap='gray'):
             ax = ax or plt.gca()

             proj = model.fit_transform(data)
             ax.plot(proj[:, 0], proj[:, 1], '.k')

             if images is not None:
                 min_dist_2 = (thumb_frac * max(proj.max(0) - proj.min(0))) ** 2
```

```
shown_images = np.array([2 * proj.max(0)])
for i in range(data.shape[0]):
    dist = np.sum((proj[i] - shown_images) ** 2, 1)
    if np.min(dist) < min_dist_2:
        # don't show points that are too close
        continue
    shown_images = np.vstack([shown_images, proj[i]])
    imagebox = offsetbox.AnnotationBbox(
        offsetbox.OffsetImage(images[i], cmap=cmap),
                              proj[i])
    ax.add_artist(imagebox)
```

Calling this function now, we see the result in Figure 46-13.

```
In [21]: fig, ax = plt.subplots(figsize=(10, 10))
         plot_components(faces.data,
                         model=Isomap(n_components=2),
                         images=faces.images[:, ::2, ::2])
```

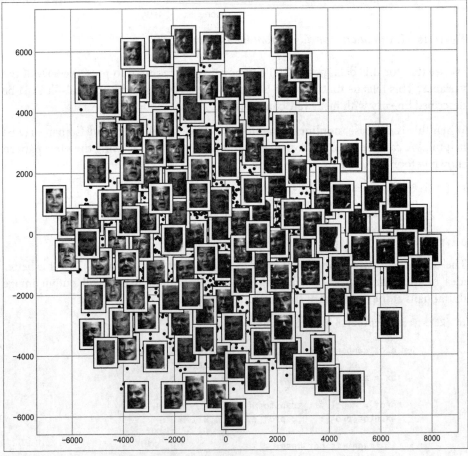

Figure 46-13. Isomap embedding of the LFW data

The result is interesting. The first two Isomap dimensions seem to describe global image features: the overall brightness of the image from left to right, and the general orientation of the face from bottom to top. This gives us a nice visual indication of some of the fundamental features in our data.

From here, we could then go on to classify this data (perhaps using manifold features as inputs to the classification algorithm) as we did in Chapter 43.

Example: Visualizing Structure in Digits

As another example of using manifold learning for visualization, let's take a look at the MNIST handwritten digits dataset. This is similar to the digits dataset we saw in Chapter 44, but with many more pixels per image. It can be downloaded from *http:// openml.org* with the Scikit-Learn utility:

```
In [22]: from sklearn.datasets import fetch_openml
         mnist = fetch_openml('mnist_784')
         mnist.data.shape
Out[22]: (70000, 784)
```

The dataset consists of 70,000 images, each with 784 pixels (i.e., the images are 28 × 28). As before, we can take a look at the first few images (see Figure 46-14).

```
In [23]: mnist_data = np.asarray(mnist.data)
         mnist_target = np.asarray(mnist.target, dtype=int)

         fig, ax = plt.subplots(6, 8, subplot_kw=dict(xticks=[], yticks=[]))
         for i, axi in enumerate(ax.flat):
             axi.imshow(mnist_data[1250 * i].reshape(28, 28), cmap='gray_r')
```

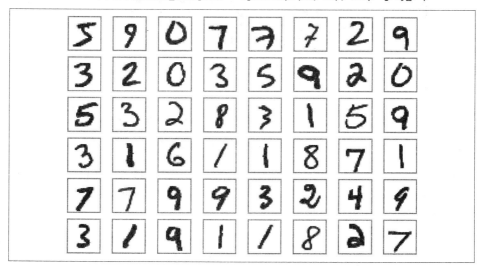

Figure 46-14. Examples of MNIST digits

This gives us an idea of the variety of handwriting styles in the dataset.

Let's compute a manifold learning projection across the data. For speed here, we'll only use 1/30 of the data, which is about ~2,000 points (because of the relatively poor scaling of manifold learning, I find that a few thousand samples is a good number to start with for relatively quick exploration before moving to a full calculation). Figure 46-15 shows the result.

```
In [24]: # Use only 1/30 of the data: full dataset takes a long time!
         data = mnist_data[::30]
         target = mnist_target[::30]

         model = Isomap(n_components=2)
         proj = model.fit_transform(data)

         plt.scatter(proj[:, 0], proj[:, 1], c=target,
                             cmap=plt.cm.get_cmap('jet', 10))
         plt.colorbar(ticks=range(10))
         plt.clim(-0.5, 9.5);
```

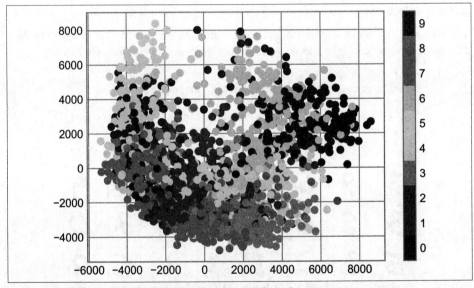

Figure 46-15. Isomap embedding of the MNIST digit data

The resulting scatter plot shows some of the relationships between the data points, but is a bit crowded. We can gain more insight by looking at just a single number at a time (see Figure 46-16).

```
In [25]: # Choose 1/4 of the "1" digits to project
         data = mnist_data[mnist_target == 1][::4]

         fig, ax = plt.subplots(figsize=(10, 10))
         model = Isomap(n_neighbors=5, n_components=2, eigen_solver='dense')
```

```
plot_components(data, model, images=data.reshape((-1, 28, 28)),
                ax=ax, thumb_frac=0.05, cmap='gray_r')
```

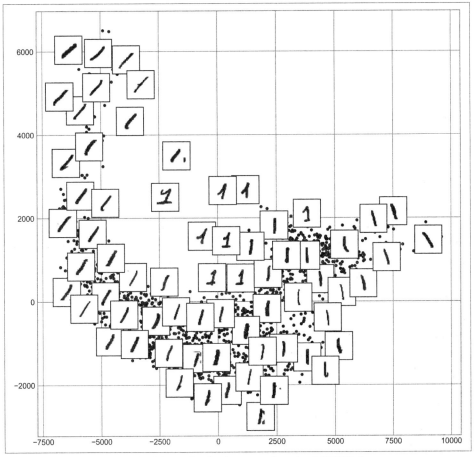

Figure 46-16. Isomap embedding of only the 1s within the MNIST dataset

The result gives you an idea of the variety of forms that the number 1 can take within the dataset. The data lies along a broad curve in the projected space, which appears to trace the orientation of the digit. As you move up the plot, you find 1s that have hats and/or bases, though these are very sparse within the dataset. The projection lets us identify outliers that have data issues: for example, pieces of the neighboring digits that snuck into the extracted images.

Now, this in itself may not be useful for the task of classifying digits, but it does help us get an understanding of the data, and may give us ideas about how to move forward—such as how we might want to preprocess the data before building a classification pipeline.

## CHAPTER 47

In Depth: k-Means Clustering

In the previous chapters we explored unsupervised machine learning models for dimensionality reduction. Now we will move on to another class of unsupervised machine learning models: clustering algorithms. Clustering algorithms seek to learn, from the properties of the data, an optimal division or discrete labeling of groups of points.

Many clustering algorithms are available in Scikit-Learn and elsewhere, but perhaps the simplest to understand is an algorithm known as *k-means clustering*, which is implemented in `sklearn.cluster.KMeans`.

We begin with the standard imports:

```
In [1]: %matplotlib inline
        import matplotlib.pyplot as plt
        plt.style.use('seaborn-whitegrid')
        import numpy as np
```

Introducing k-Means

The *k*-means algorithm searches for a predetermined number of clusters within an unlabeled multidimensional dataset. It accomplishes this using a simple conception of what the optimal clustering looks like:

- The *cluster center* is the arithmetic mean of all the points belonging to the cluster.
- Each point is closer to its own cluster center than to other cluster centers.

Those two assumptions are the basis of the *k*-means model. We will soon dive into exactly *how* the algorithm reaches this solution, but for now let's take a look at a simple dataset and see the *k*-means result.

First, let's generate a two-dimensional dataset containing four distinct blobs. To emphasize that this is an unsupervised algorithm, we will leave the labels out of the visualization (see Figure 47-1).

```
In [2]: from sklearn.datasets import make_blobs
        X, y_true = make_blobs(n_samples=300, centers=4,
                               cluster_std=0.60, random_state=0)
        plt.scatter(X[:, 0], X[:, 1], s=50);
```

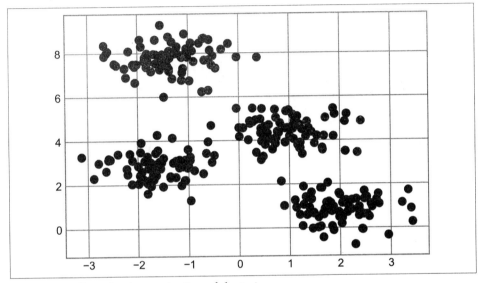

Figure 47-1. Data for demonstration of clustering

By eye, it is relatively easy to pick out the four clusters. The k-means algorithm does this automatically, and in Scikit-Learn uses the typical estimator API:

```
In [3]: from sklearn.cluster import KMeans
        kmeans = KMeans(n_clusters=4)
        kmeans.fit(X)
        y_kmeans = kmeans.predict(X)
```

Let's visualize the results by plotting the data colored by these labels (Figure 47-2). We will also plot the cluster centers as determined by the k-means estimator:

```
In [4]: plt.scatter(X[:, 0], X[:, 1], c=y_kmeans, s=50, cmap='viridis')

        centers = kmeans.cluster_centers_
        plt.scatter(centers[:, 0], centers[:, 1], c='black', s=200);
```

The good news is that the k-means algorithm (at least in this simple case) assigns the points to clusters very similarly to how we might assign them by eye. But you might wonder how this algorithm finds these clusters so quickly: after all, the number of possible combinations of cluster assignments is exponential in the number of data points—an exhaustive search would be very, very costly. Fortunately for us, such an exhaustive search is not necessary: instead, the typical approach to *k*-means involves an intuitive iterative approach known as *expectation–maximization*.

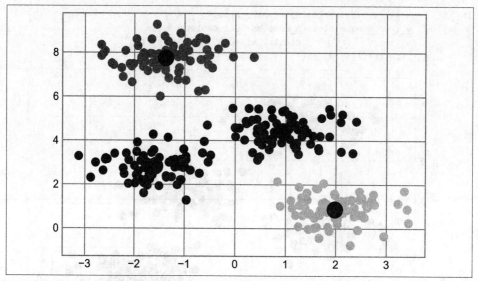

Figure 47-2. k-means cluster centers with clusters indicated by color

Expectation–Maximization

Expectation–maximization (E–M) is a powerful algorithm that comes up in a variety of contexts within data science. *k*-means is a particularly simple and easy-to-understand application of the algorithm; we'll walk through it briefly here. In short, the expectation–maximization approach here consists of the following procedure:

1. Guess some cluster centers.
2. Repeat until converged:
   a. *E-step*: Assign points to the nearest cluster center.
   b. *M-step*: Set the cluster centers to the mean of their assigned points.

Here the *E-step* or *expectation step* is so named because it involves updating our expectation of which cluster each point belongs to. The *M-step* or *maximization step* is so named because it involves maximizing some fitness function that defines the locations of the cluster centers—in this case, that maximization is accomplished by taking a simple mean of the data in each cluster.

The literature about this algorithm is vast, but can be summarized as follows: under typical circumstances, each repetition of the E-step and M-step will always result in a better estimate of the cluster characteristics.

We can visualize the algorithm as shown in Figure 47-3. For the particular initialization shown here, the clusters converge in just three iterations. (For an interactive version of this figure, refer to the code in the online appendix (*https://oreil.ly/wFnok*).)

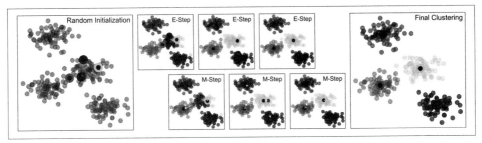

Figure 47-3. Visualization of the E–M algorithm for k-means[1]

The *k*-means algorithm is simple enough that we can write it in a few lines of code. The following is a very basic implementation (see Figure 47-4).

```
In [5]: from sklearn.metrics import pairwise_distances_argmin

        def find_clusters(X, n_clusters, rseed=2):
            # 1. Randomly choose clusters
            rng = np.random.RandomState(rseed)
            i = rng.permutation(X.shape[0])[:n_clusters]
            centers = X[i]

            while True:
                # 2a. Assign labels based on closest center
                labels = pairwise_distances_argmin(X, centers)

                # 2b. Find new centers from means of points
                new_centers = np.array([X[labels == i].mean(0)
                                        for i in range(n_clusters)])

                # 2c. Check for convergence
                if np.all(centers == new_centers):
                    break
                centers = new_centers

            return centers, labels

        centers, labels = find_clusters(X, 4)
```

---

1 Code to produce this figure can be found in the online appendix (*https://oreil.ly/yo6GV*).

```
plt.scatter(X[:, 0], X[:, 1], c=labels,
            s=50, cmap='viridis');
```

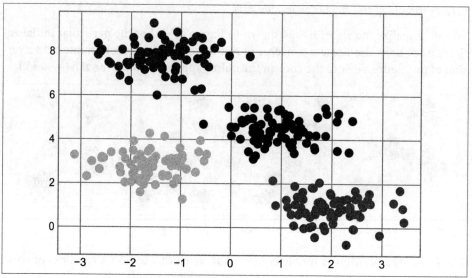

Figure 47-4. Data labeled with k-means

Most well-tested implementations will do a bit more than this under the hood, but the preceding function gives the gist of the expectation–maximization approach. There are a few caveats to be aware of when using the expectation–maximization algorithm:

The globally optimal result may not be achieved

First, although the E–M procedure is guaranteed to improve the result in each step, there is no assurance that it will lead to the *global* best solution. For example, if we use a different random seed in our simple procedure, the particular starting guesses lead to poor results (see Figure 47-5).

```
In [6]: centers, labels = find_clusters(X, 4, rseed=0)
        plt.scatter(X[:, 0], X[:, 1], c=labels,
                    s=50, cmap='viridis');
```

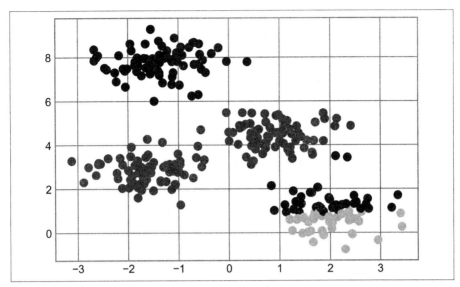

Figure 47-5. An example of poor convergence in k-*means*

Here the E–M approach has converged, but has not converged to a globally optimal configuration. For this reason, it is common for the algorithm to be run for multiple starting guesses, as indeed Scikit-Learn does by default (the number is set by the n_init parameter, which defaults to 10).

The number of clusters must be selected beforehand

Another common challenge with *k*-means is that you must tell it how many clusters you expect: it cannot learn the number of clusters from the data. For example, if we ask the algorithm to identify six clusters, it will happily proceed and find the best six clusters, as shown in Figure 40-1:

```
In [7]: labels = KMeans(6, random_state=0).fit_predict(X)
        plt.scatter(X[:, 0], X[:, 1], c=labels,
                    s=50, cmap='viridis');
```

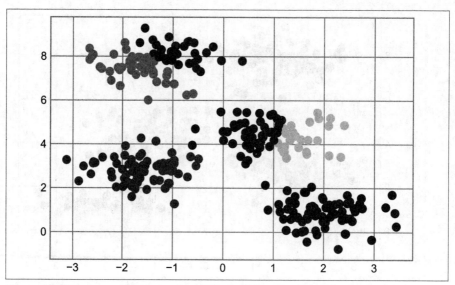

Figure 47-6. An example where the number of clusters is chosen poorly

Whether the result is meaningful is a question that is difficult to answer definitively; one approach that is rather intuitive, but that we won't discuss further here, is called silhouette analysis (*https://oreil.ly/xybmq*).

Alternatively, you might use a more complicated clustering algorithm that has a better quantitative measure of the fitness per number of clusters (e.g., Gaussian mixture models; see Chapter 48) or which *can* choose a suitable number of clusters (e.g., DBSCAN, mean-shift, or affinity propagation, all available in the `sklearn.cluster` submodule).

k-*means is limited to linear cluster boundaries*

The fundamental model assumptions of *k*-means (points will be closer to their own cluster center than to others) means that the algorithm will often be ineffective if the clusters have complicated geometries.

In particular, the boundaries between *k*-means clusters will always be linear, which means that it will fail for more complicated boundaries. Consider the following data, along with the cluster labels found by the typical *k*-means approach (see Figure 47-7).

```
In [8]: from sklearn.datasets import make_moons
        X, y = make_moons(200, noise=.05, random_state=0)

In [9]: labels = KMeans(2, random_state=0).fit_predict(X)
        plt.scatter(X[:, 0], X[:, 1], c=labels,
                    s=50, cmap='viridis');
```

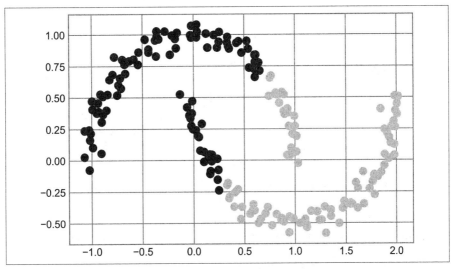

Figure 47-7. Failure of k-*means with nonlinear boundaries*

This situation is reminiscent of the discussion in Chapter 43, where we used a kernel transformation to project the data into a higher dimension where a linear separation is possible. We might imagine using the same trick to allow *k*-means to discover non-linear boundaries.

One version of this kernelized *k*-means is implemented in Scikit-Learn within the SpectralClustering estimator. It uses the graph of nearest neighbors to compute a higher-dimensional representation of the data, and then assigns labels using a *k*-means algorithm (see Figure 47-8).

```
In [10]: from sklearn.cluster import SpectralClustering
         model = SpectralClustering(n_clusters=2,
                                    affinity='nearest_neighbors',
                                    assign_labels='kmeans')
         labels = model.fit_predict(X)
         plt.scatter(X[:, 0], X[:, 1], c=labels,
                     s=50, cmap='viridis');
```

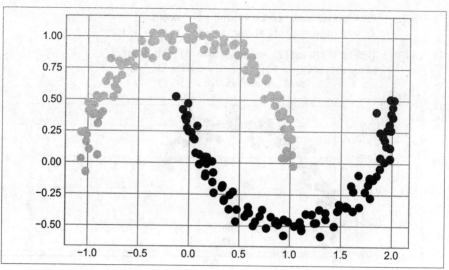

Figure 47-8. Nonlinear boundaries learned by SpectralClustering

We see that with this kernel transform approach, the kernelized *k*-means is able to find the more complicated nonlinear boundaries between clusters.

k-means can be slow for large numbers of samples

Because each iteration of *k*-means must access every point in the dataset, the algorithm can be relatively slow as the number of samples grows. You might wonder if this requirement to use all data at each iteration can be relaxed; for example, you might just use a subset of the data to update the cluster centers at each step. This is the idea behind batch-based *k*-means algorithms, one form of which is implemented in `sklearn.cluster.MiniBatchKMeans`. The interface for this is the same as for standard `KMeans`; we will see an example of its use as we continue our discussion.

Examples

Being careful about these limitations of the algorithm, we can use *k*-means to our advantage in a variety of situations. We'll now take a look at a couple of examples.

Example 1: k-Means on Digits

To start, let's take a look at applying *k*-means on the same simple digits data that we saw in Chapters 44 and 45. Here we will attempt to use *k*-means to try to identify similar digits *without using the original label information*; this might be similar to a first step in extracting meaning from a new dataset about which you don't have any *a priori* label information.

We will start by loading the dataset, then find the clusters. Recall that the digits dataset consists of 1,797 samples with 64 features, where each of the 64 features is the brightness of one pixel in an 8 × 8 image:

```
In [11]: from sklearn.datasets import load_digits
         digits = load_digits()
         digits.data.shape
Out[11]: (1797, 64)
```

The clustering can be performed as we did before:

```
In [12]: kmeans = KMeans(n_clusters=10, random_state=0)
         clusters = kmeans.fit_predict(digits.data)
         kmeans.cluster_centers_.shape
Out[12]: (10, 64)
```

The result is 10 clusters in 64 dimensions. Notice that the cluster centers themselves are 64-dimensional points, and can be interpreted as representing the "typical" digit within the cluster. Let's see what these cluster centers look like (see Figure 47-9).

```
In [13]: fig, ax = plt.subplots(2, 5, figsize=(8, 3))
         centers = kmeans.cluster_centers_.reshape(10, 8, 8)
         for axi, center in zip(ax.flat, centers):
             axi.set(xticks=[], yticks=[])
             axi.imshow(center, interpolation='nearest', cmap=plt.cm.binary)
```

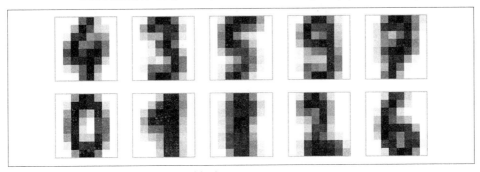

Figure 47-9. Cluster centers learned by k-means

We see that *even without the labels*, KMeans is able to find clusters whose centers are recognizable digits, with perhaps the exception of 1 and 8.

Because *k*-means knows nothing about the identities of the clusters, the 0–9 labels may be permuted. We can fix this by matching each learned cluster label with the true labels found in the clusters:

```
In [14]: from scipy.stats import mode

         labels = np.zeros_like(clusters)
         for i in range(10):
```

```
        mask = (clusters == i)
        labels[mask] = mode(digits.target[mask])[0]
```

Now we can check how accurate our unsupervised clustering was in finding similar digits within the data:

```
In [15]: from sklearn.metrics import accuracy_score
         accuracy_score(digits.target, labels)
Out[15]: 0.7935447968836951
```

With just a simple *k*-means algorithm, we discovered the correct grouping for 80% of the input digits! Let's check the confusion matrix for this, visualized in Figure 47-10.

```
In [16]: from sklearn.metrics import confusion_matrix
         import seaborn as sns
         mat = confusion_matrix(digits.target, labels)
         sns.heatmap(mat.T, square=True, annot=True, fmt='d',
                     cbar=False, cmap='Blues',
                     xticklabels=digits.target_names,
                     yticklabels=digits.target_names)
         plt.xlabel('true label')
         plt.ylabel('predicted label');
```

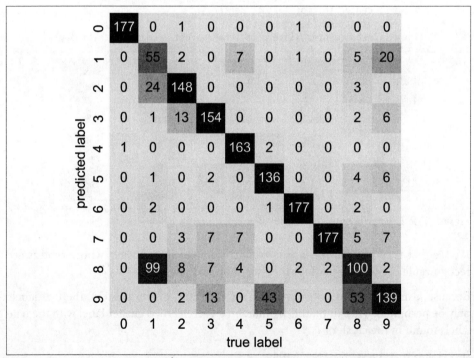

Figure 47-10. Confusion matrix for the k*-means classifier*

As we might expect from the cluster centers we visualized before, the main point of confusion is between the eights and ones. But this still shows that using *k*-means, we can essentially build a digit classifier *without reference to any known labels*!

Just for fun, let's try to push this even farther. We can use the t-distributed stochastic neighbor embedding algorithm (mentioned in Chapter 46) to preprocess the data before performing *k*-means. t-SNE is a nonlinear embedding algorithm that is particularly adept at preserving points within clusters. Let's see how it does:

```
In [17]: from sklearn.manifold import TSNE

         # Project the data: this step will take several seconds
         tsne = TSNE(n_components=2, init='random',
                     learning_rate='auto',random_state=0)
         digits_proj = tsne.fit_transform(digits.data)

         # Compute the clusters
         kmeans = KMeans(n_clusters=10, random_state=0)
         clusters = kmeans.fit_predict(digits_proj)

         # Permute the labels
         labels = np.zeros_like(clusters)
         for i in range(10):
             mask = (clusters == i)
             labels[mask] = mode(digits.target[mask])[0]

         # Compute the accuracy
         accuracy_score(digits.target, labels)
Out[17]: 0.9415692821368948
```

That's a 94% classification accuracy *without using the labels*. This is the power of unsupervised learning when used carefully: it can extract information from the dataset that it might be difficult to extract by hand or by eye.

Example 2: k-Means for Color Compression

One interesting application of clustering is in color compression within images (this example is adapted from Scikit-Learn's "Color Quantization Using K-Means" (*https://oreil.ly/TwsxU*)). For example, imagine you have an image with millions of colors. In most images, a large number of the colors will be unused, and many of the pixels in the image will have similar or even identical colors.

For example, consider the image shown in Figure 47-11, which is from the Scikit-Learn datasets module (for this to work, you'll have to have the PIL Python package installed):[2]

---

[2] For a color version of this and following images, see the online version of this book (*https://oreil.ly/PDSH_GitHub*).

```
In [18]: # Note: this requires the PIL package to be installed
         from sklearn.datasets import load_sample_image
         china = load_sample_image("china.jpg")
         ax = plt.axes(xticks=[], yticks=[])
         ax.imshow(china);
```

Figure 47-11. The input image

The image itself is stored in a three-dimensional array of size (`height`, `width`, `RGB`), containing red/blue/green contributions as integers from 0 to 255:

```
In [19]: china.shape
Out[19]: (427, 640, 3)
```

One way we can view this set of pixels is as a cloud of points in a three-dimensional color space. We will reshape the data to [`n_samples`, `n_features`] and rescale the colors so that they lie between 0 and 1:

```
In [20]: data = china / 255.0  # use 0...1 scale
         data = data.reshape(-1, 3)
         data.shape
Out[20]: (273280, 3)
```

We can visualize these pixels in this color space, using a subset of 10,000 pixels for efficiency (see Figure 47-12).

```
In [21]: def plot_pixels(data, title, colors=None, N=10000):
             if colors is None:
                 colors = data

             # choose a random subset
             rng = np.random.default_rng(0)
```

```
i = rng.permutation(data.shape[0])[:N]
colors = colors[i]
R, G, B = data[i].T

fig, ax = plt.subplots(1, 2, figsize=(16, 6))
ax[0].scatter(R, G, color=colors, marker='.')
ax[0].set(xlabel='Red', ylabel='Green', xlim=(0, 1), ylim=(0, 1))

ax[1].scatter(R, B, color=colors, marker='.')
ax[1].set(xlabel='Red', ylabel='Blue', xlim=(0, 1), ylim=(0, 1))

fig.suptitle(title, size=20);
```

```
In [22]: plot_pixels(data, title='Input color space: 16 million possible colors')
```

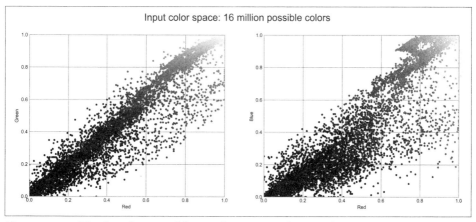

Figure 47-12. The distribution of the pixels in RGB color space[3]

Now let's reduce these 16 million colors to just 16 colors, using a *k*-means clustering across the pixel space. Because we are dealing with a very large dataset, we will use the mini-batch *k*-means, which operates on subsets of the data to compute the result (shown in Figure 47-13) much more quickly than the standard *k*-means algorithm:

```
In [23]: from sklearn.cluster import MiniBatchKMeans
         kmeans = MiniBatchKMeans(16)
         kmeans.fit(data)
         new_colors = kmeans.cluster_centers_[kmeans.predict(data)]

         plot_pixels(data, colors=new_colors,
                     title="Reduced color space: 16 colors")
```

---

3 A full-size version of this figure can be found on GitHub (*https://oreil.ly/PDSH_GitHub*).

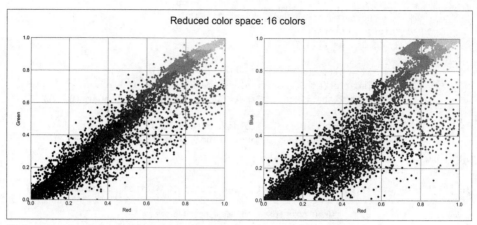

Figure 47-13. 16 clusters in RGB color space[4]

The result is a recoloring of the original pixels, where each pixel is assigned the color of its closest cluster center. Plotting these new colors in the image space rather than the pixel space shows us the effect of this (see Figure 47-14).

```
In [24]: china_recolored = new_colors.reshape(china.shape)

         fig, ax = plt.subplots(1, 2, figsize=(16, 6),
                                subplot_kw=dict(xticks=[], yticks=[]))
         fig.subplots_adjust(wspace=0.05)
         ax[0].imshow(china)
         ax[0].set_title('Original Image', size=16)
         ax[1].imshow(china_recolored)
         ax[1].set_title('16-color Image', size=16);
```

Figure 47-14. A comparison of the full-color image (left) and the 16-color image (right)

---

4 A full-size version of this figure can be found on GitHub (*https://oreil.ly/PDSH_GitHub*).

Some detail is certainly lost in the rightmost panel, but the overall image is still easily recognizable. In terms of the bytes required to store the raw data, the image on the right achieves a compression factor of around 1 million! Now, this kind of approach is not going to match the fidelity of purpose-built image compression schemes like JPEG, but the example shows the power of thinking outside of the box with unsupervised methods like k-means.

## CHAPTER 48

In Depth: Gaussian Mixture Models

The *k*-means clustering model explored in the previous chapter is simple and relatively easy to understand, but its simplicity leads to practical challenges in its application. In particular, the nonprobabilistic nature of *k*-means and its use of simple distance from cluster center to assign cluster membership leads to poor performance for many real-world situations. In this chapter we will take a look at Gaussian mixture models, which can be viewed as an extension of the ideas behind *k*-means, but can also be a powerful tool for estimation beyond simple clustering.

We begin with the standard imports:

```
In [1]: %matplotlib inline
        import matplotlib.pyplot as plt
        plt.style.use('seaborn-whitegrid')
        import numpy as np
```

Motivating Gaussian Mixtures: Weaknesses of k-Means

Let's take a look at some of the weaknesses of *k*-means and think about how we might improve the cluster model. As we saw in the previous chapter, given simple, well-separated data, *k*-means finds suitable clustering results.

For example, if we have simple blobs of data, the *k*-means algorithm can quickly label those clusters in a way that closely matches what we might do by eye (see Figure 48-1).

```
In [2]: # Generate some data
        from sklearn.datasets import make_blobs
        X, y_true = make_blobs(n_samples=400, centers=4,
                               cluster_std=0.60, random_state=0)
        X = X[:, ::-1] # flip axes for better plotting
```

```
In [3]: # Plot the data with k-means labels
        from sklearn.cluster import KMeans
        kmeans = KMeans(4, random_state=0)
        labels = kmeans.fit(X).predict(X)
        plt.scatter(X[:, 0], X[:, 1], c=labels, s=40, cmap='viridis');
```

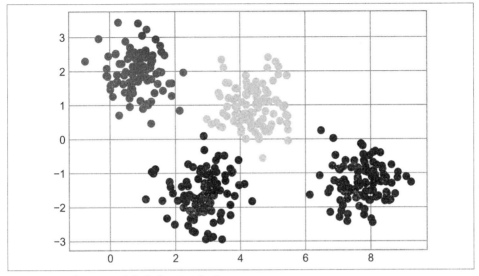

Figure 48-1. k-means labels for simple data

From an intuitive standpoint, we might expect that the clustering assignment for some points is more certain than others: for example, there appears to be a very slight overlap between the two middle clusters, such that we might not have complete confidence in the cluster assignment of points between them. Unfortunately, the *k*-means model has no intrinsic measure of probability or uncertainty of cluster assignments (although it may be possible to use a bootstrap approach to estimate this uncertainty). For this, we must think about generalizing the model.

One way to think about the *k*-means model is that it places a circle (or, in higher dimensions, a hypersphere) at the center of each cluster, with a radius defined by the most distant point in the cluster. This radius acts as a hard cutoff for cluster assignment within the training set: any point outside this circle is not considered a member of the cluster. We can visualize this cluster model with the following function (see Figure 48-2).

```
In [4]: from sklearn.cluster import KMeans
        from scipy.spatial.distance import cdist

        def plot_kmeans(kmeans, X, n_clusters=4, rseed=0, ax=None):
            labels = kmeans.fit_predict(X)

            # plot the input data
            ax = ax or plt.gca()
            ax.axis('equal')
            ax.scatter(X[:, 0], X[:, 1], c=labels, s=40, cmap='viridis', zorder=2)

            # plot the representation of the KMeans model
            centers = kmeans.cluster_centers_
            radii = [cdist(X[labels == i], [center]).max()
                        for i, center in enumerate(centers)]
            for c, r in zip(centers, radii):
                ax.add_patch(plt.Circle(c, r, ec='black', fc='lightgray',
                                        lw=3, alpha=0.5, zorder=1))

In [5]: kmeans = KMeans(n_clusters=4, random_state=0)
        plot_kmeans(kmeans, X)
```

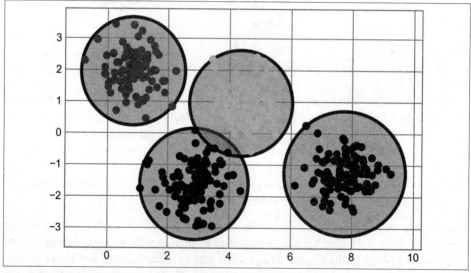

Figure 48-2. The circular clusters implied by the k-means model

An important observation for *k*-means is that these cluster models *must be circular*: *k*-means has no built-in way of accounting for oblong or elliptical clusters. So, for example, if we take the same data and transform it, the cluster assignments end up becoming muddled, as you can see in Figure 48-3.

```
In [6]: rng = np.random.RandomState(13)
        X_stretched = np.dot(X, rng.randn(2, 2))

        kmeans = KMeans(n_clusters=4, random_state=0)
        plot_kmeans(kmeans, X_stretched)
```

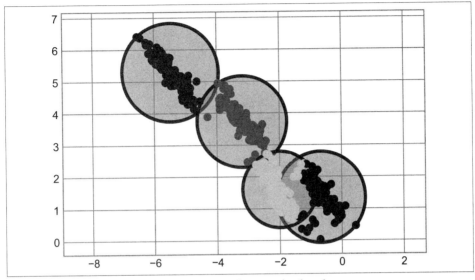

Figure 48-3. Poor performance of k-*means for noncircular clusters*

By eye, we recognize that these transformed clusters are noncircular, and thus circular clusters would be a poor fit. Nevertheless, *k*-means is not flexible enough to account for this, and tries to force-fit the data into four circular clusters. This results in a mixing of cluster assignments where the resulting circles overlap: see especially the bottom-right of this plot. One might imagine addressing this particular situation by preprocessing the data with PCA (see Chapter 45), but in practice there is no guarantee that such a global operation will circularize the individual groups.

These two disadvantages of *k*-means—its lack of flexibility in cluster shape and lack of probabilistic cluster assignment—mean that for many datasets (especially low-dimensional datasets) it may not perform as well as you might hope.

You might imagine addressing these weaknesses by generalizing the *k*-means model: for example, you could measure uncertainty in cluster assignment by comparing the distances of each point to *all* cluster centers, rather than focusing on just the closest. You might also imagine allowing the cluster boundaries to be ellipses rather than circles, so as to account for noncircular clusters. It turns out these are two essential components of a different type of clustering model, Gaussian mixture models.

Generalizing E–M: Gaussian Mixture Models

A Gaussian mixture model (GMM) attempts to find a mixture of multidimensional Gaussian probability distributions that best model any input dataset. In the simplest case, GMMs can be used for finding clusters in the same manner as *k*-means (see Figure 48-4).

```
In [7]: from sklearn.mixture import GaussianMixture
        gmm = GaussianMixture(n_components=4).fit(X)
        labels = gmm.predict(X)
        plt.scatter(X[:, 0], X[:, 1], c=labels, s=40, cmap='viridis');
```

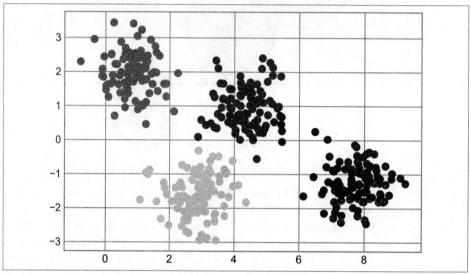

Figure 48-4. Gaussian mixture model labels for the data

But because a GMM contains a probabilistic model under the hood, it is also possible to find probabilistic cluster assignments—in Scikit-Learn this is done using the `predict_proba` method. This returns a matrix of size `[n_samples, n_clusters]` which measures the probability that any point belongs to the given cluster:

```
In [8]: probs = gmm.predict_proba(X)
        print(probs[:5].round(3))
Out[8]: [[0.    0.531 0.469 0.   ]
         [0.    0.    0.    1.   ]
         [0.    0.    0.    1.   ]
         [0.    1.    0.    0.   ]
         [0.    0.    0.    1.   ]]
```

We can visualize this uncertainty by, for example, making the size of each point proportional to the certainty of its prediction; looking at Figure 48-5, we can see that it is precisely the points at the boundaries between clusters that reflect this uncertainty of cluster assignment:

```
In [9]: size = 50 * probs.max(1) ** 2  # square emphasizes differences
        plt.scatter(X[:, 0], X[:, 1], c=labels, cmap='viridis', s=size);
```

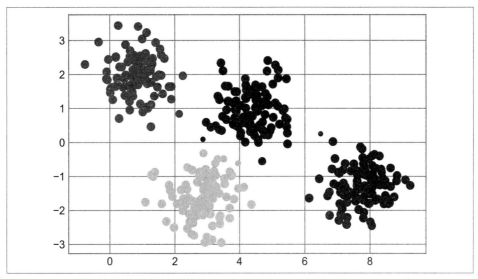

Figure 48-5. GMM probabilistic labels: probabilities are shown by the size of points

Under the hood, a Gaussian mixture model is very similar to k-means: it uses an expectation–maximization approach, which qualitatively does the following:

1. Choose starting guesses for the location and shape.
2. Repeat until converged:
   a. *E-step*: For each point, find weights encoding the probability of membership in each cluster.
   b. *M-step*: For each cluster, update its location, normalization, and shape based on *all* data points, making use of the weights.

The result of this is that each cluster is associated not with a hard-edged sphere, but with a smooth Gaussian model. Just as in the k-means expectation–maximization approach, this algorithm can sometimes miss the globally optimal solution, and thus in practice multiple random initializations are used.

Let's create a function that will help us visualize the locations and shapes of the GMM clusters by drawing ellipses based on the GMM output:

```
In [10]: from matplotlib.patches import Ellipse

         def draw_ellipse(position, covariance, ax=None, **kwargs):
             """Draw an ellipse with a given position and covariance"""
             ax = ax or plt.gca()

             # Convert covariance to principal axes
             if covariance.shape == (2, 2):
                 U, s, Vt = np.linalg.svd(covariance)
                 angle = np.degrees(np.arctan2(U[1, 0], U[0, 0]))
                 width, height = 2 * np.sqrt(s)
             else:
                 angle = 0
                 width, height = 2 * np.sqrt(covariance)

             # Draw the ellipse
             for nsig in range(1, 4):
                 ax.add_patch(Ellipse(position, nsig * width, nsig * height,
                                      angle, **kwargs))

         def plot_gmm(gmm, X, label=True, ax=None):
             ax = ax or plt.gca()
             labels = gmm.fit(X).predict(X)
             if label:
                 ax.scatter(X[:, 0], X[:, 1], c=labels, s=40, cmap='viridis',
                            zorder=2)
             else:
                 ax.scatter(X[:, 0], X[:, 1], s=40, zorder=2)
             ax.axis('equal')

             w_factor = 0.2 / gmm.weights_.max()
             for pos, covar, w in zip(gmm.means_, gmm.covariances_, gmm.weights_):
                 draw_ellipse(pos, covar, alpha=w * w_factor)
```

With this in place, we can take a look at what the four-component GMM gives us for our initial data (see Figure 48-6).

```
In [11]: gmm = GaussianMixture(n_components=4, random_state=42)
         plot_gmm(gmm, X)
```

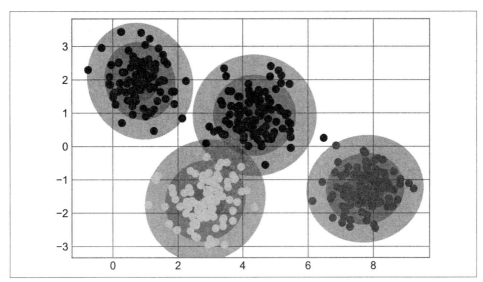

Figure 48-6. The four-component GMM in the presence of circular clusters

Similarly, we can use the GMM approach to fit our stretched dataset; allowing for a full covariance the model will fit even very oblong, stretched-out clusters, as we can see in Figure 48-7.

```
In [12]: gmm = GaussianMixture(n_components=4, covariance_type='full',
                               random_state=42)
         plot_gmm(gmm, X_stretched)
```

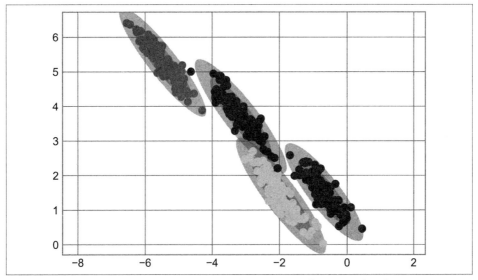

Figure 48-7. The four-component GMM in the presence of noncircular clusters

This makes clear that GMMs address the two main practical issues with *k*-means encountered before.

Choosing the Covariance Type

If you look at the details of the preceding fits, you will see that the `covariance_type` option was set differently within each. This hyperparameter controls the degrees of freedom in the shape of each cluster; it's essential to set this carefully for any given problem. The default is `covariance_type="diag"`, which means that the size of the cluster along each dimension can be set independently, with the resulting ellipse constrained to align with the axes. `covariance_type="spherical"` is a slightly simpler and faster model, which constrains the shape of the cluster such that all dimensions are equal. The resulting clustering will have similar characteristics to that of *k*-means, though it's not entirely equivalent. A more complicated and computationally expensive model (especially as the number of dimensions grows) is to use `cova riance_type="full"`, which allows each cluster to be modeled as an ellipse with arbitrary orientation. Figure 48-8 represents these three choices for a single cluster.

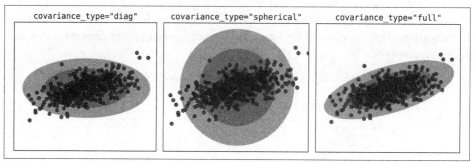

Figure 48-8. Visualization of GMM covariance types[1]

Gaussian Mixture Models as Density Estimation

Though the GMM is often categorized as a clustering algorithm, fundamentally it is an algorithm for *density estimation*. That is to say, the result of a GMM fit to some data is technically not a clustering model, but a generative probabilistic model describing the distribution of the data.

As an example, consider some data generated from Scikit-Learn's `make_moons` function, introduced in Chapter 47 (see Figure 48-9).

---

1 Code to produce this figure can be found in the online appendix (*https://oreil.ly/MLsk8*).

```
In [13]: from sklearn.datasets import make_moons
         Xmoon, ymoon = make_moons(200, noise=.05, random_state=0)
         plt.scatter(Xmoon[:, 0], Xmoon[:, 1]);
```

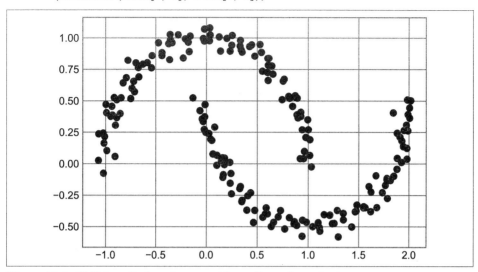

Figure 48-9. GMM applied to clusters with nonlinear boundaries

If we try to fit this with a two-component GMM viewed as a clustering model, the results are not particularly useful (see Figure 48-10).

```
In [14]: gmm2 = GaussianMixture(n_components=2, covariance_type='full',
                                random_state=0)
         plot_gmm(gmm2, Xmoon)
```

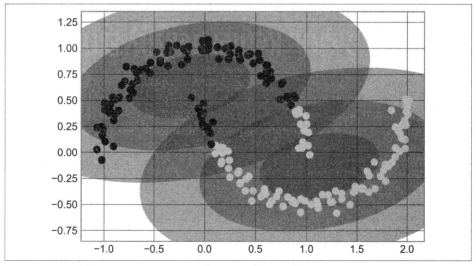

Figure 48-10. Two-component GMM fit to nonlinear clusters

But if we instead use many more components and ignore the cluster labels, we find a fit that is much closer to the input data (see Figure 48-11).

```
In [15]: gmm16 = GaussianMixture(n_components=16, covariance_type='full',
                                  random_state=0)
         plot_gmm(gmm16, Xmoon, label=False)
```

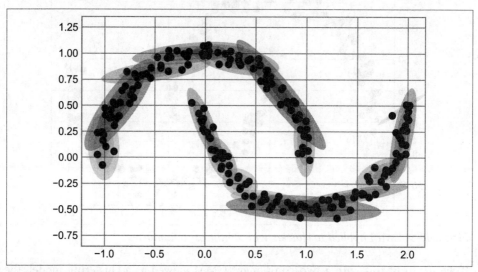

Figure 48-11. Using many GMM clusters to model the distribution of points

Here the mixture of 16 Gaussian components serves not to find separated clusters of data, but rather to model the overall *distribution* of the input data. This is a generative model of the distribution, meaning that the GMM gives us the recipe to generate new random data distributed similarly to our input. For example, here are 400 new points drawn from this 16-component GMM fit to our original data (see Figure 48-12).

```
In [16]: Xnew, ynew = gmm16.sample(400)
         plt.scatter(Xnew[:, 0], Xnew[:, 1]);
```

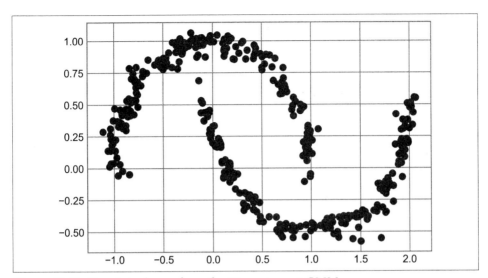

Figure 48-12. New data drawn from the 16-component GMM

A GMM is convenient as a flexible means of modeling an arbitrary multidimensional distribution of data.

The fact that a GMM is a generative model gives us a natural means of determining the optimal number of components for a given dataset. A generative model is inherently a probability distribution for the dataset, and so we can simply evaluate the *like-lihood* of the data under the model, using cross-validation to avoid overfitting. Another means of correcting for overfitting is to adjust the model likelihoods using some analytic criterion such as the Akaike information criterion (AIC) (*https://oreil.ly/BmH9X*) or the Bayesian information criterion (BIC) (*https://oreil.ly/Ewivh*). Scikit-Learn's `GaussianMixture` estimator actually includes built-in methods that compute both of these, so it is very easy to operate using this approach.

Let's look at the AIC and BIC versus the number of GMM components for our moons dataset (see Figure 48-13).

```
In [17]: n_components = np.arange(1, 21)
         models = [GaussianMixture(n, covariance_type='full',
                                   random_state=0).fit(Xmoon)
                  for n in n_components]

         plt.plot(n_components, [m.bic(Xmoon) for m in models], label='BIC')
         plt.plot(n_components, [m.aic(Xmoon) for m in models], label='AIC')
         plt.legend(loc='best')
         plt.xlabel('n_components');
```

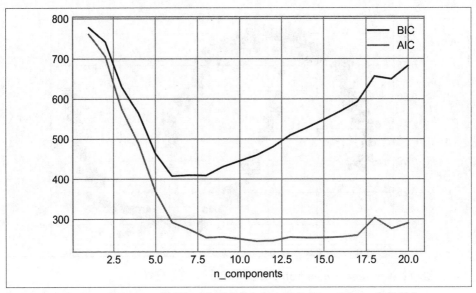

Figure 48-13. Visualization of AIC and BIC for choosing the number of GMM components

The optimal number of clusters is the value that minimizes the AIC or BIC, depending on which approximation we wish to use. The AIC tells us that our choice of 16 components earlier was probably too many: around 8–12 components would have been a better choice. As is typical with this sort of problem, the BIC recommends a simpler model.

Notice the important point: this choice of number of components measures how well a GMM works *as a density estimator*, not how well it works *as a clustering algorithm*. I'd encourage you to think of the GMM primarily as a density estimator, and use it for clustering only when warranted within simple datasets.

Example: GMMs for Generating New Data

We just saw a simple example of using a GMM as a generative model in order to create new samples from the distribution defined by the input data. Here we will run with this idea and generate *new handwritten digits* from the standard digits corpus that we have used before.

To start with, let's load the digits data using Scikit-Learn's data tools:

```
In [18]: from sklearn.datasets import load_digits
         digits = load_digits()
         digits.data.shape
Out[18]: (1797, 64)
```

Next, let's plot the first 50 of these to recall exactly what we're looking at (see Figure 48-14).

```
In [19]: def plot_digits(data):
             fig, ax = plt.subplots(5, 10, figsize=(8, 4),
                                    subplot_kw=dict(xticks=[], yticks=[]))
             fig.subplots_adjust(hspace=0.05, wspace=0.05)
             for i, axi in enumerate(ax.flat):
                 im = axi.imshow(data[i].reshape(8, 8), cmap='binary')
                 im.set_clim(0, 16)
         plot_digits(digits.data)
```

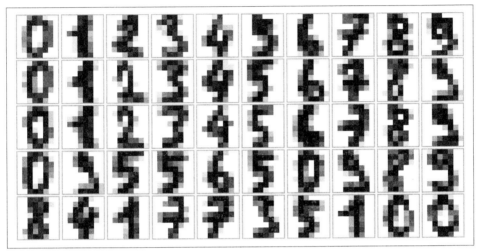

Figure 48-14. Handwritten digits input

We have nearly 1,800 digits in 64 dimensions, and we can build a GMM on top of these to generate more. GMMs can have difficulty converging in such a high-dimensional space, so we will start with an invertible dimensionality reduction algorithm on the data. Here we will use a straightforward PCA, asking it to preserve 99% of the variance in the projected data:

```
In [20]: from sklearn.decomposition import PCA
         pca = PCA(0.99, whiten=True)
         data = pca.fit_transform(digits.data)
         data.shape
Out[20]: (1797, 41)
```

The result is 41 dimensions, a reduction of nearly 1/3 with almost no information loss. Given this projected data, let's use the AIC to get a gauge for the number of GMM components we should use (see Figure 48-15).

```
In [21]: n_components = np.arange(50, 210, 10)
         models = [GaussianMixture(n, covariance_type='full', random_state=0)
                   for n in n_components]
```

```
aics = [model.fit(data).aic(data) for model in models]
plt.plot(n_components, aics);
```

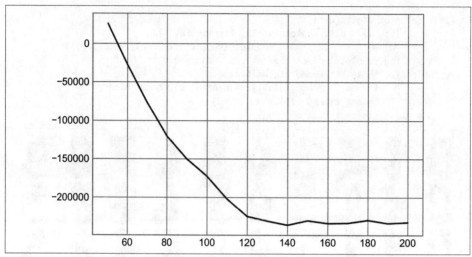

Figure 48-15. AIC curve for choosing the appropriate number of GMM components

It appears that around 140 components minimizes the AIC; we will use this model. Let's quickly fit this to the data and confirm that it has converged:

```
In [22]: gmm = GaussianMixture(140, covariance_type='full', random_state=0)
         gmm.fit(data)
         print(gmm.converged_)
Out[22]: True
```

Now we can draw samples of 100 new points within this 41-dimensional projected space, using the GMM as a generative model:

```
In [23]: data_new, label_new = gmm.sample(100)
         data_new.shape
Out[23]: (100, 41)
```

Finally, we can use the inverse transform of the PCA object to construct the new digits (see Figure 48-16).

```
In [24]: digits_new = pca.inverse_transform(data_new)
         plot_digits(digits_new)
```

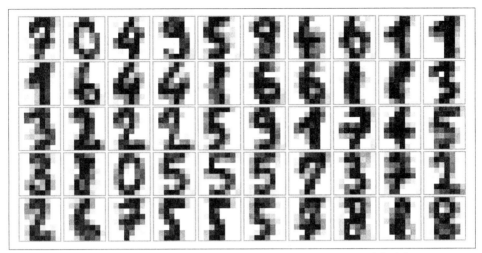

Figure 48-16. "New" digits randomly drawn from the underlying model of the GMM estimator

The results for the most part look like plausible digits from the dataset!

Consider what we've done here: given a sampling of handwritten digits, we have modeled the distribution of that data in such a way that we can generate brand new samples of digits from the data: these are "handwritten digits," which do not individually appear in the original dataset, but rather capture the general features of the input data as modeled by the mixture model. Such a generative model of digits can prove very useful as a component of a Bayesian generative classifier, as we shall see in the next chapter.

## CHAPTER 49

In Depth: Kernel Density Estimation

In Chapter 48 we covered Gaussian mixture models, which are a kind of hybrid between a clustering estimator and a density estimator. Recall that a density estimator is an algorithm that takes a D-dimensional dataset and produces an estimate of the D-dimensional probability distribution that data is drawn from. The GMM algorithm accomplishes this by representing the density as a weighted sum of Gaussian distributions. *Kernel density estimation* (KDE) is in some senses an algorithm that takes the mixture-of-Gaussians idea to its logical extreme: it uses a mixture consisting of one Gaussian component *per point*, resulting in an essentially nonparametric estimator of density. In this chapter, we will explore the motivation and uses of KDE.

We begin with the standard imports:

```
In [1]: %matplotlib inline
        import matplotlib.pyplot as plt
        plt.style.use('seaborn-whitegrid')
        import numpy as np
```

Motivating Kernel Density Estimation: Histograms

As mentioned previously, a density estimator is an algorithm that seeks to model the probability distribution that generated a dataset. For one-dimensional data, you are probably already familiar with one simple density estimator: the histogram. A histogram divides the data into discrete bins, counts the number of points that fall in each bin, and then visualizes the results in an intuitive manner.

For example, let's create some data that is drawn from two normal distributions:

```
In [2]: def make_data(N, f=0.3, rseed=1):
            rand = np.random.RandomState(rseed)
            x = rand.randn(N)
            x[int(f * N):] += 5
            return x

        x = make_data(1000)
```

We have previously seen that the standard count-based histogram can be created with the plt.hist function. By specifying the density parameter of the histogram, we end up with a normalized histogram where the height of the bins does not reflect counts, but instead reflects probability density (see Figure 49-1).

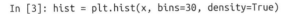

```
In [3]: hist = plt.hist(x, bins=30, density=True)
```

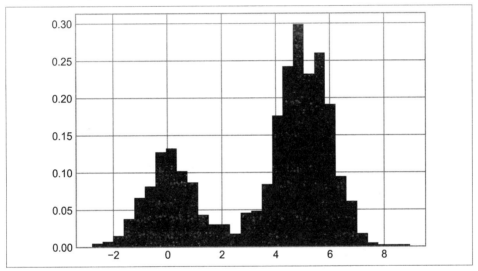

Figure 49-1. Data drawn from a combination of normal distributions

Notice that for equal binning, this normalization simply changes the scale on the y-axis, leaving the relative heights essentially the same as in a histogram built from counts. This normalization is chosen so that the total area under the histogram is equal to 1, as we can confirm by looking at the output of the histogram function:

```
In [4]: density, bins, patches = hist
        widths = bins[1:] - bins[:-1]
        (density * widths).sum()
Out[4]: 1.0
```

One of the issues with using a histogram as a density estimator is that the choice of bin size and location can lead to representations that have qualitatively different features. For example, if we look at a version of this data with only 20 points, the choice of how to draw the bins can lead to an entirely different interpretation of the data! Consider this example, visualized in Figure 49-2.

```
In [5]: x = make_data(20)
        bins = np.linspace(-5, 10, 10)
```

```
In [6]: fig, ax = plt.subplots(1, 2, figsize=(12, 4),
                               sharex=True, sharey=True,
                               subplot_kw={'xlim':(-4, 9),
                                           'ylim':(-0.02, 0.3)})
        fig.subplots_adjust(wspace=0.05)
        for i, offset in enumerate([0.0, 0.6]):
            ax[i].hist(x, bins=bins + offset, density=True)
            ax[i].plot(x, np.full_like(x, -0.01), '|k',
                       markeredgewidth=1)
```

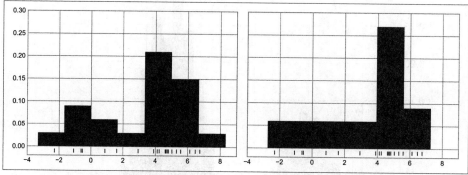

Figure 49-2. The problem with histograms: the bin locations can affect interpretation

On the left, the histogram makes clear that this is a bimodal distribution. On the right, we see a unimodal distribution with a long tail. Without seeing the preceding code, you would probably not guess that these two histograms were built from the same data. With that in mind, how can you trust the intuition that histograms confer? And how might we improve on this?

Stepping back, we can think of a histogram as a stack of blocks, where we stack one block within each bin on top of each point in the dataset. Let's view this directly (see Figure 49-3).

```
In [7]: fig, ax = plt.subplots()
        bins = np.arange(-3, 8)
        ax.plot(x, np.full_like(x, -0.1), '|k',
                markeredgewidth=1)
        for count, edge in zip(*np.histogram(x, bins)):
            for i in range(count):
                ax.add_patch(plt.Rectangle(
```

```
          (edge, i), 1, 1, ec='black', alpha=0.5))
      ax.set_xlim(-4, 8)
      ax.set_ylim(-0.2, 8)
Out[7]: (-0.2, 8.0)
```

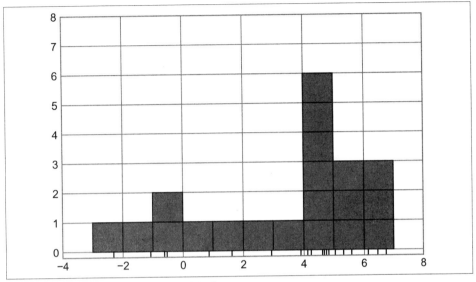

Figure 49-3. Histogram as stack of blocks

The problem with our two binnings stems from the fact that the height of the block
stack often reflects not the actual density of points nearby, but coincidences of how
the bins align with the data points. This misalignment between points and their
blocks is a potential cause of the poor histogram results seen here. But what if, instead
of stacking the blocks aligned with the *bins*, we were to stack the blocks aligned with
the *points they represent*? If we do this, the blocks won't be aligned, but we can add
their contributions at each location along the x-axis to find the result. Let's try this
(see Figure 49-4).

```
In [8]: x_d = np.linspace(-4, 8, 2000)
        density = sum((abs(xi - x_d) < 0.5) for xi in x)

        plt.fill_between(x_d, density, alpha=0.5)
        plt.plot(x, np.full_like(x, -0.1), '|k', markeredgewidth=1)

        plt.axis([-4, 8, -0.2, 8]);
```

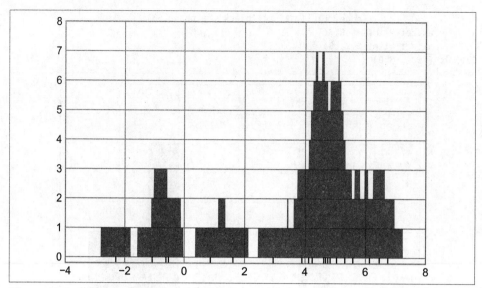

Figure 49-4. A "histogram" where blocks center on each individual point; this is an example of a kernel density estimate

The result looks a bit messy, but it's a much more robust reflection of the actual data characteristics than is the standard histogram. Still, the rough edges are not aesthetically pleasing, nor are they reflective of any true properties of the data. In order to smooth them out, we might decide to replace the blocks at each location with a smooth function, like a Gaussian. Let's use a standard normal curve at each point instead of a block (see Figure 49-5).

```
In [9]: from scipy.stats import norm
        x_d = np.linspace(-4, 8, 1000)
        density = sum(norm(xi).pdf(x_d) for xi in x)

        plt.fill_between(x_d, density, alpha=0.5)
        plt.plot(x, np.full_like(x, -0.1), '|k', markeredgewidth=1)

        plt.axis([-4, 8, -0.2, 5]);
```

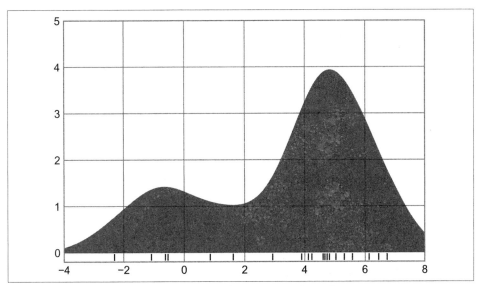

Figure 49-5. A kernel density estimate with a Gaussian kernel

This smoothed-out plot, with a Gaussian distribution contributed at the location of each input point, gives a much more accurate idea of the shape of the data distribution, and one that has much less variance (i.e., changes much less in response to differences in sampling).

What we've landed on in the last two plots is what's called kernel density estimation in one dimension: we have placed a "kernel"—a square or top hat–shaped kernel in the former, a Gaussian kernel in the latter—at the location of each point, and used their sum as an estimate of density. With this intuition in mind, we'll now explore kernel density estimation in more detail.

Kernel Density Estimation in Practice

The free parameters of kernel density estimation are the *kernel*, which specifies the shape of the distribution placed at each point, and the *kernel bandwidth*, which controls the size of the kernel at each point. In practice, there are many kernels you might use for kernel density estimation: in particular, the Scikit-Learn KDE implementation supports six kernels, which you can read about in the "Density Estimation" section (*https://oreil.ly/2Ae4a*) of the documentation.

While there are several versions of KDE implemented in Python (notably in the SciPy and statsmodels packages), I prefer to use Scikit-Learn's version because of its efficiency and flexibility. It is implemented in the sklearn.neighbors.KernelDensity estimator, which handles KDE in multiple dimensions with one of six kernels and one of a couple dozen distance metrics. Because KDE can be fairly computationally intensive, the Scikit-Learn estimator uses a tree-based algorithm under the hood and can trade off computation time for accuracy using the `atol` (absolute tolerance) and `rtol` (relative tolerance) parameters. The kernel bandwidth can be determined using Scikit-Learn's standard cross-validation tools, as we will soon see.

Let's first show a simple example of replicating the previous plot using the Scikit-Learn `KernelDensity` estimator (see Figure 49-6).

```
In [10]: from sklearn.neighbors import KernelDensity

         # instantiate and fit the KDE model
         kde = KernelDensity(bandwidth=1.0, kernel='gaussian')
         kde.fit(x[:, None])

         # score_samples returns the log of the probability density
         logprob = kde.score_samples(x_d[:, None])

         plt.fill_between(x_d, np.exp(logprob), alpha=0.5)
         plt.plot(x, np.full_like(x, -0.01), '|k', markeredgewidth=1)
         plt.ylim(-0.02, 0.22);
```

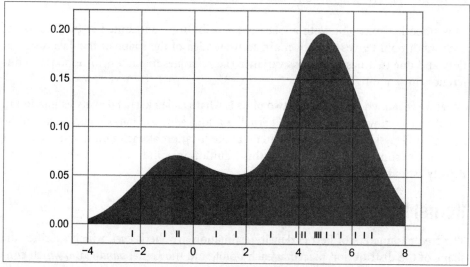

Figure 49-6. A kernel density estimate computed with Scikit-Learn

The result here is normalized such that the area under the curve is equal to 1.

Selecting the Bandwidth via Cross-Validation

The final estimate produced by a KDE procedure can be quite sensitive to the choice of bandwidth, which is the knob that controls the bias–variance trade-off in the estimate of density. Too narrow a bandwidth leads to a high-variance estimate (i.e., overfitting), where the presence or absence of a single point makes a large difference. Too wide a bandwidth leads to a high-bias estimate (i.e., underfitting), where the structure in the data is washed out by the wide kernel.

There is a long history in statistics of methods to quickly estimate the best bandwidth based on rather stringent assumptions about the data: if you look up the KDE implementations in the SciPy and `statsmodels` packages, for example, you will see implementations based on some of these rules.

In machine learning contexts, we've seen that such hyperparameter tuning often is done empirically via a cross-validation approach. With this in mind, Scikit-Learn's `KernelDensity` estimator is designed such that it can be used directly within the package's standard grid search tools. Here we will use `GridSearchCV` to optimize the bandwidth for the preceding dataset. Because we are looking at such a small dataset, we will use leave-one-out cross-validation, which minimizes the reduction in training set size for each cross-validation trial:

```
In [11]: from sklearn.model_selection import GridSearchCV
         from sklearn.model_selection import LeaveOneOut

         bandwidths = 10 ** np.linspace(-1, 1, 100)
         grid = GridSearchCV(KernelDensity(kernel='gaussian'),
                             {'bandwidth': bandwidths},
                             cv=LeaveOneOut())
         grid.fit(x[:, None]);
```

Now we can find the choice of bandwidth that maximizes the score (which in this case defaults to the log-likelihood):

```
In [12]: grid.best_params_
Out[12]: {'bandwidth': 1.1233240329780276}
```

The optimal bandwidth happens to be very close to what we used in the example plot earlier, where the bandwidth was 1.0 (i.e., the default width of `scipy.stats.norm`).

Example: Not-so-Naive Bayes

This example looks at Bayesian generative classification with KDE, and demonstrates how to use the Scikit-Learn architecture to create a custom estimator.

In Chapter 41 we explored naive Bayesian classification, in which we create a simple generative model for each class, and use these models to build a fast classifier. For Gaussian naive Bayes, the generative model is a simple axis-aligned Gaussian. With a density estimation algorithm like KDE, we can remove the "naive" element and perform the same classification with a more sophisticated generative model for each class. It's still Bayesian classification, but it's no longer naive.

The general approach for generative classification is this:

1. Split the training data by label.
2. For each set, fit a KDE to obtain a generative model of the data. This allows you, for any observation x and label y, to compute a likelihood $P(x \mid y)$.
3. From the number of examples of each class in the training set, compute the *class prior*, $P(y)$.
4. For an unknown point x, the posterior probability for each class is $P(y \mid x) \propto P(x \mid y)P(y)$. The class that maximizes this posterior is the label assigned to the point.

The algorithm is straightforward and intuitive to understand; the more difficult piece is couching it within the Scikit-Learn framework in order to make use of the grid search and cross-validation architecture.

This is the code that implements the algorithm within the Scikit-Learn framework; we will step through it following the code block:

```
In [13]: from sklearn.base import BaseEstimator, ClassifierMixin


         class KDEClassifier(BaseEstimator, ClassifierMixin):
             """Bayesian generative classification based on KDE

             Parameters
             ----------
             bandwidth : float
                 the kernel bandwidth within each class
             kernel : str
                 the kernel name, passed to KernelDensity
             """
             def __init__(self, bandwidth=1.0, kernel='gaussian'):
                 self.bandwidth = bandwidth
                 self.kernel = kernel

             def fit(self, X, y):
                 self.classes_ = np.sort(np.unique(y))
                 training_sets = [X[y == yi] for yi in self.classes_]
                 self.models_ = [KernelDensity(bandwidth=self.bandwidth,
                                               kernel=self.kernel).fit(Xi)
                                 for Xi in training_sets]
                 self.logpriors_ = [np.log(Xi.shape[0] / X.shape[0])
                                    for Xi in training_sets]
                 return self
```

```
    def predict_proba(self, X):
        logprobs = np.array([model.score_samples(X)
                             for model in self.models_]).T
        result = np.exp(logprobs + self.logpriors_)
        return result / result.sum(axis=1, keepdims=True)

    def predict(self, X):
        return self.classes_[np.argmax(self.predict_proba(X), 1)]
```

Anatomy of a Custom Estimator

Let's step through this code and discuss the essential features:

```
from sklearn.base import BaseEstimator, ClassifierMixin

class KDEClassifier(BaseEstimator, ClassifierMixin):
    """Bayesian generative classification based on KDE

    Parameters
    ----------
    bandwidth : float
        the kernel bandwidth within each class
    kernel : str
        the kernel name, passed to KernelDensity
    """
```

Each estimator in Scikit-Learn is a class, and it is most convenient for this class to inherit from the `BaseEstimator` class as well as the appropriate mixin, which provides standard functionality. For example, here the `BaseEstimator` contains (among other things) the logic necessary to clone/copy an estimator for use in a cross-validation procedure, and `ClassifierMixin` defines a default `score` method used by such routines. We also provide a docstring, which will be captured by IPython's help functionality (see Chapter 1).

Next comes the class initialization method:

```
    def __init__(self, bandwidth=1.0, kernel='gaussian'):
        self.bandwidth = bandwidth
        self.kernel = kernel
```

This is the actual code that is executed when the object is instantiated with `KDEClassifier`. In Scikit-Learn, it is important that *initialization contains no operations* other than assigning the passed values by name to `self`. This is due to the logic contained in `BaseEstimator` required for cloning and modifying estimators for cross-validation, grid search, and other functions. Similarly, all arguments to `__init__` should be explicit: i.e., `*args` or `**kwargs` should be avoided, as they will not be correctly handled within cross-validation routines.

Next comes the `fit` method, where we handle training data:

```python
def fit(self, X, y):
    self.classes_ = np.sort(np.unique(y))
    training_sets = [X[y == yi] for yi in self.classes_]
    self.models_ = [KernelDensity(bandwidth=self.bandwidth,
                                  kernel=self.kernel).fit(Xi)
                    for Xi in training_sets]
    self.logpriors_ = [np.log(Xi.shape[0] / X.shape[0])
                       for Xi in training_sets]
    return self
```

Here we find the unique classes in the training data, train a `KernelDensity` model for each class, and compute the class priors based on the number of input samples. Finally, `fit` should always return `self` so that we can chain commands. For example:

```python
label = model.fit(X, y).predict(X)
```

Notice that each persistent result of the fit is stored with a trailing underscore (e.g., `self.logpriors_`). This is a convention used in Scikit-Learn so that you can quickly scan the members of an estimator (using IPython's tab completion) and see exactly which members are fit to training data.

Finally, we have the logic for predicting labels on new data:

```python
def predict_proba(self, X):
    logprobs = np.vstack([model.score_samples(X)
                          for model in self.models_]).T
    result = np.exp(logprobs + self.logpriors_)
    return result / result.sum(axis=1, keepdims=True)

def predict(self, X):
    return self.classes_[np.argmax(self.predict_proba(X), 1)]
```

Because this is a probabilistic classifier, we first implement `predict_proba`, which returns an array of class probabilities of shape [`n_samples`, `n_classes`]. Entry [`i`, `j`] of this array is the posterior probability that sample `i` is a member of class `j`, computed by multiplying the likelihood by the class prior and normalizing.

The `predict` method uses these probabilities and simply returns the class with the largest probability.

Using Our Custom Estimator

Let's try this custom estimator on a problem we have seen before: the classification of handwritten digits. Here we will load the digits and compute the cross-validation score for a range of candidate bandwidths using the `GridSearchCV` meta-estimator (refer back to Chapter 39):

```
In [14]: from sklearn.datasets import load_digits
         from sklearn.model_selection import GridSearchCV

         digits = load_digits()

         grid = GridSearchCV(KDEClassifier(),
                             {'bandwidth': np.logspace(0, 2, 100)})
         grid.fit(digits.data, digits.target);
```

Next we can plot the cross-validation score as a function of bandwidth (see Figure 49-7).

```
In [15]: fig, ax = plt.subplots()
         ax.semilogx(np.array(grid.cv_results_['param_bandwidth']),
                     grid.cv_results_['mean_test_score'])
         ax.set(title='KDE Model Performance', ylim=(0, 1),
                xlabel='bandwidth', ylabel='accuracy')
         print(f'best param: {grid.best_params_}')
         print(f'accuracy = {grid.best_score_}')
Out[15]: best param: {'bandwidth': 6.135907273413174}
         accuracy = 0.9677298050139276
```

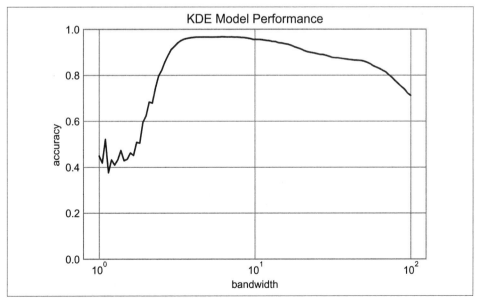

Figure 49-7. Validation curve for the KDE-based Bayesian classifier

This indicates that our KDE classifier reaches a cross-validation accuracy of over 96%, compared to around 80% for the naive Bayes classifier:

```
In [16]: from sklearn.naive_bayes import GaussianNB
         from sklearn.model_selection import cross_val_score
         cross_val_score(GaussianNB(), digits.data, digits.target).mean()
Out[16]: 0.8069281956050759
```

One benefit of such a generative classifier is interpretability of results: for each unknown sample, we not only get a probabilistic classification, but a *full model* of the distribution of points we are comparing it to! If desired, this offers an intuitive window into the reasons for a particular classification that algorithms like SVMs and random forests tend to obscure.

If you would like to take this further, here are some ideas for improvements that could be made to our KDE classifier model:

- You could allow the bandwidth in each class to vary independently.

- You could optimize these bandwidths not based on their prediction score, but on the likelihood of the training data under the generative model within each class (i.e. use the scores from `KernelDensity` itself rather than the global prediction accuracy).

Finally, if you want some practice building your own estimator, you might tackle building a similar Bayesian classifier using Gaussian mixture models instead of KDE.

## CHAPTER 50

Application: A Face Detection Pipeline

This part of the book has explored a number of the central concepts and algorithms of machine learning. But moving from these concepts to a real-world application can be a challenge. Real-world datasets are noisy and heterogeneous; they may have missing features, and data may be in a form that is difficult to map to a clean [n_samples, n_features] matrix. Before applying any of the methods discussed here, you must first extract these features from your data: there is no formula for how to do this that applies across all domains, and thus this is where you as a data scientist must exercise your own intuition and expertise.

One interesting and compelling application of machine learning is to images, and we have already seen a few examples of this where pixel-level features are used for classification. Again, the real world data is rarely so uniform, and simple pixels will not be suitable: this has led to a large literature on *feature extraction* methods for image data (see Chapter 40).

In this chapter we will take a look at one such feature extraction technique: the histogram of oriented gradients (HOG) (*https://oreil.ly/eiJ4X*), which transforms image pixels into a vector representation that is sensitive to broadly informative image features regardless of confounding factors like illumination. We will use these features to develop a simple face detection pipeline, using machine learning algorithms and concepts we've seen throughout this part of the book.

We begin with the standard imports:

```
In [1]: %matplotlib inline
        import matplotlib.pyplot as plt
        plt.style.use('seaborn-whitegrid')
        import numpy as np
```

HOG Features

HOG is a straightforward feature extraction procedure that was developed in the context of identifying pedestrians within images. It involves the following steps:

1. Optionally prenormalize the images. This leads to features that resist dependence on variations in illumination.

2. Convolve the image with two filters that are sensitive to horizontal and vertical brightness gradients. These capture edge, contour, and texture information.

3. Subdivide the image into cells of a predetermined size, and compute a histogram of the gradient orientations within each cell.

4. Normalize the histograms in each cell by comparing to the block of neighboring cells. This further suppresses the effect of illumination across the image.

5. Construct a one-dimensional feature vector from the information in each cell.

A fast HOG extractor is built into the Scikit-Image project, and we can try it out relatively quickly and visualize the oriented gradients within each cell (see Figure 50-1).

```
In [2]: from skimage import data, color, feature
        import skimage.data

        image = color.rgb2gray(data.chelsea())
        hog_vec, hog_vis = feature.hog(image, visualize=True)

        fig, ax = plt.subplots(1, 2, figsize=(12, 6),
                               subplot_kw=dict(xticks=[], yticks=[]))
        ax[0].imshow(image, cmap='gray')
        ax[0].set_title('input image')

        ax[1].imshow(hog_vis)
        ax[1].set_title('visualization of HOG features');
```

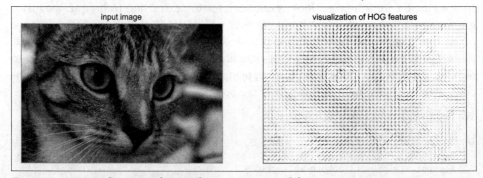

Figure 50-1. Visualization of HOG features computed from an image

HOG in Action: A Simple Face Detector

Using these HOG features, we can build up a simple facial detection algorithm with any Scikit-Learn estimator; here we will use a linear support vector machine (refer back to Chapter 43 if you need a refresher on this). The steps are as follows:

1. Obtain a set of image thumbnails of faces to constitute "positive" training samples.

2. Obtain a set of image thumbnails of non-faces to constitute "negative" training samples.

3. Extract HOG features from these training samples.

4. Train a linear SVM classifier on these samples.

5. For an "unknown" image, pass a sliding window across the image, using the model to evaluate whether that window contains a face or not.

6. If detections overlap, combine them into a single window.

Let's go through these steps and try it out.

1. Obtain a Set of Positive Training Samples

We'll start by finding some positive training samples that show a variety of faces. We have one easy set of data to work with—the Labeled Faces in the Wild dataset, which can be downloaded by Scikit-Learn:

```
In [3]: from sklearn.datasets import fetch_lfw_people
        faces = fetch_lfw_people()
        positive_patches = faces.images
        positive_patches.shape
Out[3]: (13233, 62, 47)
```

This gives us a sample of 13,000 face images to use for training.

2. Obtain a Set of Negative Training Samples

Next we need a set of similarly sized thumbnails that *do not* have a face in them. One way to obtain this is to take any corpus of input images, and extract thumbnails from them at a variety of scales. Here we'll use some of the images shipped with Scikit-Image, along with Scikit-Learn's `PatchExtractor`:

```
In [4]: data.camera().shape
Out[4]: (512, 512)

In [5]: from skimage import data, transform

        imgs_to_use = ['camera', 'text', 'coins', 'moon',
                       'page', 'clock', 'immunohistochemistry',
```

```
                            'chelsea', 'coffee', 'hubble_deep_field']
          raw_images = (getattr(data, name)() for name in imgs_to_use)
          images = [color.rgb2gray(image) if image.ndim == 3 else image
                    for image in raw_images]

In [6]: from sklearn.feature_extraction.image import PatchExtractor

        def extract_patches(img, N, scale=1.0, patch_size=positive_patches[0].shape):
            extracted_patch_size = tuple((scale * np.array(patch_size)).astype(int))
            extractor = PatchExtractor(patch_size=extracted_patch_size,
                                       max_patches=N, random_state=0)
            patches = extractor.transform(img[np.newaxis])
            if scale != 1:
                patches = np.array([transform.resize(patch, patch_size)
                                    for patch in patches])
            return patches

        negative_patches = np.vstack([extract_patches(im, 1000, scale)
                                      for im in images for scale in [0.5, 1.0, 2.0]])
        negative_patches.shape
Out[6]: (30000, 62, 47)
```

We now have 30,000 suitable image patches that do not contain faces. Let's visualize a few of them to get an idea of what they look like (see Figure 50-2).

```
In [7]: fig, ax = plt.subplots(6, 10)
        for i, axi in enumerate(ax.flat):
            axi.imshow(negative_patches[500 * i], cmap='gray')
            axi.axis('off')
```

Our hope is that these will sufficiently cover the space of "non-faces" that our algorithm is likely to see.

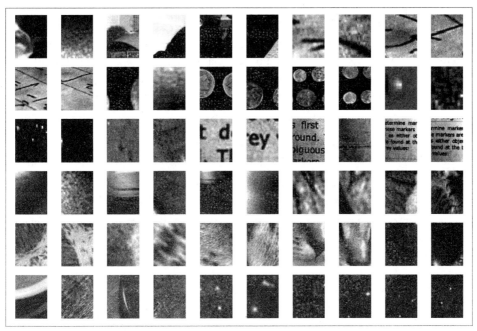

Figure 50-2. Negative image patches, which don't include faces

3. Combine Sets and Extract HOG Features

Now that we have these positive samples and negative samples, we can combine them and compute HOG features. This step takes a little while, because it involves a nontrivial computation for each image:

```
In [8]: from itertools import chain
        X_train = np.array([feature.hog(im)
                            for im in chain(positive_patches,
                                            negative_patches)])
        y_train = np.zeros(X_train.shape[0])
        y_train[:positive_patches.shape[0]] = 1

In [9]: X_train.shape
Out[9]: (43233, 1215)
```

We are left with 43,000 training samples in 1,215 dimensions, and we now have our data in a form that we can feed into Scikit-Learn!

4. Train a Support Vector Machine

Next we use the tools we have been exploring here to create a classifier of thumbnail patches. For such a high-dimensional binary classification task, a linear support vector machine is a good choice. We will use Scikit-Learn's LinearSVC, because in comparison to SVC it often has better scaling for a large number of samples.

First, though, let's use a simple Gaussian naive Bayes estimator to get a quick baseline:

```
In [10]: from sklearn.naive_bayes import GaussianNB
         from sklearn.model_selection import cross_val_score

         cross_val_score(GaussianNB(), X_train, y_train)
Out[10]: array([0.94795883, 0.97143518, 0.97224471, 0.97501735, 0.97374508])
```

We see that on our training data, even a simple naive Bayes algorithm gets us upwards of 95% accuracy. Let's try the support vector machine, with a grid search over a few choices of the C parameter:

```
In [11]: from sklearn.svm import LinearSVC
         from sklearn.model_selection import GridSearchCV
         grid = GridSearchCV(LinearSVC(), {'C': [1.0, 2.0, 4.0, 8.0]})
         grid.fit(X_train, y_train)
         grid.best_score_
Out[11]: 0.9885272620319941

In [12]: grid.best_params_
Out[12]: {'C': 1.0}
```

This pushes us up to near 99% accuracy. Let's take the best estimator and retrain it on the full dataset:

```
In [13]: model = grid.best_estimator_
         model.fit(X_train, y_train)
Out[13]: LinearSVC()
```

5. Find Faces in a New Image

Now that we have this model in place, let's grab a new image and see how the model does. We will use one portion of the astronaut image shown in Figure 50-3 for simplicity (see discussion of this in the following section, and run a sliding window over it and evaluate each patch:

```
In [14]: test_image = skimage.data.astronaut()
         test_image = skimage.color.rgb2gray(test_image)
         test_image = skimage.transform.rescale(test_image, 0.5)
         test_image = test_image[:160, 40:180]

         plt.imshow(test_image, cmap='gray')
         plt.axis('off');
```

Figure 50-3. An image in which we will attempt to locate a face

Next, let's create a window that iterates over patches of this image, and compute HOG features for each patch:

```
In [15]: def sliding_window(img, patch_size=positive_patches[0].shape,
                             istep=2, jstep=2, scale=1.0):
             Ni, Nj = (int(scale * s) for s in patch_size)
             for i in range(0, img.shape[0] - Ni, istep):
                 for j in range(0, img.shape[1] - Ni, jstep):
                     patch = img[i:i + Ni, j:j + Nj]
                     if scale != 1:
                         patch = transform.resize(patch, patch_size)
                     yield (i, j), patch

         indices, patches = zip(*sliding_window(test_image))
         patches_hog = np.array([feature.hog(patch) for patch in patches])
         patches_hog.shape
Out[15]: (1911, 1215)
```

Finally, we can take these HOG-featured patches and use our model to evaluate whether each patch contains a face:

```
In [16]: labels = model.predict(patches_hog)
         labels.sum()
Out[16]: 48.0
```

We see that out of nearly 2,000 patches, we have found 48 detections. Let's use the information we have about these patches to show where they lie on our test image, drawing them as rectangles (see Figure 50-4).

```
In [17]: fig, ax = plt.subplots()
         ax.imshow(test_image, cmap='gray')
         ax.axis('off')
```

```
Ni, Nj = positive_patches[0].shape
indices = np.array(indices)

for i, j in indices[labels == 1]:
    ax.add_patch(plt.Rectangle((j, i), Nj, Ni, edgecolor='red',
                               alpha=0.3, lw=2, facecolor='none'))
```

Figure 50-4. Windows that were determined to contain a face

All of the detected patches overlap and found the face in the image! Not bad for a few lines of Python.

Caveats and Improvements

If you dig a bit deeper into the preceding code and examples, you'll see that we still have a bit of work to do before we can claim a production-ready face detector. There are several issues with what we've done, and several improvements that could be made. In particular:

Our training set, especially for negative features, is not very complete
    The central issue is that there are many face-like textures that are not in the training set, and so our current model is very prone to false positives. You can see this if you try out the algorithm on the *full* astronaut image: the current model leads to many false detections in other regions of the image.

    We might imagine addressing this by adding a wider variety of images to the negative training set, and this would probably yield some improvement. Another option would be to use a more directed approach, such as *hard negative mining*, where we take a new set of images that our classifier has not seen, find all the patches representing false positives, and explicitly add them as negative instances in the training set before retraining the classifier.

Our current pipeline searches only at one scale

As currently written, our algorithm will miss faces that are not approximately 62 × 47 pixels. This can be straightforwardly addressed by using sliding windows of a variety of sizes, and resizing each patch using `skimage.transform.resize` before feeding it into the model. In fact, the `sliding_window` utility used here is already built with this in mind.

We should combine overlapped detection patches

For a production-ready pipeline, we would prefer not to have 30 detections of the same face, but to somehow reduce overlapping groups of detections down to a single detection. This could be done via an unsupervised clustering approach (mean shift clustering is one good candidate for this), or via a procedural approach such as *non-maximum suppression*, an algorithm common in machine vision.

The pipeline should be streamlined

Once we address the preceding issues, it would also be nice to create a more streamlined pipeline for ingesting training images and predicting sliding-window outputs. This is where Python as a data science tool really shines: with a bit of work, we could take our prototype code and package it with a well-designed object-oriented API that gives the user the ability to use it easily. I will leave this as a proverbial "exercise for the reader."

More recent advances: deep learning

Finally, I should add that in machine learning contexts, HOG and other procedural feature extraction methods are not always used. Instead, many modern object detection pipelines use variants of deep neural networks (often referred to as *deep learning*): one way to think of neural networks is as estimators that determine optimal feature extraction strategies from the data, rather than relying on the intuition of the user.

Though the field has produced fantastic results in recent years, deep learning is not all that conceptually different from the machine learning models explored in the previous chapters. The main advance is the ability to utilize modern computing hardware (often large clusters of powerful machines) to train much more flexible models on much larger corpuses of training data. But though the scale differs, the end goal is very much the same the same: building models from data.

If you're interested in going further, the list of references in the following section should provide a useful place to start!

Further Machine Learning Resources

This part of the book has been a quick tour of machine learning in Python, primarily using the tools within the Scikit-Learn library. As long as these chapters are, they are still too short to cover many interesting and important algorithms, approaches, and discussions. Here I want to suggest some resources to learn more about machine learning in Python, for those who are interested:

The Scikit-Learn website (http://scikit-learn.org)
> The Scikit-Learn website has an impressive breadth of documentation and examples covering some of the models discussed here, and much, much more. If you want a brief survey of the most important and often-used machine learning algorithms, this is a good place to start.

SciPy, PyCon, and PyData tutorial videos
> Scikit-Learn and other machine learning topics are perennial favorites in the tutorial tracks of many Python-focused conference series, in particular the PyCon, SciPy, and PyData conferences. Most of these conferences publish videos of their keynotes, talks, and tutorials for free online, and you should be able to find these easily via a suitable web search (for example, "PyCon 2022 videos").

Introduction to Machine Learning with Python, by Andreas C. Müller and Sarah Guido (O'Reilly)
> This book covers many of the machine learning fundamentals discussed in these chapters, but is particularly relevant for its coverage of more advanced features of Scikit-Learn, including additional estimators, model validation approaches, and pipelining.

Machine Learning with PyTorch and Scikit-Learn (https://oreil.ly/p268i), by Sebastian Raschka (Packt)
> Sebastian Raschka's most recent book starts with some of the fundamental topics covered in these chapters, but goes deeper and shows how those concepts apply to more sophisticated and computationally intensive deep learing and reinforcement learning models using the well-known PyTorch library (*https://pytorch.org*).

Index

explained variance ratio, 470
exponentials, 56
external code, magic commands for running, 13

F

face recognition
  Histogram of Oriented Gradients, 541-549
  Isomap, 489-493
  principal component analysis, 473-476
  support vector machines, 445-450
faceted histograms, 336
fancy indexing, 80-87
  basics, 80
  binning data, 85
  combined with other indexing schemes, 81
  modifying values with, 84
  selecting random points, 82
feature engineering, 402-409
  categorical features, 402
  derived features, 405-408
  image features, 405
  imputation of missing data, 408
  processing pipeline, 409
  text features, 404-405
feature, data point, 356
features matrix, 368
fillna() method, 129, 130
filter() method, 172
FiveThirtyEight stylesheet, 317
fixed-type arrays, 39

G

Gaussian basis functions, 424-425
Gaussian mixture models (GMMs), 512-527
  choosing covariance type, 520
  clustering with, 377
  correcting overfitting with, 523
  density estimation algorithm, 520-524
  E–M generalization, 516-520
  handwritten data generation example, 524-527
  k-means weaknesses addressed by, 512-515
  kernel density estimation and, 528
Gaussian naive Bayes classification, 375, 381, 411-414, 546
Gaussian process regression (GPR), 253
generative models, 411
get() operation, 189

get_dummies() method, 189
ggplot stylesheet, 318
GMMs (see Gaussian mixture models)
GPR (Gaussian process regression), 253
grayscale stylesheet, 319
GroupBy aggregation, 176
GroupBy object, 169-171
  aggregate() method, 172
  apply() method, 173
  column indexing, 169
  dispatch methods, 170
  filter() method, 172
  iteration over groups, 170
  transform() method, 173
groupby() operation (Pandas), 167-175
  GroupBy object and, 169-171
  grouping example, 175
  pivot tables versus, 176
  split key specification, 174
  split-apply-combine example, 167-169

H

handwritten digits, recognition of (see optical character recognition)
hard negative mining, 548
help
  IPython, 4
  magic functions, 15
help() function, 5
hexagonal binnings, 264
hierarchical indexing, 132-144
  (see also MultiIndex type)
  in one-dimensional Series, 132-135
  with Python tuples as keys, 133
  rearranging multi-indices, 141-144
  unstack() method, 134
Histogram of Oriented Gradients (HOG)
  caveats and improvements, 548-549
  for face detection pipeline, 541-549
  features, 542
  simple face detector, 543-548
histograms, 260-266
  binning data to create, 85
  faceted, 336
  kernel density estimation and, 264, 528-533
  manual customization, 312-314
  plt.hexbin() function, 264
  plt.hist2d() function, 263
  Seaborn, 333-335

About the Author

Jake VanderPlas is a software engineer at Google Research, working on tools that support data-intensive research. Jake creates and develops Python tools for use in data-intensive science, including packages like Scikit-Learn, SciPy, AstroPy, Altair, JAX, and many others. He participates in the broader data science community, developing and presenting talks and tutorials on scientific computing topics at various conferences in the data science world.

Colophon

The animal on the cover of *Python Data Science Handbook* is a Mexican beaded lizard (*Heloderma horridum*), a reptile found in Mexico and parts of Guatemala. The Greek word *heloderma* translates to "studded skin," referring to the distinctive beaded texture of the lizard's skin. These bumps are *osteoderms*, which each contain a small piece of bone and serve as protective armor.

The Mexican beaded lizard is black with yellow patches and bands. It has a broad head and a thick tail that stores fat to help it survive the hot summer months when it is inactive. On average, these lizards are 22–36 inches long, and weigh around 1.8 pounds. As with most snakes and lizards, the tongue of the Mexican beaded lizard is its primary sensory organ. It will flick it out repeatedly to gather scent particles from the environment and detect prey (or, during mating season, a potential partner).

It and the Gila monster (a close relative) are the only venomous lizards in the world. When threatened, the Mexican beaded lizard will bite and clamp down, chewing, because it cannot release a large quantity of venom at once. This bite and the aftereffects of the venom are extremely painful, though rarely fatal to humans. The beaded lizard's venom contains enzymes that have been synthesized to help treat diabetes, and further pharmacological research is in progress. It is endangered by loss of habitat, poaching for the pet trade, and locals who kill it out of fear. This animal is protected by legislation in both countries where it lives. Many of the animals on O'Reilly covers are endangered; all of them are important to the world.

The cover illustration is by Karen Montgomery, based on a black and white engraving from Wood's *Animate Creation*. The cover fonts are Gilroy Semibold and Guardian Sans. The text font is Adobe Minion Pro; the heading font is Adobe Myriad Condensed; and the code font is Dalton Maag's Ubuntu Mono.

O'Reilly Media, Inc.介绍

O'Reilly以"分享创新知识、改变世界"为己任。40多年来我们一直向企业、个人提供成功必需之技能及思想，激励他们创新并做得更好。

O'Reilly业务的核心是独特的专家及创新者网络，他们通过我们分享知识。我们的在线学习（Online Learning）平台提供独家的直播培训、图书及视频，使客户更容易获取业务成功所需的专业知识。几十年来O'Reilly图书一直被视为学习开创未来之技术的权威资料。我们全年举办的诸多会议是活跃的技术聚会场所，来自各领域的专业人士在此建立联系，讨论最佳实践并发现可能影响技术行业未来的新趋势。

我们的客户渴望作出推动世界前进的创新，我们能祝您一臂之力。

业界评论

"O'Reilly Radar博客有口皆碑。"

——Wired

"O'Reilly凭借一系列（真希望当初我也想到了）非凡想法建立了数百万美元的业务。"

——Business 2.0

"O'Reilly Conference是聚集关键思想领袖的绝对典范。"

——CRN

"一本O'Reilly的书就代表一个有用、有前途、需要学习的主题。"

——Irish Times

"Tim是位特立独行的商人，他不光放眼于最长远、最广阔的视野并且切实地按照Yogi Berra的建议去做了：'如果你在路上遇到岔路口，走小路（岔路）。'回顾过去Tim似乎每一次都选择了小路，而且有几次都是一闪即逝的机会，尽管大路也不错。"

——Linux Journal